INSTRUCTOR'S RESOURCE GUIDE

TO ACCOMPANY

Human Anatomy and Physiology

THIRD EDITION

ELAINE N. MARIEB

Holyoke Community College
Holyoke, Massachusetts

The Benjamin/Cummings Publishing Company, Inc.

Redwood City, California ■ Menlo Park, California
Reading, Massachusetts ■ New York ■ Don Mills, Ontario
Wokingham, U.K. ■ Amsterdam ■ Bonn ■ Sydney
Singapore ■ Tokyo ■ Madrid ■ San Juan

ISBN 0-8053-4282-6

1 2 3 4 5 6 7 9 10—VG—98 97 96 95 94

The Benjamin/Cummings Publishing Company, Inc.
390 Bridge Parkway
Redwood City, California 94065

Contents

UNIT 2   COVERING, SUPPORT, AND MOVEMENT OF THE BODY

UNIT 3   REGULATION AND INTEGRATION OF THE BODY

UNIT 5   CONTINUITY

Case Study: The Epilogue

Elaine N. Marieb

Holyoke Community College
Holyoke, Massachusetts

Anatomy and Physiology instructors spend a year, more or less, introducing students to their subject. Although gross, microscopic, and even submicroscopic levels of human structure and function are attended to, the "body" — the human organism and the actual objective of the students' learning—is never put back together and considered as an entirety. Consequently, even though learning necessarily takes place in bits and pieces, the finely coordinated human body is difficult for students to understand.

The Epilogue,"A Day in the Life," grew out of the above perception. My original intent was simply to provide students with the affirmation that they had actually learned a huge amount of material and to show them that, with a little help, they could make the connections which had often seemed so out of reach. However, comments from Epilogue reviewers yielded a rich lode of suggestions for how this new feature might be used creatively by instructors to encourage situational and more in-depth learning. A few of these suggestions follow, but they are by no means all-inclusive. Instructors will no doubt have still more ideas for its utilization as a learning tool.

1. Students can use the Epilogue as a study guide to help them organize their thoughts. For example, the Epilogue illustrates the importance of system interactions and helps students predict which systems' functions are closely intermeshed.

2. Before covering a system in class, have students read the related portion of the Epilogue to provide a preview of the concepts and systems interactions that relate to that system's functioning.

3. While covering a system, or immediately after its classroom completion:

    a. ask students to find all references to that system in the Epilogue and discuss the systems interactions involved.

    b. choose one example from the Epilogue and ask students to predict how the result would differ if one (or more) of the variables were changed.

    c. specify one daily activity involving the system being studied and ask the students to write out the physiological events involved. Ask students to look up and add information about the involvement of one other system not yet studied.

    d. encourage students to bounce ideas off each other for new episodes of "A Day in the Life."

    e. have students think about what they ordinarily do over the course of a day, and write about one commonplace episode in their daily life as an assignment for credit (perhaps five points of the associated test's grade). As part of the assignment, they would suggest a graphic, write a "friendly" one- to three-sentence descriptive section, and tabulate the ongoing physiological events of the body systems in computer printout form.

I hope you and your students enjoy using "A Day in the Life" as much as I enjoyed devising it. The single caution I have is that Epilogue-associated events should not become simply more tasks for the student to master. Students should be encouraged to approach the assignments from a light-hearted and even comical vantage point. Laughter is a stimulus for learning!

Case Study:
Multimedia in the Classroom

Eleanor Cauldwell

North Seattle Community College
Seattle, Washington

In the spring of 1989, I joined the faculty of North Seattle Community College to teach Human Biology: anatomy, physiology, nutrition and health. My predecessor had used cats for dissection, but I am hypersensitive to formaldehyde and many other solvents, making supervision of dissections impossible. I wanted my students to have the experience of dissection but without using preserved specimens. I discovered a series of six videodiscs that clearly presented cadaver dissections, and I was on my way to using multimedia in the classroom.

The students at North Seattle Community College represent a wide age range (the average is 29 years) and a variety of American ethnic groups and non-native groups. Some of these students grew up in the MTV generation. Others represent the pre-television generation. Some students are preparing for their first career, while others are displaced homemakers or are making career changes. Although most of the students in my classes are taking courses to fulfill the requirements of pre-healthcare programs, I have a diverse student population. To meet individual needs, I have increased my use of multimedia in the classroom. My goal is to provide a variety of approaches: visual, auditory, and kinesthetic.

Effectively incorporating multimedia in a class requires a commitment to redesign both course content and methods of presentation. Administrative support is needed to obtain the equipment, and media and computer services staff must provide the technical and creative instruction required. My multimedia classroom began with a videodisc player, two television monitors, and those six videodiscs. My system has continued to grow and now includes the following components: an InFocus video projector, a Pioneer LD-V4200 videodisc player, a Sony SVO-160 VHS VCR, a video microscope set up, a Macintosh Quadra 610 8/230 with CD-ROM drive, a SyQuest 88MB cartridge removable drive, an overhead projector, a Kodak slide projector, and a laser pointer.

The newest addition to my multimedia system is a pressure sensitive pad, with Smart Marker software from Smart Technologies, which will replace the overhead projector. Drawings can be sketched on the pad and displayed using the projector's computer channel. I use size 48 bold type on my word processing program to project type on the screen, which improves my non-native students' comprehension.

Since my lecture and laboratory sessions are both held in a very large classroom, using television monitors to show multimedia does not provide optimum viewing for the students. We decided on a ceiling-mounted InFocus video projector, which projects a four-foot-square image on a screen at the front of the room. This projector has three channels: one for projecting the computer screen, the second for the videodisc player, and the third for the VCR or video microscope. These channels can be quickly changed by remote control, allowing the use of a variety of media in the same presentation.

Several components of my system will change in the fall of 1994, reflecting the increasing need for more versatile equipment. The new additions include an InFocus Lite Pro 550 video projector, a Pioneer CLD-V2400 videodisc player, a Sony SVO-1610 VHS VCR, and a Macintosh PowerMac 6100 16/230 with CD-ROM drive and SoftWindows. The new computer system will allow the use of both Macintosh and DOS software and will project a brighter, clearer image, reducing the need to dim the lights during presentations.

When I started using videodiscs in the classroom, I selected segments using the player's remote control. This was cumbersome and I wanted a selection system that students could use themselves to review topics such as muscle structure and function, so I learned to program HyperCard to select sequences on videodiscs or videotapes. I attended an NSF-sponsored seminar, "Designing Interactive Computer-Assisted Instruction" presented by David Brooks from The University of Nebraska, Lincoln. At the seminar, I learned to create stacks that incorporate text material, illustrations, animations, and buttons to select appropriate videodisc segments. One of the first stacks was created using Voyager Videostack, a HyperCard-based videodisc interface program. Now, during our laboratory sessions and open laboratory time, students can open stacks and select sequences of the cadaver dissection videodiscs for closer review.

I continue to collect videodiscs and videotapes to provide a variety of visual and auditory experiences for the students. To select images on the videodiscs, I currently use MediaMAX software from Videodiscovery. This program can be used to select individual images, to create slide shows, and to make barcodes or buttons to be inserted in other programs such as HyperCard or Persuasion 3.0. MediaMAX is very easy to use and has a short learning curve. Many students use this program to create their own videodisc slide shows to accompany classroom projects or to review selected topics. Barcodes can be incorporated in laboratory directions or printed review materials. Students can use the barcode reader to select images from the videodiscs.

Videotapes are useful to show current applications of the principles being learned. CNN, PBS, and other stations frequently feature appropriate material. Off-air tapes that are subject to copyright rules[1] and purchased videotapes such as Bill Moyers' "Healing and the Mind" can be added to multimedia presentations. Unlike videodiscs, videotapes are a linear medium requiring time to run the tape to the desired sequence. Video Toolkit, by Abbate Video, can be used to create buttons which will locate the desired videotape sequence.

The video microscope is used to project prepared slides and living samples. For example, I show a variety of live, single-celled plants and animals to illustrate cellular structure and function. Red blood cells are used to illustrate the effects of iso-, hyper-, and hypotonic solutions. The video microscope is particularly useful for presenting histology slides and can be used to replace microscopes in a lab practical. All students see exactly the same view of each slide, and there is no opportunity for students to sabotage the test by moving the pointer on the slide.

Aldus Persuasion 3.0 is a presentation software package that facilitates the preparation of programs for direct projection from the computer to the screen. I use this program to project lecture outlines enhanced with illustrations and a variety of scanned images I have collected. One of my most successful series of modules includes pictures of individual muscles accompanied by information regarding their location, function, and origin and insertion. These modules improve my students' ability to locate and identify individual muscle groups. Persuasion 3.0 images can also be printed out as overhead transparencies, 35 mm slides, student handouts, or instructor's notes.

[1] Educational Fair-Use Copyright Rules for Video: The program (1) can only be taped by request of an instructor, not by students; (2) may be used in the classroom for ten days from the date of taping; (3) may be kept for instructor review for 45 days from date of taping; (4) must be erased after the expiration of the aforementioned permission dates.

Another way I use the computer projection system is to illustrate for the students how to use computer programs. Programs such as Anatomy Review can be demonstrated screen-by-screen to increase the confidence of less computer-literate students. In the digestion/nutrition unit, for example, students do a project using the NutriPro program. I illustrate the use of this program, comparing the nutritional value of a ham sandwich with a peanut butter and jam sandwich. Students have fewer questions and problems in the student computer labs when they have seen the software demonstrated before they begin their first project.

Thus, creative use of videodiscs, videotapes, the video microscope, and computer software in both the lecture and laboratory periods can provide experiences that complement the information found in the textbook. I believe a multimedia classroom allows me to more successfully support the intellectual growth of all of my students.

Case Study: Learning To Live with A.D.A.M.

Karen M. LaFleur

Greenville Technical College
Greenville, South Carolina

In this essay, Dr. LaFleur describes how she and her colleagues at Greenville Technical College, a two-year comprehensive institution, use A.D.A.M. software in their integrated two-year anatomy and physiology course. About 1200 students take this course each year as part of their requirements for nursing and allied health programs. There are three copies of A.D.A.M. (Comprehensive version) at Greenville Technical College: two copies in the Biology Department for use by faculty in lecture and lab and one copy for students in the Media Center.

A.D.A.M. [*A*nimated *D*issection of *A*natomy for *M*edicine] arrived at Greenville Technical College in late 1993. The installation of this program marked the beginning of change in my approach to teaching human anatomy. A.D.A.M. enables me and my students to more easily envision and comprehend complex, three-dimensional anatomical relationships. A.D.A.M. even makes it possible for me to design new and personalized teaching materials, thus incorporating contemporary technology into a traditional course.

A.D.A.M. (Comprehensive version) is a sophisticated and multi-faceted software program which can be used in a variety of ways. First, A.D.A.M. is an enormous encyclopedia of anatomical drawings superimposed one on top of the other — a 90s approach to the old textbook overlays. Based on Gray, Netter, Grant, and contemporary anatomical experts, each graphic is virtually a work of art; thousands have been merged into this interactive program. With a click of a mouse, I am given access to those graphic layers while I probe and explore some region or organ. I can point with the mouse to a structure, revealing its name and pronouncing it. I am able to print A.D.A.M.'s color graphics so I have at hand an extensive, detailed library of drawings. A.D.A.M. also allows me to mimic the mechanics of surgery, permitting me to use a "scalpel" and dissect. With the "Studio" function, I am even able to use A.D.A.M. as an instructional design tool, one which allows me to create tailored teaching materials.

I saw a prototype of A.D.A.M. in 1992 and was immediately enchanted. But at that time our department had no money for software. Fortunately, later that year the Human Anatomy and Physiology Society generously made a donation to our department in recognition of my role directing the 1991 Annual Conference. I purchased A.D.A.M. because I felt that its long-term benefits and versatility more than compensated for the price. To put it into perspective, for less than the cost of supplying cats for dissection for one semester, we purchased a permanent, reusable, versatile teaching tool, one that will not be thrown out at the end of the term.

The Biology Department is fortunate to now have two copies of A.D.A.M. in our area for use by our faculty and students. In January of 1994, a generous corporate donor gave us a third copy of the program that is installed in our college's media center.

Greenville Technical College is a comprehensive two-year college, composed of seven divisions. Two of those divisions are allied health and nursing, which account for most of our students.

We teach an integrated anatomy and physiology course. There are five full-time faculty, as well as a variable number of part-time instructors. The first semester concentrates on introductory material, skeletal, muscular, special senses, and digestive systems, while the second semester covers nervous, circulatory, respiratory, urinary, reproductive, and endocrine systems. Our anatomy labs, prior to A.D.A.M., were based on models, charts, and bone samples since, as a cost-cutting measure, we abandoned all animal dissection in the late 70s. Currently, we use an occasional fetal pig to demonstrate cavities or membranes.

Approximately 1200 students flow through our department annually. The majority of them are enrolled in one or both Anatomy and Physiology courses. Our typical full-time student fits the profile of the national average community college student: a 27-year-old white female, married, employed, with one or more children.

With heavy teaching loads, and only small amounts of time between semesters, our Biology faculty members have been learning to use A.D.A.M. along with the students. The two initial copies of A.D.A.M are used in our anatomy labs and in nearby lecture rooms. One is installed on a Macintosh Quadra with a CD player. The second station is installed on a IIvx with a built-in CD player. Both of these systems are on rolling carts which are transported to class or to lab. The systems are linked to an LCD panel and high-intensity overhead projector, which can be used to project A.D.A.M. onto a wall screen.

We are using A.D.A.M. primarily for demonstration and illustration during lecture and lab. We use it to provide students with a general orientation to the body — for example, to illustrate the thorax, its subdivisions, and its membranes. We use it to support descriptions of organ relationships—for example, how the spleen is positioned relative to the stomach or diaphragm. Once we finish using A.D.A.M. in the large group setting, we leave the program up and running to allow students access to it when we are there to offer guidance and help.

When I use A.D.A.M. in a class or lab demonstration, I involve students by eliciting from them a particular muscle's origin and insertion. If students have questions about surrounding tissues, the program allows us to move in closer or back away for more distant viewing. In both instances, the structures can be identified on the screen and even pronounced for verification and reinforcement.

Our third copy of A.D.A.M. resides in the Computer Valley. The Computer Valley consists of 24 computers in an open, supervised setting. Since these machines are available during library hours, students have open access to the program. Along with greater availability comes greater responsibility! Students are given "dissection" assignments that require them to explore some specific region. For example, I ask them to note the layers and structures that they must "remove" in order to progress through the right hypochondrium.

There are a number of specific situations where A.D.A.M. improves and refines the ability to convey anatomical relationships. Following are three situations where I feel that A.D.A.M. is an invaluable teaching tool.

1.   Musculature. While flat drawings certainly illustrate muscle attachment, the viewer is unable to move through different planes. The A.D.A.M. platform makes it possible for us to see a superficial limb or torso, and gradually to peel away structures, moving successively deeper, ultimately revealing the underlying bone. Since the program has the capacity to move from anterior, to posterior, to medial, and to lateral, we are able to see more vividly and accurately the relationship of a particular muscle to its surroundings.

2.   Joints. A.D.A.M. illustrates joint anatomy in a manner that vastly improves on what is conveyed by a series of flat drawings. With the capacity to visually dissect, the joint capsules and their relationships to tendons, ligaments, and other structures can be revealed. For my students, the locations and positions of bursae have always been difficult relationships to visualize; the multi-dimensional ability of this program allows us to probe, to see where ligaments and tendons attach, and to locate fibrous capsules.

3. Blood vessels and nerves. Tracing the course of a blood vessel on a flat page is easy. Conveying the reality of branching tributaries or the burrowing of a vessel is difficult. A.D.A.M enables us to follow the progress of a vessel through a limb, exploring interrelationships en route. We are also able to isolate a selected vessel, either by highlighting it in the natural setting or by extracting it from the body entirely.

Students seem to *really* enjoy their experiences with A.D.A.M. They use it to complete a specific assignment, to study and review required structures, to practice, and to simply "wander anatomically." At the beginning of the semester I give a minimal large-group demonstration of A.D.A.M.'s key capabilities and operations, and then pretty much turn the students loose with only a general orientation sheet in their hands. They are hesitant initially but quickly adapt to the medium. It is particularly interesting to observe their research. In my casual observations, students seem to be viewing more in this program than they do in other media. Further, they seem to ask more questions, probe relationships more often, and enjoy the whole process! Some of their candid reactions support those observations:

"I always enjoy seeing the A.D.A.M. demonstrations. I find them very helpful because they give me a good idea of things that my text and models can't show me very well." D.B.

"A.D.A.M. demonstrations make anatomy clear. I now *finally* understand the veins of the arm. I just wish that I had my own A.D.A.M.!" F.T.

While it is certainly not the answer to all pedagogical and anatomical dilemmas, A.D.A.M. software provides a captivating route around many barricades to learning. With accuracy and flair, this software makes illustrations more vivid and relationships more meaningful.

Case Study:
A.D.A.M. in the Anatomy Laboratory

Greg Smith

Saint Mary's College
Moraga, California

In this essay, Dr. Smith explains how he uses A.D.A.M. software (Comprehensive version) in his one-term anatomy course at St. Mary's College, a private four-year school. Students in biology, health sciences, nursing, and physical education use A.D.A.M. in their anatomy lab as a valuable tool to enhance their study of cadavers and models.

At Saint Mary's College, A.D.A.M. is used in the laboratory portion of our human anatomy course. This course is taken primarily by students pursuing degrees in nursing, health science, health, physical education and recreation, and biology. The career goals of these students include nursing, medicine, dentistry, physical therapy, occupational therapy, athletic training, and chiropractic. The course is designed along the lines of traditional gross anatomy, with an incorporation of histology.

My history with A.D.A.M. began about three years ago at an electronic media workshop at Stanford University, where I saw the first completed module, that of the foot. I remember thinking that comparing A.D.A.M. to other available anatomy programs was like comparing a drawing by Michelangelo to a stick figure drawing. Shortly after, I observed a demonstration sponsored by Benjamin/Cummings Publishing Company and expressed the belief that A.D.A.M. might be appropriate for undergraduate anatomy courses. I was approached by representatives from Benjamin/Cummings and A.D.A.M. Software, Inc., to serve as the director of a pilot study at Saint Mary's College to determine if, indeed, A.D.A.M. would be a valuable aid for undergraduate institutions.

The implementation of A.D.A.M. for this pilot study was relatively simple. The Biology Department had recently acquired six Macintosh computers (and other electronic audiovisual equipment) with funding from a private foundation grant. The only add-on devices needed for A.D.A.M. were a CD-ROM player and a video card.

I had evolved into the contact/resource person for computer and audiovisual use in the Biology Department and for other departments as well. I had been the motivating force behind investigations of a variety of formats for lecture and laboratory presentation. For example, we now use laser discs such as *Slice of Life* with database retrieval software, CD-ROMs such as the BioQUEST Library, and other computer simulations and demonstrations. Because of this experience, I was philosophically comfortable with the prospect of using A.D.A.M. in the lab.

I have been using A.D.A.M. for two and one-half years now, beginning with the pilot study and continuing with its incorporation into our human anatomy course. The learning curve is quite accelerated. The basic menu functions are simple to master and allow a novice computer user to easily become an independent navigator. Navigating is the operative word, as A.D.A.M. affords my students the ability to view, link, relate, and learn in a manner they choose. My educational philosophy includes providing students with as many forms of learning opportunities as possible. In the human anatomy laboratory, the students have myriad materials available, including A.D.A.M., cadavers, osteologic specimens, histologic slides, anatomic models, and laser discs. I believe these different media are necessary to reach the broad spectrum of my student's comprehension strengths and weaknesses.

A.D.A.M. was simply incorporated as an additional tool, not as a substitute for an existing format. To this end, A.D.A.M. represents a learning paradigm that is extremely effective for most students.

During the scheduled laboratory period, my students follow a general outline of objectives. How they accomplish those objectives is their decision. Within this context, the students use the media and materials that best fit their learning styles. One of the most difficult obstacles for my students is correlating what is presented in the lab with two-dimensional illustrations from which they must learn when away from the lab. A.D.A.M. helps overcome this hurdle, effectively serving as an interface between the three-dimensional nature of the cadavers and models and the two-dimensional format of illustrations. Unfortunately, the students lament the limited access to A.D.A.M. It is my desire to make it available outside the lab in other areas such as the campus computer facilities.

I also use A.D.A.M. with my honors students, who work on independent projects dissecting cadavers. Because of the independent nature of this program, A.D.A.M. has proved to be an invaluable tool for these students. They use A.D.A.M. in two ways. First, they review A.D.A.M. before dissecting a region, referencing structures to be exposed. This process minimizes the risk of removing or damaging structures that I require them to demonstrate. Second, they use A.D.A.M. to "reconstruct" regions that were dissected to expose deeper structures. This ability to add layers or structures is highly valued by all the students in the human anatomy course.

The feedback from the students is nearly all positive. They quickly recognize the value of A.D.A.M. and readily embrace it as part of their learning experience. Students express one negative point about the access speed: the time to navigate is too slow for them. I recommend using as fast a computer platform as possible. Beyond the positives and one negative, there is a wish. I am frequently asked if there is a student version they can use when studying at home or in their dormitory room.

Although I have not formally examined differences in test scores based on A.D.A.M. use, I have solicited feedback to specific questions. My students feel that A.D.A.M. has definitely helped them learn human anatomy. They like having the ability to directly access a specific structure and note its relationship to adjacent structures and regions. Being able to highlight a structure allows the students to note its true boundaries, which is important when studying the origin and insertion of muscles or tracing nerves and blood vessels. And, finally, having the different views (e.g., anterior, posterior) and links to various related information (CT scans, X rays, histologic views) provides the students with a comprehensive perspective of human anatomy.

As a general education tool, A.D.A.M. allows the student to choose the information acquisition pathway with which they feel most comfortable. This active choice is a powerful form of learning enfranchisement. Incorporating A.D.A.M. into a menagerie of laboratory materials allows students to make physical and mental connections that are so important to comprehending a visual subject like human anatomy. I personally believe that three-dimensional visualization is a form of critical thinking, and enabling this visualization is one of A.D.A.M.'s strengths.

As for future applications, using the A.D.A.M authoring function called "Studio" I would like to develop modules for use in the college's Athletic Trainers Program. These modules would demonstrate common athletic injuries, the anatomical structures involved, and the bases of specific injury evaluation protocols. Because all of the student trainers have completed human anatomy, I believe that these modules would be valuable for reviewing anatomic structures and would provide insight into functional clinical correlations. Being able to create lesson plans such as this makes A.D.A.M. a versatile tool. Human anatomic information is used in so many fields that A.D.A.M. can truly be considered an interdisciplinary resource. Even my biological anthropology students are fascinated with A.D.A.M., so much so that I expect a few of them to enroll in my human anatomy the next time it is offered.

Comments from my funding request illustrate why I believe A.D.A.M. is a valuable resource. I state that A.D.A.M. allows the student user to navigate anatomical structures utilizing interactive computer software to access a vast number of interconnected dissectable views and dynamic links. I also include my students' comments about how their appreciation of human anatomy grew as they discovered the fascinating interrelationships of various anatomic structures. During my tenure at Saint Mary's College I have striven to provide my students with materials that allow them to best learn the subject matter. The use of A.D.A.M. is consistent with my commitment to academic excellence.

1

The Human Body: An Orientation

Chapter Preview

This chapter first defines and contrasts anatomy and physiology and discusses the levels of organization complexity in the human body. The needs and functional processes common to all living organisms are reviewed. Three essential concepts — the complementarity of structure and function, hierarchy of structural organization, and homeostasis — will form the bedrock for the study of the human body and will provide the unifying threads for the topics in the book. The final section will present the language of anatomy — the terminology "tools" that are used to describe the body or its parts.

INTEGRATING THE PACKAGE

Suggested Lecture Outline

I. An Overview of Anatomy and Physiology (pp. 3-5)
    A. Introduction (p. 3)
        1. Anatomy
        2. Physiology
    B. Topics of Anatomy (p. 3)
        1. Gross Anatomy
            a. Regional Anatomy
            b. Systemic Anatomy
            c. Surface Anatomy
        2. Microscopic Anatomy
            a. Cellular Anatomy (Cytology)
            b. Histology
        3. Developmental Anatomy
            a. Embryology
        4. Pathological Anatomy
        5. Radiographic Anatomy
        6. Molecular Biology
        7. Anatomical Terminology
    C. Topics of Physiology (pp. 3-4)
        1. Renal Physiology
        2. Neurophysiology
        3. Cardiovascular Physiology
        4. Other Organ System Physiology

D. Complementarity of Structure and Function (p. 4)

II. The Hierarchy of Structural Organization (p. 5)

A. Levels of Structural Hierarchy (p. 5; Fig. 1.1, p. 4)

1. Chemical Level

a. Atoms

b. Molecules

2. Cellular Level

3. Tissue Level

4. Organ Level

5. Organ System Level

6. Organismal Level

B. Summary of Body Organ Systems (pp. 6-7; Fig. 1.2, pp. 6-7)

III. Maintaining Life (pp. 5-9)

A. Functional Characteristics (pp. 5-9)

1. Interrelationships Among Body Organ Systems (Fig. 1.3, p. 8)

2. Maintenance of Boundaries

3. Movement

4. Responsiveness (Irritability)

5. Digestion

6. Metabolism

7. Excretion

8. Reproduction

9. Growth

B. Survival Needs (p. 9)

1. Basic Goal — Maintain Life

2. Nutrients

3. Oxygen

4. Water

5. Maintenance of Body Temperature

6. Atmospheric Pressure

IV. Homeostasis (pp. 10-13)

A. Introduction (p. 10)

1. Maintenance of Stable Internal Conditions

2. Dynamic State of Equilibrium

B. General Characteristics of Control Mechanisms (p. 10; Fig. 1.4, p. 10)

1. Communication

a. Receptor

b. Control Center

c. Effector

2. Interrelationships

C. Negative Feedback Mechanisms (p. 10; Fig. 1.5, p. 11)

1. Opposite Directional Change

2. Decrease in Original Stimulus

3. Nervous System Controls

4. Endocrine System Controls

5. Other System Controls

D. Positive Feedback Mechanisms (pp. 12-13; Fig. 1.6, p. 13)

   1. Same Directional Change

   2. Increase in Original Stimulus

   3. Cascade Effect

   4. Cardiovascular System Controls

   5. Reproductive System Controls

E. Homeostatic Imbalance (p. 15)

   1. Effect of Disease

V. The Language of Anatomy (pp. 15-22)

A. Anatomical Position and Directional Terms (p. 15; Table 1.1, p. 20)

   1. Anatomical Position

      a. Human Body Erect

      b. Arms at Sides

      c. Palms Forward

      d. Feet Together

   2. Directional Terms

      a. Superior/Inferior

      b. Anterior/Posterior

      c. Medial/Lateral/Intermediate

      d. Proximal/Distal

      e. Superficial/Deep

B. Regional Terms (p. 13; Fig. 1.7, p. 14)

C. Body Planes and Sections (pp. 14-17; Fig. 1.8, p. 16)

   1. Planes

      a. Sagittal/Midsagittal/Parasagittal

      b. Frontal (Coronal)

      c. Transverse (Horizontal)

      d. Oblique

   2. Sections

D. Body Cavities and Membranes (pp. 17-21; Fig. 1.9, p. 17)

   1. Dorsal Body Cavity

      a. Cranial Cavity

      b. Vertebral (Spinal) Cavity

   2. Ventral Body Cavity

      a. Thoracic Cavity—Pleural/Pericardial

      b. Abdominopelvic Cavity—Abdominal/Pelvic

      c. Visceral Organs

      d. Diaphragm

       3. Serous Membranes (Fig. 1.10, p. 20)

          a. Parietal

          b. Visceral

          c. Serous Fluid

          d. Naming of Membranes

          e. Inflammation of Membranes

       4. Other Body Cavities

          a. Oral and Digestive Cavities

          b. Nasal Cavity

          c. Orbital Cavities

          d. Middle Ear Cavities

          e. Synovial Cavities

   E. Abdominopelvic Regions and Quadrants (pp. 21-22)

       1. Regions (Fig. 1.11, p. 21)

          a. Umbilical

          b. Epigastric

          c. Hypogastric

          d. Iliac (Right/Left)

          e. Lumbar (Right/Left)

          f. Hypochondriac (Right/Left)

       2. Quadrants (Fig. 1.12, p. 22)

          a. Right Upper

          b. Left Upper

          c. Right Lower

          d. Left Lower

Cross References

1. Basic chemical and physical principles that provide some of the groundwork for an understanding of physiological topics are presented in Chapter 2, pp. 5-53.

2. Additional information on the cellular level of structural organization is presented in Chapter 3, pp. 61-93.

3. Additional information on the tissue level of structural organization can be found in Chapter 4, pp. 103-130. The enhancement of labor contractions as an example of positive feedback is shown in Fig. 29.16, p. 1019.

5. The serous membranes of the abdominal cavity are described in Chapter 24, pp. 792-793.

6. The organs of the mediastinum are further described in Chapter 23, p. 754.

7. Hormonal control as an example of feedback regulation is examined in Chapter 17, pp. 551-553.

Transparencies Index

Bassett Atlas Figures Index

Slide Number	Figure Number	Description
14	2.3A	Sagittal section, head and neck
24	3.8A	Heart and pericardium
30	4.3	Abdominal organs in situ, omentum up
49	5.6A	Sagittal section, female pelvis
71	7.3A	Hip Joint

INSTRUCTIONAL AIDS

Lecture Hints

1. The Incredible Human Machine is an excellent videotape that offers an exciting overview of many physiological functions. With the help of sophisticated photographic techniques, the wonders of the body's internal world are revealed. The videotape is inexpensive and available from numerous vendors including Carolina Biological. Listed below are alternate methods for using the tape.

   a. Show the entire video during lecture or lab (60 minutes).

   b. Show selected sections of video during an introductory lecture or lab.

   c. Show selected sections as an introduction to each body system.

   d. Place the videotape on reserve in the library or video center and have students view it on their own. This could be required or optional (if optional, encourage viewing by adding bonus points).

2. In order to illustrate the relationship between anatomy and physiology, stress how the function of a structure is often determined by the structure itself. Bring a fork to class and ask students if they can determine its function by observing its shape. Now show the class an object they do not recognize and ask the same question (e.g., a slide or overhead of a Leeuwoenhoek simple microscope). A picture is available in the Carolina Biological catalog.

3. References can be made to levels of organization beyond the organismal level. These are not structural levels like the lower levels, but are of interest to many students. These additional levels could include populations (grouping of interacting members of the same species), communities (grouping of interacting populations), ecosystem (grouping of interacting communities), and the biosphere.

4. The body organ systems are actually an artificial grouping of structures that work toward a common goal. Stress the interrelationship between organs and systems that make the body "work" as an entire unit.

5. At times, students might substitute the term *circulatory system* for *cardiovascular system*. Explain the difference and the relationship to the lymphatic system.

6. The role of negative and positive feedback systems in maintaining or disrupting homeostasis is basic to understanding many of the physiological processes covered throughout the text. Stress the importance of feedback systems throughout the course.

7. Students often equate the term "negative" in feedback systems to something disruptive. This misunderstanding is compounded by the term "positive" also used in feedback. Stress the differences and often give examples such as a thermostat control of house temperature.

Demonstrations/Activities

1. Audio-visual materials of choice.

2. Explore the meaning of "alive."

3. Ask the students to explain how scratching an itch can be considered an example of negative feedback.

4. Assume the anatomical position and ask why this particular position is important to the study of anatomy. Then relate that any position would be acceptable as long as it was the standard for anatomical description.

5. Place a chair center stage. Ask a student to indicate how the chair would be cut in the different planes of section. The answer should include why the other options were not selected.

6. To illustrate the different degrees of protection in the dorsal and ventral cavities, ask the questions:

   a. Why do you suppose that a dog instinctively curls up to protect its abdomen?

   b. Two people have rapidly growing tumors: one in the dorsal cavity, the other in the ventral. Which one would develop symptoms first?

7. To encourage understanding of structure/function relationships, ask students to comment on the relationship between muscle and bone; the respiratory and circulatory systems.

8. Arrange for the class to attend an autopsy (after the material in Chapter 1 has been covered).

9. Use a balloon to illustrate the two layers of a serous membrane.

10. Use a torso model and/or dissected animal model to exhibit body cavities, organs, and system relationships.

11. Dissect an animal such as a preserved cat, fetal pig, large rat, etc., to illustrate anatomical relationships.

12. Use a short piece of soaker hose to illustrate a capillary as the "functional unit" of the circulatory system, noting how it is able to communicate with the interstitial spaces.

13. Use the thermostat found in the classroom (or one found in a home) to illustrate how a negative feedback system works.

Critical Thinking/Discussion Topics

1. Discuss how our intercellular environment can be described as the "sea within us."

2. List several embryonic features that form early in the developmental stages but are "lost" or converted to entirely new structures such as our "tail" (coccyx).

3. If an object were found on Mars that appeared to move and react to external stimuli, what other characteristics would be necessary to classify it as "live" and why?

4. Contrast the type of imagery obtained with X-ray machines, CT scans, DSR scans, and ultrasonics.

5. What differences are there between a free-living, single-celled organism such as a paramecium and a single human cell such as a ciliated cell of the respiratory tract?

6. Student assignment for class discussion: Bring in an article from a popular magazine (*Time, Newsweek,* etc.) that describes some environmental problem such as toxic waste disposal, pollution of the ocean, etc., that threatens human homeostasis, and be prepared to describe the problem and how it represents a threat to the body.

Library Research Topics

1. Research the historical development of anatomy and physiology.

2. Review the current definitions of death and life.

3. Develop a rationale for the chemical basis of stress and how it can affect homeostasis.

4. Explore the current research on aging and describe the effect of aging on the genetic material of the cell.

Audio-Visual Aids/Computer Software

Videotapes

1. Your Body Video Series (CA, FC010, FC 020, FC 012). Available as filmstrip or VHS. Provides overview of the human body and all systems.

2. The Incredible Human Machine (CBS, 60 min., C)

Computer Software

1. The Body Transparent (CA, DW100, Apple or IBM). An introduction to anatomy and physiology in a game format.

2. BODCAT: Human Body Simulation (CA, AT100, Apple). Student can view, interact with, and control a simulated body.

3. The Body in Focus (CA, SC76684, Apple or IBM). Self-paced exploration into the human anatomy.

4. The Human Body: An Overview (CA, BW100, Apple). Graphic view of the systems of the human body.

5. Human Systems (EI, C4350, Apple). Review of the body systems.

   See Guide to Audio-Visual Resources on p. 339 for key to AV distributors.

LECTURE ENHANCEMENT MATERIAL

Clinical and Related Terms

1. Autopsy — a medical examination of a body after death to determine the cause of death. Autopsies are also performed to aid medical understanding and to help interpret disease processes.

2. Dynamic Spatial Reconstructor (DSR) — a recently developed, specialized X-ray machine capable of creating moving, three-dimensional images of internal organs from virtually any direction. The machine can provide sectional views of organs with stop-action, enlargements, and slow-motion viewing.

3. Ectomorph — a body type in which there is an emphasis on ectodermal tissues and little fat deposition. Individuals of this type tend to be lean, tall, and delicate with long limbs and a short trunk.

4. Endomorph — a body type in which there is an emphasis on endodermal tissues and fatty deposits. Individuals of this type tend to be fat, with smooth rounded contours, short limbs, and a large trunk.

5. Fluoroscope — a device used to visualize the internal organs by means of an X-ray machine and a fluorescent screen. The object to be viewed is placed between the two devices, allowing a shadowy image to be seen on the screen.

6. Mesomorph — a body type in which there is an emphasis on mesodermal tissues and a moderate amount of fatty tissue. Individuals of this type tend to be stocky with heavy muscles, good bone structure, and prominent forearms and legs.

7. Necropsy — a synonym of autopsy.

8. Somatotype — a term used to describe a particular body type or build.

9. Thermograph — a device that detects and records the heat emanating from an area or body part.

10. Vesalius, Andreas — a famous sixteenth-century Flemish anatomist, surgeon, and medical illustrator who is regarded as the founder of modern anatomy.

11. Vital signs — basic observations made by physicians and nurses when working with patients. The signs include measuring and/or observing the respiratory rate, pulse, and temperature.

12. Vivisection — a surgical dissection performed upon a living animal (usually under anesthesia) for the purpose of physiological investigation and for the study of disease.

ANSWERS TO END-OF-CHAPTER QUESTIONS

Multiple Choice/Matching

1. c
2. a
3. e
4. a, d

5. (a) wrist; (b) hipbone; (c) nose; (d) toes; (e) scalp
6. c, d

7. (a) dorsal; (b) ventral; (c) dorsal; (d) ventral; (e) ventral
8. b

Short Answer Essay Questions

10. Since function (physiology) reflects structure, structure will determine and/or influence function. (p. 4)

11. See Fig. 1.2, pp. 10-11. Fig. 1.2 provides a summary of all the organ systems of the body.

12. Nutrients — the chemical substances used for energy and cell building; oxygen — used in the reactions that produce cellular energy; water — the liquid environment necessary for all chemical reactions; body temperature — to maintain the proper temperature for chemical reactions to procede; and atmospheric pressure — to allow gas exchange to occur. (p. 9)

13. It is the ability to maintain relatively stable internal conditions even in the face of continuous change in the outside world. (p. 10)

14. Negative feedback mechanisms operate in the opposite direction to decrease the original stimulus and/or reduce its effects, thus returning the system back to normal. Examples include regulation of body temperature and blood sugar levels. (p. 10)

   Positive feedback mechanisms operate in the same direction to enhance the original stimulus such that the activity is accelerated. Examples include regulations of blood clotting and enhancement of labor contractions. (pp. 12-13)

15. The anatomical position requires the body being erect, the arms hanging at the sides, the palms forward, the thumbs pointing away from the body, and the feet flat to the ground. It is necessary to use this standard position because most directional terms refer to the body in this position, regardless of its actual position. The use of anatomical terms saves a great deal of description and is less ambiguous. (p. 13)

16. A plane refers to an imaginary line, and a section refers to a cut along that imaginary line. (p. 14)

17. a. arm—brachial

    b. thigh—femoral

    c. chest—thoracic

    d. fingers/toes—digits

    e. anterior aspect of knee—patella (p. 14)

18. See Figs. 1.11 (p. 21), and 1.12 (p. 22). The figures and pages illustrate the regions and quadrants and list several organs for each.

Critical Thinking and Application Questions

1. a. Parietal and/or visceral pleural membranes.

    b. The membranes allow the organs to slide easily across the cavity walls and one another without friction.

    c. The organs and membranes stick together and grate against one another creating friction, heat, and pain. (p. 20)

2. a. With age, body organs and control systems become less efficient. The drop in efficiency causes the internal environment to be less and less stable. (p. 12)

    b. Examples include: bone loss with age, possibly due to lack of continued stimulation of osteoblasts by hormones (p. 169); senescent cells contain a chemical which may be responsible for cellular aging (p. 99); decrease in muscle fibers and increase in connective tissue of the skeletal muscles (p. 280); and sclerosis and thickening of heart valve flaps, possibly due to stress of blood flow (p. 630).

3. a. anterior aspect of elbow

    b. dropped his pants

    c. neck (hickey?) (p. 14)

4. Of the procedures listed, MRI would be the best choice because dense structures (e.g., the skull) do not impair the view with this technique, and it is best at producing a high-resolution view of soft tissues, particularly neural tissue. Furthermore, MRI can provide information about chemical conditions in a tissue. Thus, once the suspected tumor is localized, MRI can perform a "metabolic biopsy" to determine if it is cancerous . . . all of this without surgery. (p. 18)

5. When we take a drink, body hydration increases and thirst declines—an example of a typical negative feedback system. If it were a positive feedback system, the body's need for water (and thirst) would increase after taking a drink. (pp. 12-13)

Key Figure Questions

Figure 1.5 1. The liver. 2. Declining blood sugar levels. 3. Rising blood glucose levels. **Figure 1.6** 1. This is a positive feedback mechanism because the response to the stimulus amplifies the stimulus. 2. The cascade ends when the break in the vessel is sealed by the clot. **Figure 1.7** 1. Your armpit. 2. Your posterior skull region. 3. A finger.

Laboratory Correlations

1. Marieb, E. N. *Human Anatomy and Physiology Laboratory Manual: Cat and Fetal Pig Versions.* 3rd. ed. Benjamin/Cummings, 1989.

    Exercise 1: The Language of Anatomy

    Exercise 2: Organ Systems Overview

2. Marieb, E. N. *Human Anatomy and Physiology Laboratory Manual: Brief Version.* 3rd. ed. Benjamin/Cummings, 1992.

    Exercise 1: The Language of Anatomy

    Exercise 2: Organ Systems Overview

SUGGESTED READINGS

---

1. Chewning, E.B. *Anatomy Illustrated.* New York: Simon and Schuster, 1979.

2. Geller, S.A. "Autopsy." *Scientific American* (Mar. 1983).

3. Kennedy, D. Introduction to: "From Cell to Organism: Readings from Scientific American." *Scientific American* (1967).

4. Morris, D. *The Naked Ape.* New York: McGraw-Hill, 1968.

5. Raichle, M.E. "Visualizing the Mind." *Scientific American* 270 (April 1994): 58.

6. Smith, C.E. "Abdominal Assessment, A Blending of Science & Art." *Nursing* (Feb. 1981).

7. Snell, R.S. *Atlas of Clinical Anatomy.* Boston: Little, Brown, 1978.

8   Wilson, D.B., and J.W. Wilson. *Human Anatomy.* 2nd ed. New York: Oxford University Press, 1983.

9. Woodburne, R.T. *Essentials of Human Anatomy.* 7th ed. New York: Oxford University Press, 1983.

2

Chemistry Comes Alive

Chapter Preview

This chapter presents the basics of chemistry and biochemistry (the chemistry of living material), providing the background necessary to understand body functions. Basic chemistry, including the atom, molecules and chemical bonds, along with a description of the most important inorganic and organic molecules, will be provided.

INTEGRATING THE PACKAGE

Suggested Lecture Outline
Part 1: Basic Chemistry

I. Definition of Concepts: Matter and Energy (pp. 25-26)
    A. Matter (p. 25)
        1. States of Matter
    B. Energy (p. 25)
        1. Kinetic vs. Potential Energy
        2. Forms of Energy
            a. Chemical energy
            b. Electrical energy
            c. Mechanical energy
            d   Radiant energy

II. Composition of Matter: Atoms and Elements (pp. 26-29)
    A. Basic Terms (p. 26; Table 2.1, p. 28)
        1. Elements
        2. Atoms
        3. Atomic Symbols
    B. Atomic Structure (pp. 26-27; Fig. 2.1, p. 27)
        1. Nucleus
            a. Protons
            b. Neutrons
        2. Electrons
        3. Planetary Model
        4. Orbital Model
    C. Identifying Elements (pp. 27-28; Fig. 2.2, p. 27)
        1. Atomic Number
        2. Mass Number
            a. Isotopes (Fig. 2.3, p. 29)

3. Hydrogen Bonds (Fig. 2.9, p. 39)
 a. Surface Tension
 b. Intramolecular Bonds

V. Chemical Reactions (pp. 36-39)
 A. Chemical Equations (p. 36)
  1. Reactants
  2. Products
  3. Molecular Formula
 B. Patterns of Chemical Reactions (pp. 36-37; Fig. 2.10, p. 37)
  1. Synthesis (Combination) Reaction
  2. Decomposition Reaction
  3. Exchange (Displacement) Reaction
  4. Oxidation Reduction Reactions
 C. Energy Flow in Chemical Reactions (p. 37)
 D. Reversibility of Chemical Reactions (p. 38)
 E. Factors Influencing the Rate of Chemical Reactions (pp. 38-39)
  1. Temperature
  2. Particle Size
  3. Concentration
  4. Catalysts

Part 2: Biochemistry: The Composition and Reactions of Living Matter

I. Inorganic Compounds (pp. 39-42)
 A. Water (pp. 39-40)
  1. High Heat Capacity
  2. High Heat of Vaporization
  3. Polarity/Solvent Properties
  4. Reactivity
  5. Cushioning
 B. Salts (p. 40; Fig. 2.11, p. 39)
 C. Acids and Bases (pp. 40-42)
  1. Acids
  2. Bases
  3. pH: Acid-Base Concentration (Fig. 2.12, p. 41)
  4. Neutralization
  5. Buffers

II. Organic Compounds (pp. 42-56)
 A. Carbohydrates (pp. 43-45; Fig. 2.13, p. 44)
  1. General Characteristics
  2. Monosaccharides
  3. Disaccharides
  4. Polysaccharides
  5. Carbohydrate Functions

B. Lipids (pp. 45-47; Table 2.2, p. 46; Fig. 2.14, p. 47)

    1. General Characteristics

    2. Neutral Fats

    3. Phospholipids

    4. Steroids

    5. Eicosanoids

C. Proteins (pp. 47-53)

    1. General Characteristics

    2. Amino Acids and Peptide Bonds (Fig. 2.15, p. 48; Fig. 2.16, p. 49)

    3. Structural Levels of Proteins (Fig. 2.17, p. 49)

        a. Primary Level

        b. Secondary Level

        c. Tertiary Level

        d. Quaternary Level

    4. Fibrous and Globular Proteins (Table 2.3, p. 50)

    5. Protein Denaturation (Fig. 2.18, p. 51)

        a. Active Sites

    6. Enzymes and Enzyme Activity

        a. Characteristics of Enzymes

        b. Cofactors

        c. Activation Energy (Fig. 2.19, p. 52)

        d. Mechanism of Action (Fig. 2.20, p. 53)

    7. Stress Proteins

D. Nucleic Acids (DNA and RNA) (pp. 53-55)

    1. General Characteristics (Table 2.4, p. 55)

    2. Nucleotides

    3. DNA (Fig. 2.21, p. 54)

    4. RNA

E. Adenosine Triphosphate (ATP) (pp. 55-56; Fig. 2.22, p. 55; Fig. 2.23, p. 56)

    1. General Characteristics

    2. High-Energy Phosphate Bonds

Cross References

1. The periodic table is provided in Appendix F.

2. The oxidation-reduction reaction involving the conversion of glucose and oxygen to carbon dioxide and water is described in detail in Chapter 25, pp. 861-866.

3. Acid-base balance, electrolytes, and buffers are described in more detail in Chapter 27.

4. Phospholipids are discussed in relation to the composition and construction of membranes in Chapter 3, pp. 61-65.

5. DNA replication and the relative roles of DNA and RNA in protein synthesis are presented in Chapter 3, pp. 87-95.

6. Sodium and the sodium-potassium pump are described in detail in Chapter 27, pp. 930-939.

7. The importance of ions (minerals) in life processes is described in Chapter 25, Table 25.3, p. 855.

8. The metabolism of carbohydrates, lipids, and proteins is examined in detail in Chapter 25, pp. 858-881.

9. Examples of the basic chemistry of life processes are given in Chapter 25.

10. Digestive enzyme function is described in great detail in Chapter 24, pp. 830-837.

11. Acid function of the digestive system is presented in Chapter 24, pp. 806-810.

12. The digestion of proteins, carbohydrates, and lipids is examined in Chapter 24, pp. 830-837.

13. Acid-base balance is introduced in Chapter 23, pp. 770-772.

14. Steroid and amino-acid-based hormones are described in Chapter 17, pp. 548-553.

15. ATP, ions, and enzymes involved in the nervous impulse are explained in Chapter 11.

16. The function of ATP in muscle contraction is covered in Chapter 9, p. 269.

17. The importance of ions in generating muscle cell contraction is explained in Chapter 9.

18. Cellular ions are described in Chapter 3.

19. Enzymes and proteins in cellular structure/function is presented in Chapter 3.

20. The role of hydrogen bonding in cellular physiology is examined in Chapter 3.

21. Renal control of electrolytes is examined in great detail in Chapter 26.

Laboratory Correlations

1. Marieb, E. N. *Human Anatomy and Physiology Laboratory Manual: Cat and Fetal Pig Versions*. 3rd. ed. Benjamin/Cummings, 1989.

   None

2. Marieb, E. N. *Human Anatomy and Physiology Laboratory Manual: Brief Version*. 3rd. ed. Benjamin/Cummings, 1992.

   None

Transparencies Index

2.2/2.3	Atomic structures/Isotopes of hydrogen	2.15/2.16	Amino acids
2.5	Formation of an ionic bond	2.17	Levels of protein structure
2.6	Formation of a covalent bond	2.18	Protein denaturation
2.12	The pH scale	2.20	Enzyme action
2.13	Carbohydrates	2.21A,B	Structure of DNA
2.14	Lipids	2.22	Structure of ATP

INSTRUCTIONAL AIDS

Lecture Hints

1. *Introduction to Chemistry for Biology Students*, by George Sackheim, is an excellent aid for students who need a quick brush-up in chemistry or for those that need extra help. The book is designed as a self-paced learning guide. Most students should be able to finish a review of the essentials for Marieb Chapter 2 in about two to six hours.

2. An alternative to presenting the chemistry in Chapter 2 as a distinct block of material is to only provide the absolute minimum coverage of the topics at this time. The topics would later be expanded as areas of application were discussed.

3. Students often find the concept of isotopes confusing. A clear distinction between atomic mass and atomic weight will help clarify the topic.

4. In discussing radioisotopes it might be helpful to refer the students back to the discussion of PET scans in A Closer Look in Chapter 1.

5. Oxidation-reduction reactions involve the loss and gain of electrons. The reactant oxidized will lose electrons while the reactant reduced will gain electrons. One easy way to remember this is by using the phrase: Leo the lion goes ger. Leo stands for "loss of electrons is oxidation," and ger for "gain of electrons is reduction."

6. In biological oxidation-reduction reactions the loss and gain of electrons is often associated with the loss and gain of hydrogen atoms. Electrons are still being transferred since the hydrogen atom contains an electron.

7. The relationship between the terms *catalyst* and *enzyme* can be clarified by asking the students if all enzymes are catalysts and if all catalysts are enzymes.

8. Table 2.4, p. 58, is an excellent summary of the differences between DNA and RNA. This information will be important when discussing protein synthesis.

9. The notion that ATP is the "energy currency" of the cell should be emphasized. Students should realize that without ATP, molecules cannot be synthesized or degraded, cells can not maintain boundaries, and life processes would cease.

10. The cycling back and forth between ATP and ADP is a simple but important concept often overlooked by students.

Demonstrations/Activities

1. Audio-visual materials of choice.

2. Obtain a two-foot-long piece of thick string or cord. Slowly twist to exhibit primary, secondary, and tertiary levels of protein organization.

3. Obtain a Thompson-style vacuum tube with an internal frosted plate (to exhibit electrons), a direct current generator (Tesla coil), and bar magnet. Turn off room lights and charge one end of the tube to start electron beam. Use magnet to move electron beam up and down. This experiment helps to illustrate electrons as particles.

4. Obtain an electrolyte testing system (lightbulb setup connected to electrodes) and prepare a series of solutions such as salt, acid, base, glucose, etc. Place the electrodes into the solutions to illustrate concept of electrolytes.

5. Obtain and/or construct 3-D models of various types of biological molecules such as glucose, DNA, protein, and lipids.

6. Bring in materials or objects that are composed of common elements, e.g., a gold chain, coal, copper pipe, cast iron. Also provide examples of common compounds such as water, table salt, vinegar, and sodium bicarbonate. Solicit definitions of *atom, element,* and *compound,* and an explanation of how an atom and a molecule of compound differ.

7. Student assignment for class discussion: Find examples of the uses of radioisotopes in popular articles or nursing journals and be prepared to discuss the advantages and disadvantages of radioisotope use in medicine.

8. Prepare two true solutions (1% sodium chloride; 1% glucose) and two colloidal solutions (1% boiled starch, sol state; Jell-O, gel state). Turn off the room lights and pass a beam of light through each to demonstrate the Tyndall effect of colloids.

9. Obtain two strings of dissimilar "pop-it" beads. Put the beads together to demonstrate a synthesis reaction, and take them apart to demonstrate a decomposition reaction. Take a bead from each different chain and put them together to illustrate an exchange reaction.

10. Use a slinky to demonstrate denaturation of an enzyme. Tie colored yarn on the slinky at two sites that are widely separated, and then coil and twist the slinky upon itself to bring the two pieces of yarn next to each other. Identify the site where the yarn pieces are as the active site. Then remind them that when the hydrogen bonds holding the enzyme (or structural protein) in its specific 3-D structure are broken, the active site (or structural framework) is destroyed. Uncoil the slinky to illustrate this point.

11. Ask students to name all the foods containing saturated fats and all those containing unsaturated fats that they have eaten in the past 24 hours.

Critical Thinking/Discussion Topics

1. Discuss how two polysaccharides, starch and cellulose, each having the same subunit (glucose), have completely different properties. Why can we digest starch but not cellulose?

2. How and why can virtually all organisms—plant, animal, and bacteria—use the exact same energy molecule, ATP?

3. How could a substance such as alcohol be a solvent under one condition and a solute under another? Provide examples of solid, liquid, and gaseous solutions.

4. Describe how weak bonds can hold large macromolecules together.

5. Why can we state that most of the volume of matter, such as the tabletop you are writing on, is actually empty space?

6. When you drive up your driveway at night you see the light from the headlights on the garage door, but not in the air between the car and the door. Why? What would be observed if the night were foggy?

7. Why are water molecules at the surface of a drop of water closer together than those in the interior?

Library Research Topics

1. Explore the use of radioisotopes in the treatment of cancers.

2. Study the mechanisms by which DNA can repair itself.

3. Locate the studies of Niels Bohr concerning the structure of atoms and the location of electrons. Determine why his work with hydrogen gas provided the foundation of our knowledge about matter.

4. How can a donut provide us with so much "energy"? Find out exactly where this energy is coming from.

5. Phospholipids have been used for cell membrane construction by all members of the "cellular" world. What special properties do these molecules have to explain this phenomenon?

6. Virtually every time an amino acid chain consisting of all 20 amino acids is formed in the cell, it twists into an alpha helix, then folds upon itself into a glob. Why?

7. What is the current status of the Human Genome Project? Who is directing the project? What are the expected benefits from the study?

8. What is DNA fingerprinting? Explore the applications of this technology.

Audio-Visual Aids/Computer Software

Videotapes

1. The Chemistry of Life (EI, EP2066, VHS). Introduces the common types of biological molecules. Also available as slides.

2. The Chemistry of Proteins (EI, EP2067, VHS). Covers the general functions of proteins, amino acids, peptide bonds, and protein structure. Also available as slides.

3. The Chemistry of Carbohydrates and Lipids (EI, EP2068, VHS). The structure and role of glucose and other sugars and the various classes of lipids are illustrated. Also available as slides.

4. The Chemistry of Nucleic Acids (EI, EP2073, VHS). Illustrates the molecular structure of bases, nucleotides, base pairing, DNA and RNA and the double helix. Also available as slides.

5. DNA: Master Molecule of Life (EI, FS1255, VHS/BETA). Explains what DNA is and how it functions. An American Chemical Society Presentation. Also available as filmstrip.

6. Enzymes (CBS, 49-8755-V, 10 min., C)

Computer Software

1. Lipids (EI, C4071, Apple). Animated graphics, basic elements of lipids, and structural and functional differences are shown.

2. Proteins (EI, C4072, Apple). Animated graphics, properties of proteins, structural features, and construction of amino acids are illustrated.

3. Nucleic Acids (EI, C4073, Apple). Animated Graphics are used to highlight the distinctive features of DNA and RNA.

4. Carbohydrates (EI, C4074, Apple). Animation is used to illustrate the structure and function of the various types of carbohydrates.

5. Chemicals of Life Series (CBS, 39-8775, 39-8776, 39-8777, IBM). Highly interactive programs covering the structure of matter, carbohydrates, lipids, proteins, and nucleic acids.

See Guide to Audio-Visual Resources on p. 339 for key to AV distributors.

LECTURE ENHANCEMENT MATERIAL

Clinical and Related Terms

1. Artificial sweeteners — synthetic, low-calorie substances used as substitutes for sucrose. An earlier sweetener, Saccharin, tended to have a bitter aftertaste and was proven to be carcinogenic. The newest sweetener, Aspartame (Nutrasweet, Equal), is a synthetic compound of two amino acids (aspartic acid and phenylalanine), is 180 times as sweet as sucrose, and adds only 0.1 calories, for the equivalent of one teaspoon of sugar. Although there is no bitter aftertaste, no evidence of carcinogenesis, and fewer problems with dental caries, persons with low phenylalanine diets (phenylketonuria) or hypersensitivities to aspartame should be cautious with its use.

2. Coenzyme — an organic molecule, such as a dinucleotide, that acts as the active part of an enzyme (apoenzyme). The coenzyme and apoenzyme must combine to function as a complete enzyme (holoenzyme). Coenzymes are essential in cellular metabolism.

3. Curie — a standard unit of radioactivity used to measure the amount of radiation emitted from a radioactive source such as radon. Named after the French chemist, Marie Curie.

4. Dinucleotide — an organic molecule consisting of two nucleotides bonded together. Usually the nitrogen base of one is replaced by a vitamin such as niacin or riboflavin. The molecules act as coenzymes in cellular metabolism. Examples include nicotinamide adenine dinucleotide (NAD) and flavin adenine dinucleotide (FAD).

5. Electron-Momentum Spectrometer (EMS) — a sophisticated spectrometer that measures the final momentum and energy of a probe and target electron. It provides direct observation of specific electron orbitals and reveals detailed molecular structure and bonds.

6. Rad (rad) — radiation absorbed dose. A unit used in measuring the amount of exposure to an ionizing radiation source. The dosage may vary with the type of radiation.

7. Rem (rem) — roentgen equivalent man. The amount of ionizing radiation that has the same biological effect as one rad of X-ray. 1 rem = 1 rad X RBE (relative biological effectiveness).

8. Scanning Tunneling Microscope (STM) — an extremely sensitive scanning probe microscope that can maneuver a tiny metallic needle in three dimensions with the precision of only a few angstroms. As the probe nears a surface, the electrons "tunnel" across the gap. This device can virtually "see" individual atoms and molecules such as DNA and can study surface features such as magnetic fields and temperature.

Disorders/Homeostatic Imbalances

1. Arteriosclerosis — a condition related to the thickening, hardening, and/or loss of elasticity of the walls of arteries. The cause is unknown; however, altered lipid metabolism, excessive cholesterol and fat uptake, hypertension, and obesity are factors related to it. Treatment and/or prevention involves regular exercise, change in diet to low amounts of cholesterol and fat, moderate use of tobacco and alcohol, and a reduction in stress.

2. Atherosclerosis — a form of arteriosclerosis involving changes in the walls of arteries but not arterioles. Fatty, fibrous, cellular, and calcium deposits form plaques in the walls. The cause is unknown; however, the risk factors are the same as for arteriosclerosis. A genetic basis has been demonstrated. Treatment and prevention is the same as for arteriosclerosis.

3. Galactosemia — a genetically inherited error of metabolism seen as an inability to convert galactose found in the milk sugar, lactose, to glucose due to the absence of the enzyme galactokinase transferase or to a galactokinase deficiency. Diagnosis of the disease requires testing the child's urine for the presence of galactose and/or galactose-1-phosphate. Symptoms include the failure to thrive within a week after birth due to anorexia, vomiting, and diarrhea. Treatment involves the avoidance of all lactose- and galactose-containing foods in the diet, since the disease may result in mental retardation.

4. Kwashiorkor — a severe protein deficiency syndrome occurring in infants and young children, especially in underdeveloped countries, due to a deficiency in the quality and quantity of dietary protein. The symptoms include edema, impaired growth and development, weakness, and distention of the abdomen. Treatment requires the addition of enriched foods that contain proteins complete with all 20 amino acids.

5. Phenylketonuria, PKU — a genetically inherited disease caused by the body's inability to metabolize phenylalanine to tyrosine due to the lack of the enzyme phenylalanine hydroxylase. If untreated, severe brain damage and mental retardation could occur. The disease can be detected by testing a urine sample with Phenistix shortly before birth of the child. Symptoms include tremors, convulsions, mental deficiency, and an offensive odor of the urine and sweat. Treatment involves the reduction of phenylalanine in the diet by using preparations such as Lofenalac.

ANSWERS TO END-OF-CHAPTER QUESTIONS

Multiple Choice/Matching

1. b, d	7. c, d	13. (a) a; (2) c	19. a
2. d	8. b	14. c	20. b
3. b	9. a	15. d	21. b
4. a	10. a	16. e	22. c
5. b	11. b	17. d	
6. a	12. a, c	18. d	

Short Answer Essay Questions

23. Energy is defined as the capacity to do work, or to put matter into motion. Energy has no mass, doesn't take up space, and can be measured openly by its effects on matter. Potential energy is the energy an object has because of its position in relation to other objects. Kinetic energy is energy associated with a moving object. (p. 25)

24. According to the First Law of Thermodynamics, energy cannot be created or destroyed. Therefore, energy is not really lost, but may be released in another form such as heat or light. In this form, the energy may be partly unusable. (p. 25)

25. a. Ca   c. H   e. N   g. K   b. C   d. Fe   f. O   h. Na (p. 28)

26. a. All three are carbon with six protons. (p. 28)

    b. All possess different numbers of neutrons and therefore have a different atomic mass. (p. 29)

    c. Isotopes (p. 32)

    d. See Fig 2.4, p. 34. The figure provides a drawing of both a planetary model and an orbital model.

27. a. Add molecular weight of all atoms: 9 x 12 (C) + 8 x 1 (H) + 4 x 16 (O).

    b. Total molecular weight equals the number of grams in one mole.

    c. Divide the number of grams in the bottle by the number of grams in one mole of aspirin and that equals the total number of moles in the bottle.

    d. Answer = 2.5 moles (p. 30)

28. a. Covalent

    b. Covalent

    c. Ionic (pp. 32-35)

29. Hydrogen bonds are weak bonds that form when a hydrogen atom, already covalently linked to an electronegative atom, is attracted by another electronegative atom. Hydrogen bonding is common between water molecules, and in binding large molecules such as DNA and protein into specific three-dimensional shapes. (pp. 35-36)

30. a. The reversibility of the reaction can be indicated by double arrows of equal length.

    b. When arrows are of equal length the reaction is at equilibrium.

    c. Chemical equilibrium is reached when, for each molecule of product formed, one product molecule breaks down, releasing the same reactants. (p. 38)

31. a. Primary structure — linear molecule formed by peptide bonds; second structure — coiling of primary structure into helix; tertiary structure — folding of helical coils.

    b. Secondary level.

    c. Functional proteins achieve the tertiary level and tend to operate independently rather than in combination with others such as with structural proteins. (p. 49)

32. Dehydration refers to the joining together or synthesis of two molecules by the removal of water. Monosaccharides are joined to form disaccharides and amino acids are joined to form dipeptides by this process. Hydrolysis refers to the breakdown of a larger molecule such as a disaccharide into small molecules or monosaccharides by the addition of water. (p. 40)

33. Enzymes are highly specific biological catalysts that help to increase the rate of reactions. The exact mechanism of how enzymes decrease activation energy is not known; however, they decrease the randomness of reactions by binding specifically and temporarily to the reacting molecules. (pp. 51-53)

34. Proteins that aid the folding of other proteins into their functional three-dimensional structures. They are produced in great amounts when cells are damaged and proteins are denatured and must be replaced. (p. 53)

Critical Thinking and Application Questions

1. In a freshwater lake, there are comparatively few electrolytes (salts) to carry a current away from a swimmer's body. Hence, the chance of a severe electrical shock if lightning hit the water is real. (p. 40)

2. a. Some antibiotics compete with the substrate at the active site of the enzyme. This would tend to reduce the effectiveness of the reaction.

   b. Since the bacteria would be unable to catalyze the essential chemical reactions normally brought about by the "blocked" enzymes, the anticipated effect would be the inhibition of its metabolic activities. This would allow white blood cells to remove them from the system. However, some human cells would also be affected and could cause them to cease their functions, hopefully only temporarily. (pp. 51-53)

3. a. pH is defined as the measurement of the hydrogen ion concentration in a solution. The normal blood pH is 7.4.

   b. Severe acidosis is critical because blood comes in contact with nearly every body cell and can adversely affect the cell membranes, the function of the kidneys, muscle contraction, and neural activity. (pp. 40-42)

Key Figure Questions

Figure 2.8 In an XY covalent bond, one end of the molecule is slightly positive and one end is slightly negative. In an XX covalent bond, the charge centers coincide. **Figure 2.10** Reaction B, a decomposition reaction. **Figure 2.18** Enzyme-substrate bonding.

SUGGESTED READINGS

---

1. Cech, T.R. "RNA as an Enzyme." *Scientific American* 255 (Nov. 1986): 64-75.

2. Doolittle, R.F. "Proteins." *Scientific American* 253 (Oct. 1985): 88-99.

3. Horgan, J. "In the Beginning." *Scientific American* 264 (Feb. 1991): 116-125.

4. Karplus, M., and J.A. McCammon. "The Dynamics of Proteins." *Scientific American* 254 (Apr. 1986): 42-51.

5. Koshland, D.E., Jr. "Protein Shape and Biological Control." *Scientific American* 229 (Oct. 1973): 52-64.

6. Miller, J.A. "The Gene Idea." *Science News* 121 (Mar. 1983): 180-182.

7. Mistry, N.B., et al. "Particles with Naked Beauty." *Scientific American* 249 (July 1983): 106-115.

8. Preuss, P. "The Shape of Things to Come." *Science* 83 (Dec. 1983): 81-87.

9. Sochurek, H. "Medicine's New Vision." *National Geographic* 171 (Jan. 1987): 2-40.

10. Trefil, J. "The Search for Truth." *Discover* 10 (Dec. 1989): 56-61.

11. Weinberg, R.A. "The Molecules of Life." *Scientific American* 254 (Apr. 1986): 48-57.

12. Welch, W.J. "How Cells Respond to Stress." *Scientific American* 268 (May 1993): 56.

3

Cells: The Living Units

Chapter Preview

This chapter focuses on structural characteristics and functional processes common to all of our cells. The plasma membrane's structure and function, the cytoplasm, the cellular organelles, and the nucleus are analyzed in detail. Specialized cells and their unique functions are considered in detail in later chapters.

INTEGRATING THE PACKAGE

Suggested Lecture Outline

I. Overview of the Cellular Basis of Life (pp. 60-61)
    A. Historical Perspectives (p. 60)
    B. Basic Concepts (p. 61)
    C. Cellular Properties (p. 61)
    D. The Generalized Cell (p. 61; Fig. 3.1, p. 62)

II. The Plasma Membrane: Structure (pp. 61-65)
    A. The Fluid Mosaic Model (pp. 61-63; Fig. 3.2, p. 63)
        1. Basic Features
        2. Phospholipid Bilayer
        3. Integral Proteins (Fig. 3.3, p. 64)
        4. Peripheral Proteins
        5. Glycocalyx
    B. Specializations of the Plasma Membrane (p. 64)
        1. Microvilli
        2. Membrane Junctions (Fig. 3.4, p. 65)
            a. Basic Characteristics
            b. Tight Junctions
            c. Desmosomes
            d. Gap Junctions

III. The Plasma Membrane: Functions (pp. 65-75)
    A. Membrane Transport (pp. 65-73; Table 3.1, p. 70)
        1. Basic Characteristics
        2. Passive Processes
            a. Basic Characteristics
            b. Simple Diffusion (Fig. 3.6, p. 67)
            c. Osmosis (Fig. 3.7, p. 68)

Review Items

9. Comparison of DNA and RNA (Table 2.4, p. 55)

10. Hydrogen bond (Chapter 2, p. 35; Fig. 2.9, p. 36)

Cross References

1. A detailed discussion of aerobic cellular respiration is presented in Chapter 25.

2. The role of smooth ER in calcium ion storage and release is discussed in Chapter 9.

3. Lysosomal rupture (autolysis) and the self-digestion of cells is further discussed in relation to autoimmune diseases in Chapter 8.

4. mRNA codons and the amino acids they specify are listed in Appendix B.

5. Cell division in relation to hereditary process is presented in Chapter 30, pp. 1029-1030.

6. Reproductive cell division and gamete production is detailed in Chapter 28.

7. Tight junctions and the blood/testis barrier is presented in Chapter 28, p. 963.

8. The functions of flagella and cilia are illustrated in Chapter 28, p. 963.

9. Mitochondria and energy production in sperm cells is mentioned in Chapter 28, p. 963.

10. Membrane transport related to electrolyte and water balance is described in Chapter 27.

11. Hydrostatic pressure and the movement of fluid through membranes is described in Chapter 26, p. 905.

12. Examples of membrane transport are presented in Chapter 25.

13. Microvilli and the greatly increased absorptive surface area in the epithelial cells of the small intestine are described in Chapter 24, p. 814.

14. Membrane transport related to the absorption of digested substances is examined in Chapter 24, pp. 834-837.

15. The function of lysozyme in protection of the body is described in Chapter 22, p. 709.

16. The diffusion of respiratory gases is described in great detail in Chapter 23, pp. 763-767.

17. The function of cilia in nonspecific defense of the body is presented in Chapter 22, p. 709.

18. Cell junctions and the movement of substances through capillary walls is further explained in Chapter 20, p. 664.

19. Cell junctions and cardiac function are examined in Chapter 19.

20. Membrane receptors and functions related to the autonomic nervous system are described in Chapter 14.

21. Specialized forms of cytoskeletal elements are described in Chapter 11, p. 343

22. Nervous system membrane potentials are described in detail in Chapter 11, pp. 351-370.

23. Microfilaments as contractile elements are covered in Chapter 9, p. 253.

Laboratory Correlations

1. Marieb, E. N. *Human Anatomy and Physiology Laboratory Manual: Cat and Fetal Pig Versions.* 3rd. ed. Benjamin/Cummings, 1989.

   Exercise 3: The Microscope

   Exercise 4: The Cell — Anatomy and Division

   Exercise 5: The Cell — Transport Mechanisms and Cell Permeability

2. Marieb, E. N. *Human Anatomy and Physiology Laboratory Manual: Brief Version*. 3rd. ed. Benjamin/Cummings, 1992.

    Exercise 3: The Microscope

    Exercise 4: The Cell — Anatomy and Division

    Exercise 5: The Cell — Transport Mechanisms and Cell Permeability

Transparencies Index

3.1	Structure of the generalized cell	3.17	Protein packaging in Golgi apparatus
3.2	Structure of the plasma membrane	3.20	Cytoskeleton
3.6	Diffusion through the plasma membrane	3.22	Cilia structure and function
		3.25	Chromosome structure
3.9	Sodium-potassium pump	3.27	Protein synthesis
3.10/3.11	Exocytosis and endocytosis	3.28	Mitosis
3.13	G protein–linked receptor mechanism	3.29	Information flow from DNA to protein

---

INSTRUCTIONAL AIDS

Lecture Hints

1. It is important to stress the distinction between passive and active processes.

2. Students will often equate pinocytosis with "cell drinking" and forget that the importance of the process is taking in the dissolved solutes in the fluid rather than the solvent itself.

3. The generation and maintenance of a resting membrane potential can be covered now or postponed until the contraction of a skeletal muscle fiber is presented in Chapter 9.

4. Explain that the nuclear membrane is a double membrane and that each membrane is a phospholipid bilayer. Some students will equate phospholipid bilayer and double membrane.

5. As a method of review, the cellular organelles can be grouped as being either membranous, microtubular, or "other."

6. Clarify the distinction between centrioles and centromeres.

7. Clearly distinguish between mitosis and cytokinesis.

8. In order to reinforce the idea of complementary base pairing, point out that cytosine and thymine are pyrimidines (single ring structures) and guanine and adenine are purines (double ring structures). For proper spacing it is necessary to combine a purine with a pyrimidine for each step in the DNA "ladder" (a three ring wide step). Furthermore, point out that adenine-thymine form two hydrogen bonds and cytosine-guanine form three hydrogen bonds.

9. Students can readily remember the complementary base pairing of cytosine-guanine by pointing out the similarity of G and C.

Demonstrations/Activities

1. Audio-visual materials of choice.

2. Use an animal cell model to demonstrate the various organelles and cell parts.

3. Use a set of transparencies or slides of electron micrographs to illustrate subcellular organelles.

4. Use a chocolate chip cookie to illustrate the composition and characteristics of the cell membrane. Hint: chocolate chips = proteins, cookie dough = lipids. Any blond hairs coming out of the cookies could even illustrate glycolipids! Be sure to point out that the membrane is not actually rigid, but that the "chocolate chips should really be floating freely in the fluid cookie dough."

5. Set up models of DNA and RNA to illustrate complementary base pairing.

6. Extract DNA from a beaker of lysed bacterial cells using a glass rod to illustrate the fibrillar nature of the molecule.

7. Use models of chromosomes with detachable chromatids to illustrate mitotic phases.

8. Ask students to name examples of diffusion, osmosis, and filtration commonly found in daily life.

9. Secure a glass funnel containing filter paper over a beaker. Illustrate how greater fluid pressure (provided by more fluid in the funnel) leads to faster filtration.

10. Set up one or more of the following simple diffusion demonstrations:

    a. Place a large histological dye crystal on the center of an agar plate a few hours before lecture. A ring of color will appear radiating from the crystal. The plate can be displayed on an overhead projector.

    b. Place a crystal of dye in a beaker or water and display on an overhead projector. A dramatic demonstration that allows students to easily relate to the process involved.

    c. Use a bottle of perfume (or other substance) to illustrate diffusion in the classroom. Don't announce its use until it has diffused.

11. A simple osmometer: Place a glucose solution in a dialysis sac and tie securely to a length of glass tubing. Secure the tubing with a stand and clamp so that the dialysis bag is immersed in distilled water. Have students observe the fluid level in the tube over time.

12. If a microscope/TV camera system is available (or a microprojector), set it up to show the effects of: (a) physiologic saline, (b) hypertonic saline, and (c) distilled water on red blood cells.

Critical Thinking/Discussion Topics

1. Cells tend to have a relatively small and uniform size. Why aren't cells larger? Discuss your answer.

2. What are the advantages and disadvantages of asexual reproduction? Is mitosis an asexual reproductive method?

3. What is the value of start and stop signals in mRNA?

4. Why have certain cells of the body, such as muscle and nerve cells, "lost" their ability to divide?

5. Why must each daughter cell produced by mitosis have mitochondria?

6. Procaryotic cells (bacteria) do not have mitochondria. Where are the enzymes of ATP production? How does this support the theory that mitochondria arose from ancient bacteria that "infected" the ancestors of animal cells?

7. Use the mathematical equations for surface area and volume determination to show that volume increases faster than surface area.

8. Why is damage to the heart more serious than damage to the liver (or other organ)?

9. Start with a cell containing 24 (or any hypothetical number you wish) chromosomes, and in each stage of mitosis predict the number of chromosomes and chromatids present.

10. Why is precise division of chromosomes during mitosis so important?

11. What could be the evolutionary advantage of genetically programming cellular aging?

Library Research Topics

1. Receptor mediated endocytosis is a highly selective mechanism of ingesting molecules. How could it be used to kill cancer cells?

2. Why do we age? What appears to initiate the aging process and do we have any cellular mechanisms that control or facilitate this process?

3. Are all cancers caused by carcinogens? What other substances can cause cancer?

4. How can hybridomas aid research techniques and facilitate in our understanding of the immune system?

5. Can protein molecules move within the cell membrane? What research supports your finding?

6. Many genetic diseases are caused by mutations that change the sequence of the nitrogen bases in the DNA. How many codons are changed in the genetic disease sickle-cell anemia? What amino acid is substituted in the hemoglobin because of this mutation?

7. During the past few years experimental implants of fetal tissue have been used for treatment of brain disorders such as Parkinson's disease. What is the current status of such experimentation? What are some of he moral, ethical, and legal concerns involving such experimentation?

8. How has the advent of recombinant DNA techniques aided in our understanding of proteins such as interferon, insulin, and interleukins?

9. Compare and contrast procaryotic and eucaryotic cells.

10. Compare and contrast plant and animal cells.

11. Summarize the history of the discovery of the structure of DNA by reviewing the two following books:

    Sayre, A. *Rosalind Franklin and DNA*. New York: Norton and Company, 1975.

    Watson, J.D. *The Double Helix*. New York: Norton and Company, 1980.

Audio-Visual/Computer Software

Slides

1. Chromosome Movement Set (CBS, 48-1147)

2. Chromosome Structure Set (CBS, 48-1145)

3. Inside the Cell: Microstructures, Mechanisms, and Molecules Set (CBS 48-1127b)

4. Membrane Set (CBS, 48-1135)

5. Onion Mitosis Set (CBS, 48-1135)

6. Whitefish Mitosis Set (CBS, 48-1132)

Videotapes

1. Aging (FHS, QB-845, VHS/BETA). Program covers the physical processes of aging and examines changes in various body systems.

2. Cancer: Fundamental Ideas (CBS, 22 min., 49-8419-V, VHS)

3. The Cell Cycle, Mitosis, and Cell Division (CBS, 21 min., 49-8108-V, VHS)

4. Cell Motility (CBS, 46 min., 49-8109-V, VHS)

5. Cell Structure and Function (EI, EP-2069, VHS). Compares procaryotic and eucaryotic cells and organelles.

6. Membranes (CBS, 41 min., 49-8121-V, VHS)

7. Origin of Cellular Life (EI, EP-2070, VHS). Introduces current hypotheses regarding the steps in the origin of life.

Computer Software

1. Cell Functions: Growth and Mitosis (CCMI, IBM). Tutorial program reviewing cell functions, growth, and mitosis.

2. DNA: The Master Molecule (CBS, 39-9070, Apple, IBM). Presentation of the structure and function of DNA by means of interactive tutorials and simulations.

3. Mitosis (EI, C-3025, Apple). Graphics and text combine to guide student through phases of mitosis.

4. Modern Genetics: Chromosomes and Coding (CCMI, IBM). Explains the replication of DNA, protein synthesis, chemical makeup of a chromosome, and inheritance of mutations.

5. Nucleic Acids (CBS, 39-8792, Apple). Presents the distinct features of nucleic acids and compares DNA and RNA.

6. Osmosis (CBS, 39-8758, Apple). Uses graphics to simulate the flow of solute/solvent across a semipermeable membrane.

7. Osmotic Pressure (COND, Apple). Simulates the development of osmotic pressure due to varying types of solutes and solute concentration.

8. Protein Synthesis (CBS, 39-8791A, Apple). Proteins and Nucleic Acids are briefly discussed followed by a complete presentation of protein synthesis.

9. Protein Synthesis (EI, C-4552, Apple). Animated graphics used to explain complex process of protein synthesis.

See Guide to Audio-Visual Resources on page 339 for key to AV distributors.

LECTURE ENHANCEMENT MATERIAL

Clinical and Related Terms

1. DNA helicase — a DNA replication enzyme that functions in the unwinding of the DNA helix at the DNA replication fork.

2. DNA ligase — a DNA repair enzyme that seals any nicks that may have formed in the DNA helix during replication or other cellular activity.

3. Geriatrics — the study of the problems of aging. It is related to the field of gerontology.

4. Helix-destabilizing protein — a DNA replication enzyme that functions by binding to the unwound single strand of the replicating DNA. Molecules of helix-destabilizing protein form long rows on the single stranded DNA and thus expose the strand for base-pairing under the control of the DNA polymerase enzyme.

5. Hybridoma — a hybrid cell formed by the fusion of a normal cell such as a lymphocyte and a tumor cell. Most hybridomas are formed by fusion with antibody-producing cells that can then form large quantities of highly specific monoclonal antibodies for indefinite periods of time.

6. Metaplasia — a progressive change in a group of cells to an abnormal form for that tissue.

7. Microtomography — a technique that combines the high-magnification capabilities of the electron microscope with the three-dimensional imagery of computed tomography (CT) to obtain high-quality images of living cells. Applications include studies of cellular movement, cancerous cell growth, and the effects of antibiotics and antitumor drugs on cells.

8. Mutagen — any substance that can induce mutations in the DNA of cells. Mutagenic agents include X-rays, UV light, nicotine, and various petroleum products.

9. Recombinant DNA — genetic units of DNA that are removed from one organism and inserted into the DNA of another. Common techniques involve the use of fragments of human or animal DNA that is spliced into bacterial chromosomes, usually in units called plasmids. Upon appropriate stimulation, the bacterium can produce large quantities of the spliced DNA gene product.

10. Ringer's Solution — a sterile, isotonic solution of sodium chloride, calcium chloride, and potassium chloride in water. Used as a physiological salt solution for topical use for burns and surgical wound cleansing.

Disorders/Homeostatic Imbalances

1. Progeria — premature senility occurring in early childhood. The syndrome is characterized by small size, absence of facial and pubic hair, wrinkled and dry skin, gray hair, and atherosclerosis. There is no known etiology or treatment.

2. Tay Sach's Disease — a genetically inherited metabolic disorder involving a specific population group that is characterized by marked degeneration of brain tissue, dementia, blindness, and death. The disorder involves a deficiency of the enzyme hexosaminidase A that results in the accumulation of lipids in certain brain cells. A test is available to detect the disease in fetuses. No treatment or therapy is currently available.

ANSWERS TO END-OF-CHAPTER QUESTIONS

Multiple Choice/Matching

1. d	7. c	13. c
2. a, c	8. d	14. b
3. b	9. a	15. d
4. d	10. a	16. a
5. b	11. b	17. b
6. e	12. d	18. d

Short Answer Essay Questions

19. a. Mitochondria. (p. 77)

    b. Ribosomes, endoplasmic reticulum, and golgi bodies. (pp. 77-79)

    c. Lysosomes — digestive bag of hydrolytic enzymes for intracellular digestion. (p. 80)

    d. Peroxisomes — an organelle that detoxifies harmful and/or toxic substances produced during metabolism. (p. 81)

20. Each daughter cell produced following mitosis is genetically identical to the mother cell. Each cell then, contains part of the original cell, thus a portion of the very first original cell will always be found in each and every daughter cell. In terms of immortality, a cancer cell is truly immortal since it is not bounded by normal cellular constraints and does not actually "age." (p. 89)

21. Red blood cells, when mature, actually eject their nucleus without genetic material to guide repair processes (protein synthesis and metabolism). These cells cannot reproduce and will ultimately die. (p. 84)

22. The "sugar-coated" proteins are called glycoproteins and act as highly specific biological markers that aid cellular interactions. (p. 74) Other markers, the glycolipids, may serve similar functions as surface markers.

23. a. Body fluids, such as interstitial fluids, blood plasma, and cerebrospinal fluid, are important as transport and dissolving media. (p. 95)

    b. Cellular secretions, such as gastric and pancreatic fluids, aid in digestion, while others, such as saliva and mucous, act as lubricants. (p. 95)

    c. Extracellular matrix represents a jellylike substance that acts as a tissue "glue" in all tissues and helps to determine the characteristics of connective tissues. (p. 95)

24. The sodium-potassium pump acts to maintain a polarized state of the membrane by maintaining the diffusion gradient of sodium and potassium ions. The pump couples the transport of sodium and potassium ions so that with each "turn" of the pump, three sodium ions are ejected out of the cell and two potassium ions are carried back into the cell. (p. 71)

Critical Thinking and Application Questions

1. In each case, living cells have been immersed in a hypotonic solution which will result in water entry into the cells. In the case of celery, where the cells are also bounded by cell walls of cellulose, water entry makes the cell "stiff" due to hydrostatic pressure. In the case of skin cells, as water is absorbed, the cells swell causing the skin to take an undulating course to accommodate greater cell volume. (p. 69).

2. By interfering with normal digestion and absorption of food material, the infectious agents are causing the intestinal cell membrane to become impermeable to solute (food) molecules in the intestines and the solute molecules within the cells. As a result of this situation, the effect of the difference between the intestinal cells' content osmolarity (compartment 1) and the intestinal content osmolarity (compartment 2) will not only prevent water reabsorption by the intestinal cell but will cause water to move rapidly from compartment 1 into compartment 2, resulting in the diarrhea. (pp. 67-70)

3. a. By damaging the mitotic spindle, vincristine will inhibit the proper formation of the microtubules used in pushing the centrioles toward the opposite poles of the cell. Failure to do this will result in the cell being unable to complete its mitotic division process, thus killing the cell. (p. 89)

    b. By binding to the DNA and blocking mRNA synthesis, adriamycin effectively inhibits protein synthesis. Cessation of this process inhibits the cell from replacing enzymes and other proteins required for cellular survival. (pp. 89-95)

Key Figure Questions

Figure 3.1 Events occurring within membrane boundaries can take place unhampered by events occurring outside those membrane boundaries. **Figure 3.2** Filaments of the cytoskeleton. **Figure 3.11** Lungs, open to the external environment, collect dust and other airborne debris. In the lungs of smokers, carbon particles are added to this debris. **Figure 3.12** 1. Both ions are being pumped against their concentration gradient. 2. Sodium ions are being pumped against their electrical gradient.

SUGGESTED READINGS

1. Alberts, B., et al. *Molecular Biology of the Cell.* New York: Garland, 1983.

2. Beardsley, T. "Aging Comes of Age." *Scientific American* 260 (May 1988): 17, 21.

3. Bredt, D.S., and S.H. Snyder. "Biological Roles of Nitric Oxide." *Scientific American* (1993): 22.

4. Bretscher, M.S. "The Molecules of the Cell Membrane." *Scientific American* 253 (Oct. 1985): 100-108.

5. Cech, T.R. "RNA as an Enzyme." *Scientific American* 255 (Nov. 1986): 64-75.

6. Dickerson, R.E. "The DNA Helix and How It Is Read." *Scientific American* 249 (Dec. 1983): 94-111.

7. Feldman, M., and L. Eisenbach. "What Makes a Tumor Cell Metastatic?" *Scientific American* 259 (Nov. 1988): 60-87.

8. Gilman, A.G., and M.E. Linder. "G Proteins." *Scientific American* (1993): 70.

9. Gore, R., B. Dale, and D. Meltzer. "The Awesome Worlds Within a Cell." *National Geographic* 150 (Sept. 1976): 355-395.

10. Grunstein, M. "Histones as Regulators of Genes." *Scientific American* 267 (Oct. 1992): 68.

11. Hakomori, S. "Glycosphingolipids." *Scientific American* 254 (May 1986): 44-53.

12. Kornberg, R.D., and A. Klug. "The Nucleosome." *Scientific American* 244 (Feb. 1981): 52-64.

13. Lancaster, J.R. "Nitric Oxide in Cells." *American Scientist* 80 (May-June 1992): 248.

14. Liotta, L.A. "Cancer Cell Invasion and Metastasis." *Scientific American* 266 (Feb. 1992): 54.

15. Lis, L., and N. Sharon. "Carbohydrates in Cell Recognition." *Scientific American* 268 (Jan. 1993): 82.

16. Loscalzo, J., D.J. Singel, and J.S. Stamler. "Biochemistry of Nitric Oxide and Its Redox-Activated Forms." *Science* 258 (Dec. 1992): 1898.

17. Marx, J.L. "Oxygen-Free Radicals Linked to Many Diseases." *Science* 235 (Jan. 1987): 529-531.

18. McIntosh, J.R., and K.L. McDonald. "The Mitotic Spindle." *Science* 235 (Jan. 1987): 48-56.

19. Miller, J.A. "Between the Cells: Control by Glue." *Science News* 128 (July 1985).

20. Normura, M. "The Control of Ribosome Synthesis." *Scientific American* 250 (Jan. 1984): 102-144.

21. Ostro, M.J. "Liposomes." *Scientific American* 256 (Jan. 1987): 102-111.

22. Pennisi, E. "DNA Repeats Tied to Neuromuscular Diseases." *Science News* 144 (Feb. 1993): 20.

23. Pennisi, E. "New Gene Ties Cancer, Cell Cycle." *Science News* 144 (Nov. 1993): 356.

24. Radman, M., and R. Wagner. "The High Fidelity of DNA Duplication." *Scientific American* 259 (Aug. 1988): 40-47.

25. Rennie, J. "DNA's New Twists." *Scientific American* 266 (March 1993): 122.

26. Ross, J. "The Turnover of Messenger RNA." *Scientific American* 260 (Apr. 1988): 48-55.

27. Rothman, J.E. "The Compartmental Organization of the Golgi Apparatus," *Scientific American* 253 (Sept. 1985): 74-89.

28. Rubenstein, E. "Diseases Caused by Impaired Communication Between Cells." *Scientific American* 242 (Mar. 1980): 102-107.

29. Sloboda, R.D. "The Role of Microtubules in Cell Structure and Cell Division." *American Scientist* 68 (May-June 1980): 290-298.

30. Snyder, S.H. "Nitric Oxide: First in a New Class of Neurotransmitters?" *Science* 257 (July 1992): 494.

31. Steintz, J.A. "snurps." *Scientific American* 258 (June 1988): 56-63.

32. Unwin, N., and R. Henderson. "The Structure of Proteins in Biological Membranes." *Scientific American* 250 (Feb. 1984): 78-94.

33. Weaver, R.F. "The Cancer Puzzle." *National Geographic* 150 (Sept. 1976): 396-399.

34. Weber, K., and M. Osborn. "The Molecules of the Cell Matrix." *Scientific American* 253 (Oct. 1985): 110-120.

35. Welch, W.J. "How Cells Respond to Stress." *Scientific American* 268 (May 1993): 56.

36. Weinberg, R.A. "Finding the Anti-Oncogene." *Scientific American* 259 (Sept. 1988): 44-50.

37. Weintraub, H.M. "Antisense RNA and DNA." *Scientific American* 262 (Jan. 1990): 40-46.

4

Tissues: The Living Fabric

Chapter Preview

This chapter describes the functions, locations, and characteristics of the four main tissue types of the body: epithelial, connective, muscular, and nervous tissues. The correlation between structure and function, and the interrelationships that tissues have with each other will be emphasized.

INTEGRATING THE PACKAGE

Suggested Lecture Outline

I. Introduction to Tissues (p. 102)
    A. Basic Characteristics (p. 102)
    B. Definitions (p. 102)
II. Epithelial Tissue (pp. 103-112)
    A. Features of Epithelia (p. 103)
    B. Special Characteristics of Epithelia (p. 104)
        1. Cellularity
        2. Specialized Contacts
        3. Polarity
        4. Basement Membrane
        5. Avascularity
        6. Regeneration
    C. Classification of Epithelia (pp. 103-112; Fig. 4.1, p. 104)
        1. Criteria
            a. Cellular Shape
            b. Cellular Arrangement (Layers)
            c. Major Classes
        2. Simple Epithelia (Fig. 4.2, pp. 105-106)
            a. Simple Squamous
            b. Simple Cuboidal
            c. Simple Columnar
            d. Pseudostratified Columnar
        3. Stratified Epithelia (Fig. 4.2, pp. 107-109)
            a. Stratified Squamous
            b. Stratified Cuboidal
            c. Stratified Columnar
            d. Transitional

D. Glandular Epithelia (pp. 110-112)
1. Basic Characteristics
2. Endocrine Glands
3. Exocrine Glands
a. Basic Characteristics
b. Unicellular Glands (Fig. 4.3, p. 110)
c. Multicellular Glands (Fig. 4.4, p. 111; Fig. 4.5, p. 112)
III. Connective Tissue (pp. 112-124)
A. Functions of Connective Tissue (p. 112)
B. Common Characteristics of Connective Tissue (p. 113)
1. Common Origin (Fig. 4.6, p. 113)
2. Degrees of Vascularity
3. Extracellular Matrix
C. Structural Elements of Connective Tissue (pp. 113-115)
1. Basic Characteristics
2. Areolar Connective Tissue (Fig. 4.7, p. 114)
3. Ground Substance
4. Fibers
a. Collagen
b. Elastic
c. Reticular
5. Cells
a. Basic Characteristics
b. Primary Blast Cell Types
c. Macrophages
D. Types of Connective Tissue (pp. 115-124; Fig. 4.8, pp. 116-121)
1. Embryonic Connective Tissue: Mesenchyme
a. Mesenchymal Tissue
b. Mucous Connective Tissue
2. Connective Tissue Proper
a. Subclasses of Connective Tissue Proper
b. Areolar
c. Adipose
d. Reticular
e. Dense Regular
f. Dense Irregular
3. Cartilage (Fig. 4.7, pp. 119-120)
a. Basic Characteristics
b. Hyaline
c. Elastic
d. Fibrocartilage

4. Bone (Osseous Tissue) (Fig. 4.7, p. 121)
    a. Basic Characteristics
    b. Cell Types
5. Blood (Vascular Tissue) (Fig. 4.7, p. 121)

IV. Epithelial Membranes (p. 124; Fig. 4.9, p. 125)
    A. Cutaneous Membranes
    B. Mucous Membranes
    C. Serous Membranes

V. Muscle Tissue (pp. 124-126; Fig. 4.10, pp. 126-127)
    A. Basic Characteristics (p. 124)
    B. Types of Muscle Tissue (pp. 124-126)
        1. Skeletal Muscle
        2. Cardiac Muscle
        3. Smooth Muscle

VI. Nervous Tissue (p. 126; Fig. 4.11, p. 128)
    A. Basic Characteristics (p. 126)
    B. Types of Cells (p. 126)
        1. Neurons
        2. Supporting Cells

VII. Tissue Repair (pp. 125-131; Fig. 4.12, p. 129)
    A. General Mechanisms (p. 125)
    B. Steps of Tissue Repair (pp. 128-129)
        1. Inflammation
        2. Organization
        3. Regeneration and/or Fibrosis
    C. Factors Affecting the Repair Process (pp. 129-131)
        1. General Factors
        2. Tissue Type

VIII. Developmental Aspects of Tissues (pp. 131-132)
    A. Embryonic and Fetal Development of Tissues (p. 131; Fig. 4.13, p. 131)
        1. Primary Germ Layer Formation
            a. Ectoderm
            b. Mesoderm
            c. Endoderm
        2. Specialization of Germ Layers
    B. Effect of Aging on Tissues (p. 132)

Review Items

1. The hierarchy of structural organization (Chapter 1, p. 67)
2. Divisions of the ventral body cavity (Chapter 1, p. 21)

Cross References

1. The function of keratin in keratinized stratified squamous epithelium is further explained in Chapter 5.

2. Additional information on ductless (endocrine) glands can be found in Chapter 17.

3. A complete discussion of osseous tissue and the structure and growth of bone is in Chapter 6, pp. 159-172.

4. A complete discussion of cutaneous membrane (skin) is presented in Chapter 5.

5. Skeletal and smooth muscle are presented in greater detail in Chapter 9; cardiac muscle is in Chapter 19.

6. A complete discussion of nervous tissue is given in Chapter 11.

7. The inflammatory and immune responses are considered in detail in Chapter 22.

8. Epithelial cell characteristics related to filtration, secretion, and absorption is described in detail in Chapter 26.

9. Epithelial cells of the absorptive and secretory cells of the digestive tract are described in Chapter 24.

10. Cartilagenous support of respiratory structures is covered in Chapter 23, pp. 747-751.

11. The pseudostratified epithelium in the lining of the trachea is mentioned in Chapter 23, p. 750.

12. Interstitial fluid (generation and removal) is further explained in Chapter 21, pp. 694-697.

13 Reticular connective tissue support of lymphatic tissue is mentioned in Chapter 21, p. 697.

14. The epithelial and connective tissue components of the blood vessels are described in Chapter 20, pp. 644-645.

15. The serous coverings of the heart, the epithelium of the heart, and the use of connective tissue in cardiac valves is examined in detail in Chapter 19.

16. Blood is further described in Chapter 18, pp. 585-598.

17. The function of nervous tissue is presented in detail in Chapters 13 and 19.

18. The connective tissue coverings of the muscles are described in Chapter 9, p. 249.

19. The role of connective tissues in ligaments and tendons is discussed in Chapter 8, pp. 225, 226-227.

20. The formation of osseous tissue is further described in Chapter 6, pp. 163-166.

21. Chondrocytes and cartilage as related to bone formation is presented in Chapter 6, p. 165.

22. The function of the basement membrane in skin is covered in Chapter 5, pp. 136-139.

23. The role of connective tissues in providing strength for the integument is presented in Chapter 5, p. 139.

24. Different exocrine glands found in the skin are examined in Chapter 5, pp. 144-145.

25. The role of cartilage in joint formation is examined in Chapter 8, pp. 227-228.

Laboratory Correlations

1. Marieb, E. N. *Human Anatomy and Physiology Laboratory Manual: Cat and Fetal Pig Versions.* 3rd. ed. Benjamin/Cummings, 1989.

   Exercise 6: Classification of Tissues

2.   Marieb, E. N. *Human Anatomy and Physiology Laboratory Manual: Brief Version*. 3rd. ed.
     Benjamin/Cummings, 1992.

     Exercise 6: Classification of Tissues

Transparencies Index

INSTRUCTIONAL AIDS

Lecture Hints

1.   The relationship between structure and function is important and can be readily illustrated by examples of epithelial tissues. Stress how the multilayered structure of stratified squamous epithelium is much better adapted for surfaces exposed to wear and tear, while simple squamous epithelium is better adapted for filtration.

2.   Stratified squamous epithelium is usually the first of the multilayered epithelial tissues presented. Emphasize that only the surface cells are flattened. The student's first conception is often that the tissue is composed of multiple layers of thin flat cells.

3.   Stress the difference in structure and function between keratinized and nonkeratinized stratified squamous epithelium.

4.   Emphasize the uniqueness of the matrix when explaining the classification of the connective tissues. This will be helpful since this group of tissues seems so diverse to the students. Students may often lose sight of how and why such a diverse group is classified together. Also relate the type of matrix to the specific function of the tissue.

5.   Students are sometimes confused about why collagen and elastic fibers are called white and yellow fibers respectively, even though under microscopic observation they appear to be pink and black respectively. This is because prepared specimens are stained, not natural colors.

6.   Point out that hyaline cartilage contains large numbers of collagen fibers even though they will not be visible on the slides observed in lab.

7.   Emphasize that cartilage is avascular and that this results in a slow repair or healing rate. Also point out that the ears are very prone to frostbite since the entire blood supply of the structure is limited to the skin covering the cartilage.

8.   While presenting the information on bone (osseous tissue) stress that this is a living tissue that has a direct blood supply. Often the student conception of bone is that it is nonliving material (due to observations in the lab).

9.   Mention that the "fibers" in blood are unique in that they are composed of a soluble protein that only becomes insoluble during the process of clot formation.

10. Epithelial membranes are composed of epithelial and connective tissues. The best example to illustrate this is the skin (cutaneous membrane).

11. Stress that regeneration is not the same as repair. Point out that in order for a salamander to regenerate a tail, segments of the chromosome involved in tail formation must be activated. Repair involves "standard" mitotic processes.

12. Compare and contrast regeneration in a variety of animals (planaria, salamanders, lizards) and humans.

Demonstrations/Activities

1. Audio-visual materials of choice.

2. Use 3-D models, such as a cube (for cuboidal), a fried egg (for squamous), a drinking glass (for columnar), to illustrate the various types of epithelial tissues.

3. Obtain or prepare 2 x 2 slides of all the tissues used during the lecture presentation of histology.

4. Illustrate how tissues are sectioned to show how thin sections are made. Remind students that slides possess only a small, thin slice of tissue and that the slide may have more than one type of tissue.

5. Obtain slides to illustrate different types of stains used in staining tissues, such as hematoxyli-neosin, Sudan black B, Wright's stain, etc.

6. Use models of epithelial, connective tissue, muscle cells, and a neuron to illustrate how the cells of the different tissue types are similar and dissimilar.

7. Ask the students to make a list of all the things the body could not do if connective tissue were absent.

8. Cover your fist with a collapsed balloon to demonstrate the relationship between parietal and visceral layers of a serous membrane.

9. Use a human torso model to indicate the locations of mucous and serous membranes.

10. Use models of skeletal, cardiac, and smooth muscle to compare and contrast these tissue types.

11. In lab, cut a planaria transversely and follow the regeneration of the missing body region over the next few weeks.

12. Use moderate pressure to scrape a fingernail along the anterior surface of the forearm to demonstrate the beginnings of the inflammatory response (redness, swelling).

Critical Thinking/Discussion Topics

1. What types of inflammations are there, and how do they differ from each other?

2. How are tissues prepared and sectioned to produce the various tissue slides seen in this textbook?

3. How can macrophages detect what is foreign and what is self (not foreign) in the body?

4. Of what medical significance is the entrance into the tissue spaces of the body of a microorganism that could degrade collagen? Name an example and describe the disease it causes.

5. If all cells of the body arise from the same embryonic cell (zygote), how can each cell take on specific roles? Could any of these differentiated cells revert to a different cell type?

6. In some cysts and tumors, bone, hair, and even teeth can be found. How can this happen?

7. Since cartilage is avascular, how is it supplied with the essentials of life?

8. Other than to reduce bleeding and prevent microbial invasion, why are wounds sutured?

9. There appears to be an inverse relationship between potential regeneration and level of specialization of tissues. Why might this be so?

Library Research Topics

1. Basement membranes provide the interface between epithelium and connective tissue. What is the chemical composition of this layer and why is this area of great interest to cell biologists?

2. What is the current status of cloning? Is it feasible for human cells?

3. What is suction lipectomy? Who can perform this procedure?

4. Why can some cells regenerate and others not? What advantages and disadvantages are there for either case?

5. Some nutritionists suggest that obesity later in life results from overfeeding during infancy and childhood. Review some of the articles that support or dispute this theory.

Audio-Visual Aids/Computer Software

Slides

1. Preparing Tissues for Microscopy (LPI)

2. Animal Tissues—Epithelium, Cartilage (NTA, #50, Microslide Viewer)

3. Animal Tissues—Muscle, Bone, Connective (NTA, #51, Microslide Viewer)

4. Animal Cells and Tissues Set 1 (BM). Examines the structure and specialization of epithelial cells and tissues, connective tissue, adipose tissue, cartilage, and bone.

5. Basic Human Histology Set (CBS), 482200). Set of 100 transparencies that covers a broad spectrum of human histology.

6. Basic Mammalian Tissue Type Set (CBS, 481754). 33 frames showing the four basic tissues.

7. Connective Tissue Proper (CBS, 482173)

8. Connective Tissue Set (CBS, 482116)

9. Epithelial Tissue Set (CBS, 482114, 482172)

10. Histology of Basic Tissue Types (EI, #611)

11. Introduction to Mammalian Histology Set (CBS, 481755M)

12. Skeletal Tissue Set (CBS, 482118)

13. Striated Muscle Tissue Set (CBS, 482134)

Videotapes

1. Cytology and Histology (EI, SS0035V, VHS/BETA). Series also available as slide set.

2. Cells, Tissues, and Organs (SU, 11 min. VHS). Examines cell differentiation from single cells to tissues to organs.

Computer Software

1. Human Life Processes II: Systems Level (CCMI, IBM). Demonstrates how cells and tissues function together in organs.

See Guide to Audio-Visual Resources on p. 339 for key to AV distributors.

LECTURE ENHANCEMENT MATERIAL

Clinical and Related Terms

1. Biopsy — the removal and microscopic examination of living tissue for diagnostic purposes. Often used to determine if tissue is malignant or benign. Types of biopsies include aspiration (to obtain tissue by means of a suction needle), sternal (to obtain bone marrow), and cone (to obtain tissue from the uterus).

2. Exfoliative cytology — the removal and microscopic examination of cells from a lesion or from cells that have sloughed off the body. Used to determine malignancy, microbiological changes, and homeostatic changes in vaginal secretions, urine, sputum, abdominal fluids, etc. Procedures include washing, aspiration, smear (Pap smear), and scraping.

3. Hyperplasia — abnormal increase in the size of a tissue or organ due to the excessive growth or proliferation of normal cells. May occur as a result of inflammation, infection, trauma, or may be of unknown etiology.

4. Ischemia — a deficiency in the blood supply to a tissue or organ due to a blockage or constriction of blood vessels supplying the area. May be localized and temporary. Myocardial ischemia or deficiency of the blood supply to the heart muscle due to blockage of coronary arteries is a common example.

5. Microtome — a specialized "slicing" machine used to prepare very thin slices of tissues for microscopic observation. Tissues used for microscopic work are sliced less than ten micrometers thick, using either a rotary or sliding microtome.

6. Papanicolaou test — a simple, painless exfoliative test used most often for the early detection of cancer of the cervix and uterus. Also known as the Pap smear. The test is based on the observation that cancerous cells close to the epithelial surface will slough off into the surrounding secretions. The material can then be placed on a slide, stained, and examined under the microscope. The test can be applied to other areas, such as the vagina, lung, stomach, and bladder.

Disorders/Homeostatic Imbalances

1. Carcinoma — a malignant growth of epithelial cells that may infiltrate any other tissue and can metastasize through the blood stream to other tissues. The cause of this form of cancer is unknown, but may result from excessive exposure to viruses, chemicals, sunlight, or from hormonal or genetic factors. Most cancers of the breasts, uterus, skin, tongue, and GI tract arise as carcinomas. Examples include basal cell, chorionic, epidermoid, glandular, nasopharyngeal, and squamous cell carcinomas.

2. Edema — a generalized or localized abnormal accumulation of fluids within the body tissue (intercellular) spaces. Localized factors include insect bites, bacterial infections, inflammation, trauma, and poor lymphatic drainage. Generalized factors may involve a decrease in cardiac output, liver failure, decreased renal function, a decrease in blood plasma proteins, electrolyte imbalances, and malnutrition. Treatment usually requires bed rest, decreased sodium intake, and correction of any physiological malfunctions such as neutralizing any inflammation, restoring cardiac output, and correcting any electrolyte imbalances. Examples are related to areas affected, such as brain, lungs, and feet.

3. Fistula — an abnormal tubelike opening extending from within the body to the outside or between two organs or cavities. The openings may be formed as a result of surgery, infection, injury, or congenital abnormalities. Examples of fistulas include anal, branchial (embryonic fistulas), cervical, biliary, parotid, and umbilical fistulas. Most fistulas can be treated by surgery.

4. Pleurisy — an inflammation of the serous lining of the lung caused by infection, injury, trauma, or tumor, or as a complication of any of a number of infectious agents. Symptoms include fever,

cough, chills, sharp pain, and rapid, shallow breathing. It may be chronic or acute, fibrinous or purulent, dry or wet. In dry pleurisy, the amount of serous fluid separating the two membranous layers remains unchanged. However, the two layers may rub against each other and become swollen and painfully inflamed. In wet pleurisy, the fluid level increases abnormally and interferes with normal breathing by compressing the lung. Usually there is little pain; however, if the fluid becomes infected and pusfilled (empyema), the condition can change. Treatment usually involves bed rest, application of heat, and increased fluid intake. Antibiotics may be needed in cases of infection.

5. Sarcoma — a malignant growth of connective tissue cells such as bone, cartilage, blood, or lymph, that may develop rapidly and metastasize through lymph vessels to other tissues. The cause is unknown. Most sarcomas are named for the tissue they affect, such as fibrosarcoma (fibroid connective tissue), osteosarcoma (bone tissue), lymphosarcoma (lymphatic tissue), and chondrosarcoma (cartilage tissue). Kaposi's sarcoma, a malignant vascular cancer that is characterized by the development of bluish-red nodules on the skin, is often associated with immunosuppressed patients, such as AIDS patients.

ANSWERS TO END-OF-CHAPTER QUESTIONS

Multiple Choice/Matching Questions

1. a, c, d, b   3. b, f, a, d, g, d   5. c
2. c, e   4. b   6. b

Short Answer Essay Questions

7. Groups of closely associated cells that are similar in structure and perform a common function. (p. 102)

8. Protection—stratified squamous; absorption—simple columnar; filtration—simple squamous; secretion—simple cuboidal. (p. 103)

9. The covering and lining epithelia are classified on the basis of the shape of the cells and the number of cell layers present. The three common shapes are squamous, cuboidal, and columnar. The classes in terms of cell number are: simple (single layer) or stratified (multiple layers). In some cases, such as with endothelium, it is important to indicate their special location in the body. (pp. 104-105)

10. Merocrine glands (sweat glands) secrete their products by exocytosis; holocrine glands (oil glands) release their products by lysis of the entire cell; apocrine (mammary glands) release their products by pinching off parts of the cell contents. (p. 112)

11. Binding—areolar; support—cartilage; protection—bone; insulation—adipose; and transportation —blood. (pp. 116-123)

12. Fibroblast; chondroblast; osteoblast. (pp. 116-123)

13. Ground substance—interstitial fluid, proteioglycans, and glycosaminoglycans; fibers—collagen, elastic, reticular. (pp. 113-114)

14. The matrix gets to its position due to secretion by the undifferentiated cells located throughout the matrix. (p. 114)

15. a. areolar (p. 114)

    b. elastic cartilage (p. 123)

    c. elastic tissue (p. 123)

    d. mesenchyme (p. 116)

    e. fibrocartilage (p. 123)

    f. hyaline cartilage (p. 123)

16. The macrophage system is involved in overall body defenses. (p. 115)

17. See Fig. 4.10, pp. 126-127. The figure illustrates the location, function, and description of the three muscle types.

18. Neurons are highly specialized cells that generate and conduct nerve impulses, whereas neuroglial cells are nonconducting cells that support, insulate, and protect the neurons. (p. 126)

19. Tissue repair begins with organization, during which the blood clot is replaced by granulation tissue. If the wound is small and the damaged tissue is actively mitotic, the tissue will regenerate and cover the fibrous tissue. When a wound is extensive or the damaged tissue amitotic, it is repaired only by using fibrous connective tissue. (pp. 126-129)

20. Ectoderm—epithelium and nervous; mesoderm—connective, muscle, and epithelium; endoderm—epithelium. (p. 131)

Critical Thinking and Application Questions

1. No. Cartilage heals slowly because it lacks the blood supply necessary for the healing process. (p. 123)

2. The skin is subjected to almost constant friction, which wears away the surface cells, and is charged with preventing the entry of damaging agents and with preventing water loss from the body. A stratified squamous epithelium with its many layers is much better adapted to stand up to abrasion than is simple epithelium (single layer cells); also the stratified epithelia regenerate more efficiently than simple epithelia. Finally, keratin is a tough waterproofing protein that fills the bill for preventing dessication and acting as a physical barrier to injurious agents. Since a mucosa is a wet membrane, it would be ineffective in preventing water loss from the deeper tissues of the body. (p. 103)

3. If ligaments contained more elastic fibers they would be more stretchy, thus joints would be more flexible. However, the function of the ligaments is to bond bones together securely so proper controlled joint movement can occur. Ligaments that are more elastic would result in floppy joints in which the bones involved in the joint would be prone to misalignment and dislocation. (p. 123)

4. Epithelium, because epithelia remain mitotic throughout life. This is not the case for nervous and muscle tissue, and some forms of connective tissue. (p. 129)

Key Figure Questions

Figure 4.5 The holocrine gland, because cells that are fragmented and secreted have to be replaced.
Figure 4.12 1. Budding capillaries and soft connective tissue. 2. They bud from uninjured blood vessels in the area. 3. The epidermis would not be continuous. The injured region would be totally replaced by scar tissue.

SUGGESTED READINGS

---

1. Caplan, A.I. "Cartilage." *Scientific American* 251(4) (Oct. 1984): 84-94.

2. Cormack, D.H. *Ham's Histology.* 9th ed. New York: J.B. Lippincott Company, 1987.

3. DiFiore, M.S. *An Atlas of Human Histology.* 5th ed. Philadelphia: Lea and Febiger, 1981.

4. Eyre, D.R. "Collagen: Molecular Diversity in the Body's Protein Scaffold." *Science* 207 (1980): 1315.

5. Fawcett, D.W. *A Textbook of Histology*. 11th ed. Philadelphia: W.B. Saunders, 1986.

6. Kessel, R.G., and R.H. Kardon. *Tissues and Organs: A TextAtlas of Scanning Electron Microscopy*. San Francisco: W.H. Freeman & Company, 1979.

7. Leeson, T.S., and C.R. Leeson. *Histology*. 4th ed. Philadelphia: W.B. Saunders, 1981.

8. Miller J.A. "The Connective Tissue Perspective." *Science News* 121(8) (Feb. 1982): 124.

9. Ross, M.H., and E.J. Reith. *Histology: A Text and Atlas*. New York: Harper & Row, 1985.

10. Weiss, L., and R.O. Greep. *Histology*. 5th ed. New York: McGraw-Hill, 1983.

5

The Integumentary System

Chapter Preview

This chapter describes the body covering (skin) and its contribution to the body integrity as a whole. The skin and its derivatives (sweat and oil glands, hair, and nails), make up a very complex set of organs that serves a number of functions including protection, excretion, body temperature regulation, sensation, and vitamin D synthesis.

INTEGRATING THE PACKAGE

Suggested Lecture Outline

I.   The Skin (pp. 135-140; Fig. 5.1, p. 136)
   A. Basic Characteristics (pp. 135-136)
      1. Epidermis
      2. Dermis
   B. Epidermis (pp. 136-139)
      1. Basic Characteristics
      2. Cells of the Epidermis
         a. Keratinocytes
         b. Melanocytes
         c. Langerhans' Cells
         d. Merkel Cells
      3. Layers of the Epidermis (Fig. 5.2, p. 137; Fig. 5.3, p. 138)
         a. Stratum Basale
         b. Stratum Spinosum
         c. Stratum Granulosum
         d. Stratum Lucidum
         e. Stratum Corneum
   C. Dermis (p. 139)
      1. Basic Characteristics
      2. Layers of Dermis
         a. Papillary Layer
         b. Reticular Layer
   D. Skin Color (pp. 142-143)
      1. Melanin
      2. Carotene
      3. Hemoglobin

       4. Skin Color Variations
          a. Redness (Erythema)
          b. Pallor
          c. Jaundice
          d. Bronzing
          e. Black and Blue Marks

II. Appendages of the Skin (pp. 140-145)

    A. Hairs and Hair Follicles (pp. 140-143; Fig. 5.4, p. 141)
       1. Basic Characteristics
       2. Structure of a Hair
          a. Basic Features
          b. Regions
          c. Layers
          d. Pigment
       3. Structure of a Hair Follicle (Fig. 5.5, p. 142; Fig. 5.6, p. 143)
          a. Basic Features
          b. Associated Structures
       4. Distribution, Types, and Growth of Hair
          a. Basic Features
          b. Types of Hair
          c. Hair Growth and Density
       5. Hair Thinning and Baldness
          a. Basic Features
          b. Male Pattern Baldness
          c. Hair Thinning

    B. Nails (pp. 143-144)
       1. Basic Characteristics
       2. Structure of a Nail (Fig. 5.7, p. 144)
          a. Free Edge
          b. Body
          c. Root
          d. Nail Bed
          e. Nail Matrix
          f. Lunula
          g. Nail Folds
          h. Eponychium (Cuticle)

    C. Sweat (Sudoriferous) Glands (p. 144)
       1. Basic Characteristics
       2. Eccrine Sweat Glands
       3. Apocrine Sweat Glands
       4. Ceruminous Glands
       5. Mammary Glands

Review Items

5. Dense irregular connective tissue (Chapter 4, p. 122; Fig. 4.8F, p. 119)

6. Fibers in matrix of connective tissue (Chapter 4, p. 115)

7. Simple coiled tubular glands (Chapter 4, p. 111; Fig. 4.4, p. 111)

8. Simple branched alveolar glands (Chapter 4, p. 111; Fig. 4.4, p. 111)

9. Merocrine glands (Chapter 4, p. 112; Fig. 4.5, p. 112)

10. Holocrine glands (Chapter 4, p. 112; Fig. 4.5, p. 112)

Cross References

1. Additional information on jaundice and the buildup of bilirubin can be found in Chapter 18, p. 591 and Chapter 24, p. 821.

2. The effects of androgens are explained in Chapter 29.

3. Body temperature regulation is discussed in Chapter 25, pp. 885-890.

4. Organ and tissue transplants, and prevention of rejection are discussed in Chapter 22, pp. 731-734.

5. Mechanical and chemical nonspecific defense mechanisms are described in great detail in Chapter 22, pp. 708-715.

6. Sebaceous and sudoriferous glands of the ear canal are examined in Chapter 16, p. 527.

7. Cutaneous sensation and reflex activity is covered in Chapter 13, p. 456.

Laboratory Correlations

1. Marieb, E. N. *Human Anatomy and Physiology Laboratory Manual: Cat and Fetal Pig Versions.* 3rd. ed. Benjamin/Cummings, 1989.

    Exercise 7: The Integumentary System

    Exercise 8: Classification of Membranes

2. Marieb, E. N. *Human Anatomy and Physiology Laboratory Manual: Brief Version.* 3rd. ed. Benjamin/Cummings, 1992.

    Exercise 7: The Skin

Transparencies Index

5.1   Skin Structure
5.5   Structure of a hair and follicle
5.8   Estimating the extent of burns

Bassett Atlas Figures Index

Slide Number	Figure Number	Description
59	6.7A	Shoulder joint

INSTRUCTIONAL AIDS

Lecture Hints

1. The strata basale and spinosum are often referred to collectively as the growing layers (stratum germinativum). Some authors consider the stratum germinativum to be only the stratum basale. Students are easily confused if terminology is not consistent between lecture, text, and lab test.

2. By pointing out that the hypodermis is also called the superficial fascia, there will be less confusion with the deep fascia covered in Chapter 9.

3. Stress that the hypodermis is not a skin layer, but is actually beneath the skin.

4. The hypodermis (superficial fascia) is an important location of fat storage that insulates the body. This layer is more prominent in females than males, resulting in a softer feel to the touch. This softer skin is considered a secondary sex characteristic of the female.

5. Ultraviolet radiation damages the viable cells of the epidermis by causing formation of thymine dimers in DNA.

6. Dark-skinned races developed close to the equator where ultraviolet radiation exposure is the highest. Fair-skinned races are more prevalent away from the equator. As a group, the Eskimos have darker skin even though they are far from the equator because of secondary exposure due to reflection of sunlight from the snow.

7. The stratified squamous epithelium in the mouth of some desert lizards (e.g., the Gila monster) is black due to the high activity of melanocytes. In other desert lizards (e.g., the horned lizard) the peritoneum is black. This melanin is thought to provide additional protection from the high degree of ultraviolet exposure in the desert.

8. During lab, students often try to locate and identify the stratum lucidum in all skin slides. Stress that this layer of the epidermis is present only in thick skin.

9. Explain that the skin plays a role in regulating body temperature by evaporation of sweat and by controlling blood flow through dermal blood vessels.

10. Some sebaceous glands are not associated with hair follicles and open directly onto the skin surface. Examples include the sebaceous glands of the skin, lips, and eyelids (tarsal glands).

11. Actual contact with the environment is through a layer of dead rather than living cells. This specialization was critical for the evolution of life forms to evolve that could survive in a terrestrial environment.

Demonstrations/Activities

1. Audio-visual materials of choice.

2. Use 3-D models of skin to illustrate layers and strata.

3. Use a microprojector and microscope slides of skin to illustrate layers. Use slides of skin from the scalp and palm to contrast the differences in the layers.

4. Have a small fan operating. As students file into the classroom spray their arm or hand with water. Ask them to describe the sensation as the water evaporates from the skin, and to explain why evaporation of water (or sweat) is important to temperature homeostasis. Repeat the demonstration with alcohol and ask why the cooling effect is greater.

5. Provide small glass plates and instruct students to observe the change in the color of their skin while pressing the heel of their hand firmly against the glass. Ask them to explain the color change, and what would happen to the skin if the pressure were prolonged.

6. Show the students a picture of a heavily wrinkled person. Ask them to list all the factors that have contributed to the skin deterioration seen.

Critical Thinking/Discussion Topics

1. What role does the skin play in the regulation of body temperature?

2. Why exactly can animals with thick fur, such as Alaskan huskies, resist extremely cold temperatures?

3. Humans are often called the "naked apes." Since we have extensive hair follicles all over our body, why do you suppose we lack body hair?

4. If the skin acts as a barrier to most substances, how can it initiate an allergic response to such things as poison ivy?

5. Many organisms such as snakes, insects, and lobsters shed their "skin" periodically. How does this compare to the process taking place in humans?

6. The air is 80°F and the lake temperature is 70°F. Why do you first feel cold when you enter the water? Why do you feel chilled when exiting the water?

7. Why does axillary hair not grow as long as hair on the scalp? How long would scalp hair grow if it were not cut?

8. Discuss the advantages and disadvantages of using the drug minoxidil for stimulating hair regrowth in bald men.

9. Which structures located in the dermis are of epidermal origin?

10. When fair-skinned individuals go outside on a cold windy day, their skin turns "white" and after a time turns "red." Explain.

11. Nancy has a dry skin condition and prefers to take her bath in the evening. Would it be more effective for her to apply a skin care lotion such as Keri lotion in the morning or in the evening after taking a bath? Why?

12. Why is it more difficult to get a suntan during the winter months even though the sun is closer to the earth during this season?

13. Individuals living in Ohio may be able to go out into the sun for three hours and not burn, but if they go to Florida during spring break, they may get a sunburn after only two hours. Why?

14. Describe the difference between the A and B types of ultraviolet rays relative to skin damage.

15. Why does a suntan eventually fade?

16. Other than to reduce bleeding, why are wounds sutured together?

Library Research Topics

1. Explore the literature on the latest techniques and materials such as test-tube skin, synthetic skin, and heterograft skin used in skin grafting.

2. The long-term effects of sunburn seem to include severe wrinkling of the skin and skin cancer. What are the latest statistics on this problem and what has been done to correct it?

3. What are the latest therapies for baldness?

4. Although our skin is a "barrier" to microbes, prepare a list of microbes, such as bacteria, yeast, fungi, protozoans, and arthropods, that may reside on or in our skin.

5. Accutane (Isotretinoin) is a prescriptive drug approved in the early 1980s for treatment of severe cystic acne. Is this drug safe to use during pregnancy?

6. Student assignment: Look up the signs and symptoms of basal cell carcinoma, squamous cell carcinoma, and malignant melanoma for class discussion.

Audio-Visual Aids/Computer Software

Slides

1. Skin (NTA, #57, Microslide Viewer)

2. Skin and Its Function (EI, SS-0335F, Filmstrip or Slides). Integumentary system and its parts.

3. First Aid: Newest Techniques, Burns (SC)

4. Skin Cancer: The Sun and You (CBS, 52-7655)

Videotapes

1. Skin Deep (FHS, QB-822, VHS/BETA). Reviews sense receptors, taste buds, touch sensors, and olfactory cells. Also available as 16 mm film.

2. Surgical Skin Preparation (TF, 13 min., C, VHS/BETA). Demonstrates three phases of surgical skin preparation.

3. Nova: The Wonders of Plastic Surgery (CBS, 49-3654-V, 60 min., C, VHS)

Computer Software

1. Dynamics of Human Skin (EI, C-3059, 1988, Apple or IBM)

    See Guide to Audio-Visual Resources on p. 339 for key to AV distributors.

LECTURE ENHANCEMENT MATERIAL

Clinical and Related Terms

1. Allograft — a tissue or skin graft that is transferred from a genetically nonidentical donor of the same species.

2. Antiperspirant — an agent that inhibits or prevents perspiration. Also called an anhidrotic agent.

3. Autograft — a tissue or skin graft that is transferred from one part of the patient's body to another part.

4. Collagen implant — a surgical procedure involving the implantation of collagen (usually from an animal) under the skin to improve or eliminate scars or tissue damage due to acne, burns, diseases, and wrinkles. The collagen acts as a structural framework.

5. Comedo — a collection of oily material and keratin that blocks up the excretory duct of the sebaceous gland. Often discolors to black (blackhead) and may be associated with acne vulgaris and seborrheic dermatitis.

6. Corn — a circumscribed painful, horny induration and thickening of the skin, caused by friction or pressure from poorly fitting shoes or hose. May be treated by eliminating the irritants or by using a keratolytic agent for removal of the corn.

7. Dandruff — normal or excessive shedding of dry, white, scaly material from the scalp. May be aggravated by conditions such as seborrheic dermatitis.

8. Debridement — the removal of all dead, damaged, contaminated, and foreign material from an infected or burned wound until all surrounding healthy tissue is exposed.

9. Decubitus ulcer — a severe bedsore or ulcer that forms due to the obstruction of blood flow. May occur especially on the elbow, hip, and leg.

10. Deodorant — an agent used to destroy or mask odors. Often used in conjunction with an antiperspirant.

11. Depilatory — an agent capable of removing or destroying hair.

12. Dermabrasion — a surgical procedure for the removal of acne scars, tattoos, wrinkles, or other deformities from the skin. May require the use of sandpaper-like material or mechanical methods.

13. Dermatome — an instrument used to carefully cut thin sections of skin for skin transplantation.

14. Desquamation — the normal or abnormal shedding of skin or other epithelial tissue in scales or sheets. May be seen in newborn infants during the first weeks of life.

15. Detritus — any broken-down, degenerative, or particulated material that is produced or remains after the wearing away or disintegration of a substance or tissue, such as the skin.

16. Diaphoresis — excessive or profuse sweating.

17. Heterograft — a tissue or skin graft that is transferred from one individual to another individual of a completely different species. Also known as a xenograft.

18. Hypothermia — an accidental or induced condition of lowered body temperature.

19. Impacted cerumen — a condition in which an abnormal amount of ear wax or cerumen is produced and becomes impacted in the external auditory canal. Treatment may require irrigation of the ear with a warm glycerine solution or the mechanical removal of the plug.

20. Intradermal — within the skin or dermis. An intracutaneous injection.

21. Keratosis — any abnormal horny or keratinized growth such as a wart or callus.

22. Lund-Browder Classification — a method for estimating the extent of burns similar to the rule of nines. Most often used for children to more accurately determine the proportion of body surface affected.

23. Papule — a small, solid, elevated, and circumscribed area or lesion on the skin. Usually red in color. May precede vesicular or pustular eruptions such as from syphilis, measles, or smallpox.

24. Petechia — a small, purplish, round, nonraised, hemorrhagic spot on the skin or mucous membrane that may indicate a severe fever or vascular abnormality.

25. Pimple — a papule or pustule often seen on the skin of adolescents with acne.

26. Pruritus — a severe itching. Common in skin disorders, allergic inflammation, and parasitic infestations.

27. Pustule — a small, raised area of the skin that is filled with pus or lymph. Often associated with acne vulgaris, varicella, syphilis, and other types of dermatitis.

28. Pyrogen — any substance or agent that causes fever.

29. Ringworm — a popular term for infections of the skin due to various species of fungi.

30. Subcutaneous — beneath the skin; injections or infusions that are given into the subcutaneous tissue. Also called hypodermic.

31. Topical — a term referring to a particular area; an agent that is applied to a specific area and that affects only that area.

Disorders/Homeostatic Imbalances

Congenital Disorders

1. Anhidrosis — an abnormal reduction in or absence of the secretion of sweat. May be temporary or permanent, generalized or localized, or genetically related. Treatment may require soothing ointments, lubricants, and reduction of any irritants. Air conditioning may reduce discomfort.

2. Ichthyosis congenital — a hereditary skin disorder characterized by severe drying of the skin and excessive growth of keratinocytes and keratin. The skin appears rough and scaly, often resembling fish scales.

Immunologic/Inflammatory Disorders

1. Actinic dermatitis — an inflammation and erythema of the skin caused by over-exposure to radiant energy such as sunlight, UV light, and X-rays.

2. Eczema — a general term used to describe any superficial inflammatory process involving the skin. Generally marked by redness, itching, tiny papules, and weeping, oozing vesicles that usually crust over. Common in children due to allergic reactions to foods or inhalants.

3. Pemphigus — a group of uncommon but severe and often lethal skin diseases characterized by vesicles and blisters that become infected. Pemphigus vulgaris, the most common type, begins with eruptions on the oral or nasal mucosa and spreads to other parts of the body.

4. Urticaria — a vascular reaction of the skin exhibiting transient, elevated, reddish patches with severe itching. Also known as hives. May be caused by reactions to foods, infections, or stress.

Infestations and Infectious Dermatitis

1. Athlete's foot — a fungal infection of the skin of the foot caused by a number of mycotic agents, including Candida and Trichophyton. Also known as tinea pedis. Prevention includes keeping the feet dry and open to the air. Treatment includes the use of antifungal agents, such as clotrimazole and griseofulvin.

2. Leprosy (Hansen's Disease) — an inflammatory but rarely contagious bacterial disease caused by Mycobacterium leprae. The severity and/or event of disease appears to be related to the host's immune status. Sulfone medication is very effective in therapy, but in untreated cases the infection may result in peripheral sensory nerve damage and/or the formation of chronic granulomatous lesions of the skin.

3. Molluscum contagiosum — a common, benign, contagious, usually self-limited viral disease of the skin, exhibiting firm, rounded papules. Treatment, if needed, involves cauterization.

4. Pediculosis — an infection of lice affecting the head and body (Pediculus humanis), or genitalia (Phthirus pubis). The disease causes intense itching and may lead to secondary infections. The eggs are usually visible and are attached to the hair shafts. Treatment with a shampoo containing lindane (Kwell) appears to control the lice.

5. Scalded skin syndrome — a form of impetigo caused by Staphylococcus aureus that results in the peeling of skin layers in sheets due to a reaction to the staphylococcal toxins. Treatment requires extensive antibiotic and antihistamine therapy.

6. Scabies — a contagious, itchy dermatosis caused by the itch mite, Sarcoptes scabiei. The mites form burrows in the folds of the skin in the groin and chest and cause extensive edema. The disease is considered a sexually transmitted disease, as it is transmitted by direct contact. A solution of lindane (Kwell) controls the mites.

Proliferative/Neoplastic Disorders

1. Basal cell carcinoma (BCC) — a malignancy of the skin that begins in the stratum basale. The tumor rarely metastasized and in 90% of the cases is located between the hairline and the upper lip. The primary cause is excessive exposure to sunlight.

2. Malignant melanoma — a cancer of the melanocytes that may begin wherever there is pigment. Melanomas may develop from melanocytes in normal skin or from a pigmented skin blemish such as a mole or birthmark.

3. Nevus — a circumscribed, often congenital discoloration of the skin or mucous membrane. Also called a mole. May involve epidermal, connective, nervous, and vascular elements.

4. Polyp — a tumor-like growth or mass that protrudes from the surface of a mucous membrane, usually on a stalk or pedicle. Commonly found as benign tumors in the nose, uterus, and rectum. If they become malignant, surgical removal is necessary.

5. Seborrheic keratosis — a common benign tumor of basal cell-like cells that arise spontaneously and rapidly on the trunk, face, and extremities. Usually occurs after middle age. The lesions may appear as plaques with greasy surfaces and slight pigmentation. Lesions may require surgical removal.

6. Squamous cell carcinoma — a malignancy of the skin that arises from keratinocytes of the stratum spinosum. Most appear on areas of the body exposed to sun, but may develop elsewhere on the body.

7. Wart — a circumscribed, elevated, cutaneous tumor, usually caused by papilloma viruses. May refer to benign, nonviral tumors. Seems to develop on exposed parts of the fingers, hands, face, and other areas. Viral-induced warts may be spread by rubbing or scratching and may heal spontaneously. Only effective treatment appears to be electrocautery and/or cryosurgery.

ANSWERS TO END-OF-CHAPTER QUESTIONS

Multiple Choice/Matching

1.	a	6.	b	10.	b
2.	c	7.	c	11.	a
3.	d	8.	c	12.	d
4.	d	9.	b	13.	b
5.	b				

Short Answer Essay Questions

14. Cells of the stratum spinosum are called prickle cells because of their spiky shape in fixed tissues; granules of keratohyalin and lamellated granules appear in the cells of the stratum granulosum. (p. 138)

15. Generally not. Most "bald" men have fine vellus hairs that look like peach fuzz in the "bald" areas. (p. 143)

16. Due to the lack of adipose tissue, there is a decrease in the skin's ability to act as a shock absorber and insulator. (p. 139)

17. The skin acts as a mechanical barrier to water and infectious agents and as a chemical barrier to absorb the UV light. (p. 145)

18. First-degree burns affect only the epidermis; second-degree burns affect down to the dermis; and third-degree burns affect down to the subcutaneous tissue and muscle. (pp. 146-147)

19. Hair formation begins with an active growth phase, followed by a resting phase. After the resting phase a new hair forms to replace the old one. Factors that affect growth cycles include nutrition, hormones, local dermal blood flow, body region, gender, age, genetic factors, physical or emotional trauma, excessive radiation, and certain drugs. Factors that affect hair texture include hormones, body region, genetic factors, and age. (pp. 141-142)

20. Cyanosis is a condition in which the skin of Caucasians turns blue due to improperly oxygenated hemoglobin. (p. 140)

21. Wrinkling is due to the loss of elasticity of the skin, along with the loss of the subcutaneous tissue and is hastened by prolonged exposure to wind and sun. (p. 149)

22. a. A whitehead is formed by blockage of the duct of a sebaceous gland with sebum. When this sebum oxidizes, it produces a blackhead. When a blocked sebaceous gland becomes infected, it produces a pimple. (p. 145)

    b. Noninfectious dandruff is the normal shedding of the stratum corneum of the scalp. (p. 139)

    c. Greasy hair and a shiny nose both result from the secretion of sebum onto the skin. (p. 145)

    d. Stretch marks represent small tears in the dermis, as the skin is stretched by obesity or pregnancy. (p. 139)

    e. A freckle is a small area of pigmentation in the epidermis, caused by an accumulation of melanin. (p. 139)

    f. Fingerprints are films of sweat derived from sweat glands that open along the epidermal ridges of the palm. (p. 139)

23. (a) Porphyria. Porphyria victims lack the ability to form the heme of Hb. Buildup of intermediate byproducts (porphyrins) in the blood cause lesions in sun-exposed skin. Dracula was said to have drunk blood and to have shunned the daylight. (p. 148)

24. Stratum corneum cells are dead. By definition, cancer cells are rapidly dividing cells. (p. 138)

Critical Thinking and Application Questions

1. His long-term overexposure to ultraviolet radiation in sunlight is considered to be a risk factor for the development of skin cancer. In addition, moles or pigmented spots that show asymmetry (A), border irregularity (B), color variation (C), and a diameter greater than 6 mm (D) are all signs of a possible malignant melanoma. He should seek immediate medical attention. If it is a malignant melanoma, the chance for survival is not high, but early detection increases the survival rate. (p. 149)

2. The two most important problems encountered clinically with a victim of third-degree burns are a loss of body fluids resulting in dehydration and an electrolyte imbalance, and the risk of infection. Intact skin effectively blocks not only the diffusion of water and water-soluble substances out of the body, but acts as a barrier limiting the invasion of various microorganisms. (p. 147)

3. Chronic physical irritation or inflammation can lead to excessive hair growth in the region affected due to an increase in blood flow to the area. (p. 143)

Key Figure Questions

Figure 5.1 1. The stratum basale or basal layer of the epidermis. 2. Adipose cells. **Figure 5.2** The desmosomes connecting the keratinocytes are important to maintain continuity of the epidermis.

SUGGESTED READINGS

---

1. Cotsarelis, G., T. Sun, and R.M. Lavker. "Label-Retaining Cells Reside in the Bulge Area of Pilosebaceous Unit: Implications for Follicle Stem Cells, Hair Cycle, and Skin Carcinogenesis." *Cell* 61 (June 1990): 1329-1337.

2. Cowen, R. "Speeding Up Wound Healing the EGF Way." *Science News* 136 (July 1989): 39.

3. Dahl, M.V. "Acne: How It Happens and How It's Treated." *Modern Medicine* (Sept. 1982).

4. Edelson, L.E., and J.M. Fink. "The Immunologic Function of Skin." *Scientific American* 252 (June 1985): 46-53.

5. Elden, H.R. *Biophysical Properties of the Skin.* New York: Wiley, 1971.

6. Franklin, D. "Skin by the Yard Covers Massive Burns." *Science News* 126 (Aug. 1984): 102.

7. Jaret, P. "White Fingernails—Golden Eyes." *In Health* 4 (Jan.-Feb. 1990): 47-90.

8. Langone, J. "Gone Today, Hair Tomorrow." *Time* (Aug. 1988): 78.

9. Maugh, T.H. "Hair: A Diagnostic Tool to Compliment Blood Serum and Urine." *Science* 202 (1978): 1271.

10. Montagna, W. "The Skin." *Scientific American* (Feb. 1965).

11. Pawelek, J.M., and A.M. Korner. "The Biosynthesis of Mammalian Melanin." *American Scientist* 70 (Mar.-Apr. 1982): 136-137.

12. Raloff, J. "Hairy Portals for Toxic Chemicals." *Science News* 133 (June 1988): 407.

13. Rosenfels, A. "Some of a Body's Crucial Functions Are Only Skin Deep." *Smithsonian* 19 (May 1988): 159-180.

14. Ross, R. "Wound Healing." *Scientific American* (June 1969).

15. Satterberg, F. "Battle of the Bald." *Hippocrates* 3 (Sept.-Oct. 1989): 28-30.

16. Sobel, D. "Serious Face Saving." *Health* (Oct. 1989): 65-67, 82-84.

17. Vargo, N.L. "The Skin Cancer Success Story." *RN* (July 1987): 50-57.

18. Weisburd, S. "Skin Reborn From Muscle." *Science News* 132 (Sept. 1987): 164.

19. Weiss, R. "Wrestling With Wrinkles." *Science News* 134 (Sept. 1988): 200-202.

20. Wurtman, R.J. "The Effects of Light on the Human Body." *Scientific American* (1975).

6

Bones and Bone Tissue

Chapter Preview

This chapter focuses on the chemistry, histology, general structure, and function of bone tissue. In addition, the dynamics of formation and mechanisms for remodeling throughout life are illustrated. Various disorders that affect the bone tissue, such as osteoporosis, osteomalacia, and Paget's disease will also be described.

INTEGRATING THE PACKAGE

Suggested Lecture Outline

I. Skeletal Cartilages (pp. 155-156; Fig. 6.1, p. 157)
   A. Basic Structure, Types, Location (p. 156)
   B. Growth of Cartilage (p. 156)
II. Functions of the Bones (p. 156)
   A. Support (p. 156)
   B. Protection (p. 156)
   C. Movement (p. 156)
   D. Storage (p. 156)
   E. Blood Cell Formation (p. 156)
III. Classification of Bones (pp. 156-158; Fig. 6.2, p. 158)
   A. Shape (p. 157)
   B. Types of Tissue (p. 157)
      1. Compact
      2. Spongy
IV. Bone Structure (pp. 158-162)
   A. Gross Anatomy (pp. 158-160)
      1. Typical Long Bone (Fig. 6.3, p. 159)
         a. Diaphysis
         b. Epiphysis
         c. Epiphyseal Line
         d. Periosteum
         e. Endosteum
         f. Articular Cartilage
      2. Short, Irregular, and Flat Bone (Fig. 6.4, p. 160)
         a. Compact and Spongy Bone
         b. Diploe

          c. Periosteum

          d. Endosteum

      3. Location of Hematopoietic Tissue in Bones

          a. Red Marrow

          b. Red Marrow Cavities

          c. Yellow Marrow

      4. Bone Markings (Table 6.1, p. 162)

  B. Microscopic Structure of Bone (pp. 160-161; Fig. 6.5, p. 161; Fig. 6.6, p. 162)

      1. Compact Bone

          a. Haversian System (Osteon)

          b. Lamellar Bone

          c. Haversian Canal

          d. Volkmann's Canals

          e. Osteocytes

          f. Lacunae

          g. Canaliculi

          h. Interstitial Lamellae

          i. Circumferential Lamellae

      2. Spongy Bone

          a. Trabeculae

          b. Lamellae

          c. Osteocytes

  C. Chemical Composition of Bone (pp. 162-163)

      1. Organic Components

          a. Cells

          b. Osteoid

      2. Inorganic Components

V. Bone Development (Osteogenesis) (pp. 163-166)

  A. Formation of the Bony Skeleton (pp. 163-165)

      1. Intramembranous Ossification (Fig. 6.7, p. 163)

      2. Endochondral Ossification (Fig. 6.8, p. 164)

  B. Bone Growth (pp. 165-166)

      1. Growth in Length of Long Bones (Fig. 6.9, p. 165)

      2. Appositional growth

      3. Hormonal Regulation of Bone Growth During Youth (Fig. 6.10, p. 166)

VI. Bone Homeostasis: Remodeling and Repair (pp. 166-171)

  A. Bone Remodeling (pp. 168-170)

      1. Basic Processes

          a. Bone Remodeling Units

          b. Bone Deposition

          c. Bone Reabsorption

   2. Control of Remodeling
       a. Negative Hormonal Feedback (Fig. 6.11, p. 168)
       b. Mechanical Feedback (Fig. 6.12, p. 169)
   B. Repair of Fractures (pp. 170-171)
       1. Types of Fractures (Table 6.2, p. 171)
       2. Treatment
       3. Phases of Repair (Fig. 6.13, p. 171)
           a. Hematoma Formation
           b. Callus Formation
           c. Bony Callus Formation
           d. Remodeling
VII. Homeostatic Imbalances of Bone (p. 172)
   A. Osteoporosis (p. 172)
   B. Osteomalacia (p. 172)
   C. Paget's Disease (p. 172)
VIII. Developmental Aspects of Bones: Timing of Events (pp. 172-173)
   A. Embryonic and Fetal Development of Bone Tissue (p. 172)
   B. Development of Bone Tissue Through Adolescence (p. 173)
   C. Effect of Aging on Bone Tissue (p. 173)
IX. Making Connections: Interrelationships Between the Integumentary System and Other Body Organ Systems (pp. 174-175)

Review Items

1. Bone (osseous tissue) (Chapter 4, p. 123; Fig. 4.8J, p. 121)
2. Calcium salts (Chapter 2, p. 40)
3. Chondroblasts (Chapter 4, p. 115)
4. Collagen fibers (Chapter 4, p. 115) .
5. Fibroblasts (Chapter 4, p. 115)
6. Fibrocartilage (Chapter 4, p. 123; Fig. 4.8I, p. 120)
7. Hyaline cartilage (Chapter 4, p. 123; Fig. 4.8G, p. 119)
8. Proteoglycans (Chapter 4, p. 115)

Cross References

1. Individual bones that make up the skeleton are presented in Chapter 7.
2. Articular cartilage and joint structure is examined in Chapter 8, p. 226.
3. Identifying marks of individual bones are given in Chapter 7.
4. Gigantism and dwarfism as related to bone growth and length are discussed in Chapter 17, pp. 556-558.
5. Devices used to speed up repair and healing of fractures are discussed in A Closer Look, Chapter 6, p. 167.
6. The effects of parathyroid hormone and calcitonin on bone homeostasis are further explained in Chapter 17, p. 565.
7. Hematopoietic tissue is further described in Chapter 18, p. 588.

Laboratory Correlations

1. Marieb, E. N. *Human Anatomy and Physiology Laboratory Manual: Cat and Fetal Pig Versions.* 3rd. ed. Benjamin/Cummings, 1989.

   Exercise 9: Bone Classification, Structure, and Relationships: An Overview

2. Marieb, E. N. *Human Anatomy and Physiology Laboratory Manual: Brief Version.* 3rd. ed. Benjamin/Cummings, 1992.

   Exercise 8: Bone Classification, Structure, and Relationships: An Overview

Transparencies Index

6.1	Cartilages in the adult skeleton and body	6.9	Growth in length of a long bone
6.2	Classification of bones on the basis of shape	6.10	Long bone growth and remodeling during youth
6.3	Structure of a long bone	6.11	Hormonal controls of Ca^{2+} in blood
6.5	Microscopic structure of compact bone		

Bassett Atlas Figures Index

Slide Number	Figure Number	Description
68	6.16A	Wrist bones

INSTRUCTIONAL AIDS

Lecture Hints

1. Students often erroneously distinguish between long and short bones on the basis of size. Stress that the distinction is based on shape, not size.

2. Emphasize the difference between the epiphyseal plate and epiphyseal line.

3. Point out that the perichondrium does not cover the articular cartilages.

4. Emphasize the difference between red and yellow marrow.

5. Point out that osteocytes in Haversian systems are not isolated from each other, but tied together by canaliculi.

6. Compare and contrast location and function of osteocytes, osteoblasts, and osteoclasts.

7. Point out that long bone growth ends sooner in females (18 yrs) than males (21 yrs).

8. Emphasize that bones can be remodeled or grow appositionally, even after longitudinal growth has ceased.

9. Point out that greenstick fractures are more common in children, since their bones contain a higher proportion of organic matrix and are more flexible.

10. Distinguish between a simple (closed) and a compound (open) fracture.

11. Emphasize that bones must be mechanically stressed to remain healthy. Physical activity pulls on bones, resulting in increased structure. Inactivity results in bone atrophy.

Demonstrations/Activities

1. Obtain a sectioned long bone, such as a femur, to illustrate major parts of a bone. A fresh sectioned bone could be used to illustrate the periosteum and difference between red and yellow marrow.

2. Obtain a fetal skeleton to illustrate early stages of bone development.

3. Illustrate the chemical nature of bone tissue by placing one chicken bone in nitric acid and one in the oven. The nitric acid will leach out the calcium salts and the oven will break down the organic matter.

4. Obtain X-rays of young children, teenagers and adults, to illustrate changes in the epiphyseal plate.

5. Obtain X-rays of various types of fractures. If possible, obtain X-rays that illustrate healing stages following the fracture.

6. Obtain a 3-D model of a Haversian system to illustrate the microscopic characteristics of bone.

7. As an analogy, hold a bundle of uncooked spaghetti to illustrate the arrangement of osteons within compact bone.

8. Obtain a cleared and stained pig embryo to show the development of osseous tissue (Carolina Biological).

9. Break a green twig to illustrate a greenstick fracture. Then contrast with a dry twig.

Critical Thinking/Discussion Topics

1. Explore the statement, "Multiple pregnancies will result in the mother losing all the enamel from her teeth and calcium from her bones." Is this all true, all false, or only partly true?

2. Prepare a list of the hormonal abnormalities that can affect the growth of bones, both in children and in adults.

3. If air pollution becomes much worse, could it have an effect on bone development? Why?

4. Calcium plays an important role in bone formation. What other roles does calcium play in the body?

5. Full-contact sports seem to be a part of the curriculum for primary-school-age children. In the view of bone development, is this wise?

6. Prehistoric remains of animals consist almost exclusively of bones and teeth. Why?

7. If bone tissue is so hard, how can we move teeth from one location in the jaw to another?

8. Why are infections more common with compound fractures than simple fractures?

9. What would be the effect of extended weightlessness on the skeletal system? How can these effects be minimized or at least reduced?

10. Why does the top of a cooked chicken drumstick often come off while it is being eaten?

11. Why are greenstick fractures more common in children?

Library Research Topics

1. Research the latest technique, such as the Ilizarov procedure, used to lengthen bones damaged in accidents or illnesses.

2. What is involved in a bone marrow transplant, and is it a risky and difficult procedure?

3. What drugs or treatments are available to help correct conditions of gigantism and dwarfism and how do they work?

4. Explore the procedures used in bone tissue transplants where pieces of bone are removed from one part of the body and implanted into another.

5. What effect would steroid use have on the bone tissue and bone marrow?

6. How are electrical fields being used to stimulate bone growth and repair?

Audio-Visual Aids/Computer Software

Slides

1. Specialized Connective Tissue: Cartilage and Bone Set (CBS, 48-2174)

2. Systems of the Human Body—The Skeletal System and Its Function Set (CBS), 48-2065C)

Videotapes

1. The Development of Bone (TF, 19 min., C, 1988). Step-by-step development, from cartilage to bone.

   See Guide to Audio-Visual Resources on page 339 for key to AV distributors.

LECTURE ENHANCEMENT MATERIAL

Clinical and Related Terms

1. Demineralized bone matter (DBM) — demineralized bone tissue derived from bones of cadavers that can be surgically implanted into the body to replace lost sections of bone.

2. Dysostosis — a term referring to defective ossification. Often a genetic defect in the normal calcification of fetal cartilage.

3. Exostosis — a benign horny growth that projects from the surface of a bone; usually capped with cartilage.

4. Heterotrophic ossification — a nonmalignant bony growth that may occur after a fracture.

5. Marrow biopsy — the removal of a sample of red bone marrow by introducing a needle into the marrow cavity of the sternum.

6. Osteoma — a benign or malignant bony tumor that develops on a bone or other structure.

7. Pulsating electromagnet fields — a procedure involving electrical stimulation to facilitate bone repair.

8. Tenodesis — the surgical connection of a tendon to a bone. Used to restore muscle balance to a joint, to restore lost muscular function, or to increase the effectiveness of a joint area.

Disorders/Homeostatic Imbalances

Congenital Malformations

1. Craniofacial Dysostosis — a hereditary defect in the ossification of fetal cranial and facial bones, resulting in acrocephaly, exophthalmic, strabismus, and a parrot-beaked nose. Also known as Crouzon's disease.

2. Multiple Osteocartilagenous Exostosis — an exostosis characterized by multiple cartilage-capped, mushroom-shaped, bony projections that extend from the surface of bones. Most commonly found on bones of the shoulder, knee, and ankle. Also known as an osteochondroma.

3. Osteogenesis Imperfecti — also known as brittle bone. One of the most common hereditary bone diseases, characterized by a defective synthesis of the organic and inorganic matrix of bone. The disease is clinically diagnosed by finding thin, poorly formed bones, multiple fractures, loose-jointedness, and discoloration of the teeth.

Tumors of Bone

1. Osteoid osteoma — a small, benign, painful, neoplasm, most commonly found in the diaphyses of the tibia and femur, and arising within the cortical bone. Tends to occur in young adult males with no known etiology. Treatment requires surgical removal.

2. Osteosarcoma (osteogenic sarcoma) — a malignant, primary tumor of bone tissue, characterized by malignant, mesenchymal cell stroma with the formation of osteoid, bone, and/or cartilage by the tumor cells. May be classified as an osteoblastic, fibroblastic, or chondroblastic sarcoma, depending on the primary cell type in the tumor. May be hereditary or caused by radiation or oncogenic viruses. Primarily arises in the metaphyseal ends of the long bones of the extremities. Symptoms include pain, tenderness, and swelling of the affected part; however, many tumors remain silent. The prognosis, although improving, remains poor.

Tumors of Cartilage

1. Chondrosarcoma — a malignant tumor derived from chondroblasts that represent the second most common type of bone cancer. They are easier to manage then osteosarcomas and more amenable to surgical removal and cure. They tend to be large, bulky, gel-like, and grayish-white, with lobular growths. The tumors arise slowly, are painful, and cause swelling in the affected area.

2. Osteochondroma (Exostosis) — a benign bone tumor consisting of bone that projects from the surface of a bone and is capped by cartilage. May be hereditary or may arise spontaneously. Tumors have a tendency toward malignancy.

Infectious Disorders

1. Osteomyelitis — a pyogenic infection of the bone and especially bone marrow. Most cases begin as an acute infection but may become chronic if appropriate antibiotic therapy is not applied promptly. Organisms reach the bone marrow from the blood stream, a focal infection, or as a consequence of trauma. Symptoms include malaise, fever, chills, leukocytosis, and intense, throbbing pain in the affected area.

Hormonal Disorders

1. Acromegaly — an abnormal enlargement of the extremities of middle-aged persons, especially the cranial and facial bones, and the bones of the hands and feet. Usually results from hyper-secretion of the growth hormone (GH) from the anterior pituitary. Treatment requires either surgical intervention (transphenoidal hypophysectomy), pituitary radiation, or both.

2. Cushing's Disease (Syndrome) — a syndrome characterized by the excessive elaboration of cortisol, resulting in truncal obesity, moon face, hypertension, acne, weakness, and osteoporosis.

3. Osteitis Fibrosa Cystica (Von Recklinghausin's Disease) — a condition of prolonged or severe hyperparathyroidism causing chronic bone absorption and destruction. Bone changes are similar to osteomalacia and osteoporosis. Due to the increased parathormone production, continued demineralization of the bone tissue occurs. Osteoclastic action increases, resulting in generalized bone resorption, with thinning of the cancellous and compact tissues.

ANSWERS TO END-OF-CHAPTER QUESTIONS

Multiple Choice/Matching

1.	e	6.	b	11.	b
2.	b	7.	c	12.	c
3.	c	8.	b	13.	b
4.	d	9.	e	14.	b
5.	e	10.	c		

Short Answer Essay Questions

15. Cartilage has greater resilience because its matrix lacks bone salts, but its cells receive nutrients via diffusion from blood vessels that lie external to the cartilage. By contrast, bone has a beautifully engineered system of canaliculi for nutrient delivery, and for that reason its regeneration is much faster and more complete. (pp. 156, 160, 170)

16. For this answer, please refer to pp. 164-165 of *Human Anatomy and Physiology*, Third Edition.

17. Macroscopic appearance of compact bone indicates it is dense, while the appearance of spongy bone indicates it is porous; microscopic appearance of compact bone indicates it possesses Haversian systems, while spongy bone lacks osteons; compact bone is located along the shaft, while spongy bone is located at the ends or within the shaft. (pp. 160-161)

18. The increase in thickness of compact bone is counteracted by the resorption of bone by osteoclasts. (p. 168)

19. An osteoid seam is an unmineralized band of bony matrix that is about 10-12 micrometers wide. Between this seam and older bone is an abrupt transition called a calcification front. The osteoid seam always stays the same width, indicating that osteoid must mature before it can be calcified. This area then changes quickly from an unmineralized matrix to a mineralized matrix. (p. 163)

20. Two control loops regulate bone remodeling. One is a negative feedback hormonal mechanism involving mechanical and gravitational forces involving muscle pull that services the needs of the skeleton itself. (p. 168)

21. a. First decade is fastest, fourth decade is slowest. (p. 173)

    b. Elderly usually experience bone loss and osteoporosis and lack of blood supply. (p. 172)

    c. Children have proportionally more organic matrix. (p. 172)

Critical Thinking and Application Questions

1. A bony callus is the conversion of the fibrocartilagenous callus to a calcified callus indicating healing. (p. 170)

2. Vitamin D and milk provide dietary calcium and the vitamin is needed for its uptake; the sun helps to activate the vitamin D. Thick epiphyseal plates indicate poor calcification of the growing area. Because of this lack of sufficient calcium, the bones will be more pliable, and weight-bearing bones, like those in the leg, will bend. (p. 172)

3 The compact lamellar structure of dense bone produces structural units designed to resist twisting and other mechanical stresses placed on bones. In contrast, spongy bone is made up of trabeculae only a few cell layers thick containing irregularly arranged lamellae. (p. 162)

4. According to Wolff's law, bone growth and remodeling occurs in response to stress placed on such bones. With disuse, the bones in the limbs not being used will begin to atrophy. (p. 169)

5. Presumably the epiphyseal plant-bone junction has separated. The same would not happen to the boy's 23-year-old sister because by this age, epiphyseal plates have been replaced by bone and are no longer present. (p. 166)

Key Figure Questions

Figure 6.9 1. Cartilage 2. Because their avascular matrix is hardened by deposit of bone salts. **Figure 6.10** Its shaft would be the original length and it would have two very elongated and enlarged ends. **Figure 6.11** Calcium salts are deposited in bone and blood calcium levels fall.

SUGGESTED READINGS

1. Arehart-Treichel, J. "Boning Up On Osteoporosis." *Science News* 124 (Aug. 1983): 140.

2. Bienenstock, H. "Diagnosis: Arthritis." *Hospital Medicine* (Oct. 1984).

3. Bourne, G. W. *The Biochemistry and Physiology of Bone.* 2nd ed. New York: Academic Press, 1976.

4. Chisolm, J.J. "Lead Poisoning." *Scientific American* (Feb. 1971).

5. Edward, D.D. "Arthritis: Looking for Immunotherapy." *Science News* 131 (April 1987): 228.

6. Fackelmann, K.A. "Fluoride-Calcium Combo Builds Better Bones." *Science News* 135 (Jan. 1987): 36.

7. Fackelman, K.A. "Prolonged Nursing and the Risk of Bone Loss." *Science News* 143 (June 1993): 407.

8. Frame, B., and M.J. McKenna. "Osteoporosis: Postmenopausal or Secondary." *Hospital Practice* 20 (Oct. 1985): 36-47.

9. Hall, B.K. "Cellular Differentiation in Skeletal Tissue." *Biological Reviews* 45 (1970): 455.

10. Hall, B.K. "The Embryonic Development of Bone." *American Scientist* 76 (Mar. 1988): 174-175.

11. Harris, W.H., and R.P. Heaney. *Skeletal Renewal and Metabolic Bone Disease.* Boston: Little, Brown and Co., 1970.

12. Hendricks, M. "Estrogen Receptors Detected in Bone Cells." *Science News* 134 (July 1988): 5.

13. Johnson, G.T. "Arthroscopic Surgery on the Knee." *The Harvard Medical School Health Letter* (Mar. 1982).

14. Raloff, J. "Reasons for Boning Up On Manganese." *Science News* 130 (Sept. 1986): 199.

15. Riggs, B. L. "A New Option For Treating Osteoporosis." *New England Journal of Medicine* 327 (July 1990): 124-125.

16. Silberner, J. "Osteoporosis: Most Answers Yet to Come." *Science News* 131 (Feb. 1987): 116.

17. Thomsen, D.E. "Electrifying Biology." *Science News* 127 (Apr. 1985): 268.

18. Vaughn, J.M. *The Physiology of Bone.* 3rd ed. New York: Oxford University Press, 1981.

19. Wallis, C. "Making Bones as Good as New." *Time* 125 (14 Jan. 1985): 62.

20. Weisburd, S. "New Bone-Loss Risk Factors in Young Women." *Science News* 132 (Nov. 1987): 347.

21. Weiss, R. "Bone Density Drops With Thyroid Therapy." *Science News* 133 (June 1988): 359.

7

The Skeleton

Chapter Preview

This chapter describes and identifies all the major bones, parts, and processes of the axial and appendicular skeleton. The axial skeleton consists of the skull, vertebral column, and bony thorax. The appendicular skeleton consists of the pectoral girdle, upper and lower limbs, and pelvic girdle. Developmental changes in facial and cranial contours, spinal curvatures, and upper- to lower-body ratios will also be presented.

INTEGRATING THE PACKAGE

Suggested Lecture Outline

Part 1: The Axial Skeleton

I.   Basic Regions (p. 179; Fig. 7.1, p. 180)

II. The Skull (pp. 179-194); Table 7.1, pp. 188-189)

    A.   General Characteristics

    B.   Overview of Skull Geography (pp. 179; Figs. 7.2 - 7.4, pp. 181-183)

    C.   Cranium (pp. 181-190)

        1.   Frontal Bone (Fig. 7.2, p. 181; Fig. 7.3A, p. 182; Fig. 7.4B, p. 183)

        2.   Parietal Bones and the Major Sutures (Figs. 7.2A and 7.2B, p. 181)

        3.   Occipital Bone (Fig. 7.4, p. 183)

        4.   Temporal Bones (Fig. 7.5, p. 185)

        5.   Sphenoid Bone (Fig. 7.6, p. 186)

        6.   Ethmoid Bone (Fig. 7.7, p. 187)

        7.   Sutural Bones (Fig. 7.2B, p. 181)

    D.   Facial Bones (pp. 190-192; Fig. 7.8, p. 190)

        1.   Mandible

        2.   Maxillary Bones

        3.   Zygomatic Bones

        4.   Nasal Bones

        5.   Lacrimal Bones

        6.   Palatine Bones

        7.   Vomer

        8.   Inferior Nasal Conchae

    E.   Special Characteristics of the Orbits and Nasal Cavity (pp. 191-192)

        1.   The Orbits (Fig. 7.9, p. 192)

        2.   The Nasal Cavity (Fig. 7.10, p. 193)

Review Items

1. Bone markings (Chapter 6, p. 163; Table 6.1, p. 163)

2. Classification of bones (Chapter 6, p. 158; Fig. 6.1, p. 158)

3. Fibrocartilage (Chapter 4, p. 124; Fig. 4.7J, p. 121)

4. Hyaline Cartilage (Chapter 4, p. 124; Fig. 4.7H, p. 120)

Cross References

1. Joints are discussed in Chapter 8.

2. Bones of the middle ear cavity are discussed in detail in Chapter 16, p. 527.

3. Sutures are discussed in Chapter 8, p. 223.

4. The reinforcing muscles for the vertebral column are discussed in Chapter 10, pp. 302-305.

5. Table 7.2, p. 199, compares and contrasts the regional characteristics of the cervical, thoracic, and lumbar vertebrae.

6. The bones of the skull that function in the respiratory system are described in Chapter 23, pp. 744-751.

8. Muscles of the face are covered in Chapter 10, p. 296.

9. Muscles of the thorax are presented in Chapter 10, p. 306.

10. Muscles of the upper extremity are presented in Chapter 10, p. 312.

11. Muscles of the pelvic girdle are presented in Chapter 10, p. 310.

12. Muscles of the lower extremity are presented in Chapter 10, p. 324.

Laboratory Correlations

1. Marieb, E. N. *Human Anatomy and Physiology Laboratory Manual: Cat and Fetal Pig Versions.* 3rd. ed. Benjamin/Cummings, 1989.

   Exercise 10: The Axial Skeleton

   Exercise 11: The Appendicular Skeleton

   Exercise 12: The Fetal Skeleton

2. Marieb, E. N. *Human Anatomy and Physiology Laboratory Manual: Brief Version.* 3rd. ed. Benjamin/Cummings, 1992.

   Exercise 9: The Axial Skeleton

   Exercise 10: The Appendicular Skeleton

Transparencies Index

7.1	The human skeleton	7.10	Special anatomical characteristics of the nasal cavity
7.2	Anatomy of anterior skull	7.11	Paranasal sinuses
7.3A	Anatomy of lateral skull	7.14	Ligaments and fibrocartilage discs
7.3B	Anatomy of lateral skull (sagittal view)	7.17	Posterolateral views of articulated vertebrae
7.4A	Anatomy of interior portion of skull		
7.4B	Anatomy of interior portion of skull (superior view)	7.18	The sacrum and coccyx
		7.19	The bony thorax
7.9	Special anatomical characteristics of the orbits	7.20	A typical true rib and its articulations

Bassett Atlas Figures Index

Slide Number	Figure Number	Description
1	1.1A	Brain surface and vessels
7	1.7	Posterior view of brain in situ
13	2.2	Sagittal section, oral and nasal cavities
14	2.3A	Sagittal section, head and neck
52	5.9A	Pelvic girdle ligaments
63	6.11	Elbow joint
68	6.16A	Wrist bones
86	7.18A	Ankle joints

INSTRUCTIONAL AIDS

Lecture Hints

1. A good indicator of student comprehension of the spatial relationship among facial bones is the ability to list the bones making up the eye orbit.

2. Point out during the lecture that the styloid process of the temporal bone is often damaged during the preparation of a skeleton. As a result, skulls available in the lab may be lacking this fragile structure.

3. Students often have difficulty in identification of sphenoid and ethmoid bones. It would be helpful to show disarticulated specimens during lecture.

4. Point out that all facial bones (except mandible) articulate with the maxillae.

5. Remembering common meal times, 7 a.m., 12 noon, and 5 p.m., may help students to recall the number of bones in the three regions of the vertebral column.

6. "Atlas supports the world" can be used to help students remember that the atlas is first and axis second.

7. Point out that correct anatomical terminology specifies "arm" as the portion between shoulder and elbow; and "leg" refers to the portion between knee and ankle.

8. Although the obturator foramen is large, it is really closed by fibrous membrane in life.

9. In anatomical position the radius/ulna and fibula/tibia are in alphabetical order from the outside.

10. Dysplasia of the hip is common in some breeds of dog (German shepherd, English sheepdog).

Demonstrations/Activities

1. Audio-visual materials of choice.

2. Use an articulated skeleton to:

   a. indicate its protective and support aspects

   b. identify individual bones

3. Obtain a skull with its calvarium cut and a vertebral column to illustrate how these bones provide protection for the delicate neural tissues.

4. Obtain a skull that shows Wormian bones.

5. Use a Beauchene (disarticulated) skull to demonstrate the individual skull bones, and to show the fragile internal structure of bones containing sinuses.

6. The cranium is remarkably strong for its weight and thinness of cranial bones. This is in part due to the curvature of the cranium. To demonstrate this "self-bracing" effect, attempt to break an egg by squeezing it in the palm of your (or a student's) hand.

7. Use a disarticulated vertebral column to illustrate similarities/differences between vertebrae.

8. Point out that the superior articular surface of the atlas is elongated, matching the surface of the occipital condyle.

9. Tie different colors of string around the lamina and pedicle of a vertebra and pass it around class.

10. Obtain X-rays that exhibit abnormal curvatures (scoliosis, lordosis, kyphosis).

11. Obtain different ribs and indicate how each is similar and different.

12. Give a group of students a thoracic vertebra and a rib and ask them to articulate the two together.

13. Point out differences between the male and female pelvis.

14. Obtain a sacrum to show fusion of the vertebrae.

15. Use a fetal skeleton to emphasize the changes in skull and body proportions that occur after birth, and to point out the fact that initially the skeleton is formed (mostly) of hyaline cartilage rather than bone.

Critical Thinking/Discussion Topics

1. List several skeletal landmarks that can be used to guide a nurse or physician in giving injections, locating areas for surgery, and assisting in the diagnosis of internal conditions.

2. What effect would exaggerated exercise or the complete lack of exercise have on bones such as the tibia, femur, and humerus if it occurred during childhood, or during adulthood?

3. Numerous children are born with a congenital hip defect. Why is this area affected so often and what can be done to correct the defect?

4. Years ago, students used to carry large, heavy books on one arm or the other. Today, most students are using knapsacks. What difference, if any, could be detected in the spinal column between then and now?

5. Various religious writings have suggested that a rib was taken from man to create woman. Are any ribs missing from the male rib cage? What could explain this discrepancy?

6. Humans have a short neck, but giraffes have a long one. Does the giraffe have more neck vertebrae to accommodate this extra length? What other similarities or variations can be found between human bone structure and other animals?

7. How is it possible to "taste" eye drops shortly after they are placed in the eye?

8. When a dentist injects novocaine near the mandibular foramen, why does the lower lip become numb?

9. At birth only two of the four spinal curvatures are present. The secondary curvatures (cervical and lumbar) do not develop until after birth. What is the role of these secondary curvatures, and why do they develop?

Library Research Topics

1. There is a new technique, known as percutaneous automated discectomy, that involves back surgery without stitches. How safe is it, and when can it be employed?

2. Temporomandibular-joint disorders are very common and painful. What methods of treatment are there and how successful are they? (See *Time*, April 25, 1988.)

3. Spinal deviations such as scoliosis are very difficult to repair. What are the current methods of treatment, both invasive and noninvasive?

4. Paleontologists and archaeologists have unearthed many prehistoric skulls and bones of human-like creatures and animals. How can they reconstruct the soft features and tissues of these animals from only their skeletal remains?

5. Trace the origin of genetic disorders such as spina bifida and cleft palate starting with the human embryo. What is the explanation for these defects?

6. In the surgical repair of a herniated disc it may be necessary to remove some of the nucleus pulposus. The removal may be by conventional surgery or by chemonucleosus. What are the advantages of the latter procedure?

Audio-Visual Aids/Computer Software

Slides

1. The Skeletal System and Its Function (EA, SS-03260F, Filmstrips or Slides). Skeletal substructure reviewed.

2. Introductory Physiology Series: Bones and Muscles (McGraw-Hill)

Videotapes

1. Osteology of the Skull (TF, C, VHS/BETA, 1988). An eight-part series covering the osteology of the skull.

2. Osteology of the Upper Limb (TF, 23 min., C, VHS/BETA, 1988)

3. Skeleton: An Introduction (TF, 46 min., C, VHS/BETA)

4. Skull Anatomy Series (TF, C, VHS/BETA, 1988). A nine-part series that offers a complete approach to teaching and review of the skull.

5. Skeletal and Topographic Anatomy Series (TF, C, VHS/ BETA, 1988). A 30-part series that illustrates the anatomy of the head, neck, thorax, extremities, etc.

6. Anatomy of the Skull I: Superior, Posterior, and Lateral Views (BM, 28 min.)

7. Anatomy of the Skull II: Frontal and Lateral Aspects (BM, 25 min.)

8. Anatomy of the Skull III: Basal Aspects (BM, 40 min.)

9. Anatomy of the Skull IV: Interior of the Cranial Cavity (BM, 36 min.)

10. Anatomy of the Thorax (BM, 36 min.)

11. Anatomy of the Upper Limbs (BM, 52 min.)

12. Anatomy of the Vertebral Column (BM, 30 min.)

13. Bodywatch: Beyond Calcium (SU, 30 min.)

Computer Software

1. Dynamics of the Human Skeletal System (EI, C-3052, 1988, Apple or IBM)

2. Bone Probe (PLP, CH-175027, Apple 64K)

   See Guide to Audio-Visual Resources on p. 339 for key to AV distributors.

LECTURE ENHANCEMENT MATERIAL

Clinical and Related Terms

1. Acrocephaly — a condition also called oxycephaly in which the top of the skull is pointed or conical. Usually due to the premature closure of the coronal and lambdoidal sutures.

2. Bowleg — an outward bending of the legs, sometimes seen in cases of rickets.

3. Bunion — an abnormal thickening, swelling, and inflammation of the bursa of the great toe, resulting in a lateral displacement. Usually caused by excessively tight shoes.

4. Bursectomy — an excision and drainage of a bursa.

5. Caudal anesthesia — a procedure in which anesthetics are injected through the sacral hiatus to anesthetize the sacral and coccygeal nerves.

6. Clawfoot — a deformity, sometimes hereditary, of the foot, where the longitudinal arch is extremely high and there is a flexion of the distal joints.

7. Craniotomy — a surgical operation on the cranium. Sometimes used to decrease the size of the fetal skull to facilitate a difficult delivery.

8. Deviated nasal septum — a laterally deflected nasal septum that, if severe, may block a nasal passageway, thus requiring surgery.

9. Gravity inversion — a physical therapy traction technique, using the body's own weight to decompress the backbone by inversion.

10. Knock-knee — a deformity in which the knees rub or knock against each other while walking. Usually occurs in children as a result of irregular growth of the leg bones, injury to the ligaments, or injury to the ends of the bones.

11. Pelvimetry — the measurement of the opening and capacity of the pelvis, usually to determine if there is enough room for a normal delivery.

12. Rachiotomy — the surgical incision of a vertebra or parts of the vertebral column.

13. Sinusitis — an inflammation of the sinuses, usually the paranasal sinuses.

14. Trephone — a saw used for removing a small circular piece of bone from the skull. Also, in ancient Peru, a religious practice (trephining) of drilling holes into the skulls of victims to release "evil" spirits.

15. Whiplash — a term used for injury to the cervical vertebrae caused by a sudden jerking or backward-forward motion. In severe cases, the dens of the axis may be driven into the medulla, resulting in death.

16. Wormian bone — an irregular-shaped bone of variable size that forms in suture lines of the skull.

Disorders/Homeostatic Imbalances

Congenital Disorders

1. Cleft Lip/Palate — a congenital defect in the fusion of the embryonic maxillary and/or median nasal processes leading to a minor or major fissure of the soft and/or bony tissues of the lip and palate. May be unilateral or bilateral, complete or incomplete. Surgical repair is usually necessary.

2. Chondromalacia Patellae — an abnormal softening of the articular cartilage of the patella. Usually of hereditary influence.

3. Congenital Dislocation of Hip — a common congenital defect of the hip-joint socket (acetabulum) where the affected socket is shallow, the rim improperly developed, and the ligaments of the joint relatively lax. The head of the femur fails to maintain its proper position and is easily dislocated.

4. Mandibular Cleft — a congenital defect in the fusion of the embryonic mandibular processes leading to a minor or major fissure of the soft and/or bony tissues of the mandible.

5. Macrocephaly — an abnormal enlargement of the cranium. May be due to improper fusion or development of the fontanels or to continued production of growth hormones.

6. Microcephaly — an abnormally small cranium. Often due to premature closure of the fontanels.

7. Polydactyly — a congenital disorder usually resulting in the presence of supernumerary fingers or toes.

8. Syndactyly — a common disorder, usually congenital, resulting in the persistence of webbing between adjacent fingers or toes, resulting in digits that are more or less completely attached.

Infectious/Inflammatory Disorders

1. Brodie's Abscess — an abscess, usually caused by tuberculosis or some other infectious agent, that affects the head of a bone. Symptoms include pain, swelling, and tenderness of the area.

2. Saber Shin — a condition seen in congenital syphilis that is characterized by a sharp anterior convexity of the tibia (shin).

3. Pott's Disease — an inflammation of the spine, resulting in carious lesions or osteitis of the vertebrae, usually caused by systemic tuberculosis. If untreated, the disease may cause complete destruction of the vertebrae and intervertebral discs.

4. Spondylosis — a serious, usually chronic, inflammation of the vertebrae, commonly associated with tuberculosis. Continued erosion of the vertebrae may result in hunchback.

ANSWERS TO END-OF-CHAPTER QUESTIONS

Multiple Choice/Matching

1. (1) b,g; (2) h; (3) d; (4) d,f; (5) e; (6) c; (7) a,b,d,h; (8) i
2. (1) g; (2) f; (3) b; (4) a; (5) b; (6) c; (7) d; (8) e
3. (1) b; (2) f; (3) e; (4) a; (5) h; (6) e; (7) f.

Short Answer Essay Questions

4. Cranial bones: parietal, temporal, frontal, occipital, sphenoid, and ethmoid. The facial bones: mandible, vomer, maxillae, zygomatics, nasals, lacrimals, palatines, and inferior conchae. Cranial bones provide sites for attachment, and enclose and protect the brain. The facial bones form the framework of the face, hold eyes in position, provide cavities for organs of taste and smell, secure the teeth, and anchor facial muscles. (pp. 179-193)

5. At birth, the skull is huge relative to the face and body. During childhood and adolescence, the face grows out from the skull. By adulthood, the cranial and facial skeletons are the appropriate proportional size. (p. 219)

6. Normal curves are: cervical, thoracic, lumbar, and sacral. The thoracic and sacral are primary; the cervical and lumbar are secondary. (p. 194)

7. Cervical vertebrae possess transverse foramina, have small bodies and bifid spinous processes; thoracic vertebrae possess facets for the ribs and have circular vertebral foramina; lumbar vertebrae have massive bodies and blunt spines. (p. 199)

8. a. The discs act as shock absorbers and allow the spine to flex and extend.

   b. The annulus fibrosis composed of fibrocartilage is more external and contains the nucleus pulposus. The nucleus is the semifluid substance enclosed by the annulus.

   c. The annulus provides strength and durability.

   d. The nucleus provides resilience.

   e. Disc herniation involves protrusion of the nucleus pulposus through the annulus. (p. 195)

9. The bony thorax includes the thoracic vertebrae, ribs, sternum, and costal cartilages. (p. 200)

10. a. True rib: attached at both ends directly; false rib: attached at vertebrae directly, sternum indirectly.

    b. A floating rib is a false rib.

    c. Floating ribs lack a costal cartilage connection. (p. 202)

11. The pelvic girdle functions to attach and transfer the weight of the body to the lower limbs. The bones are large, strong, and securely attach the bones of the thigh to the axial skeleton. The pectoral girdle bones are light and quite mobile to provide flexibility at the expense of strength and stability. (p. 208)

12. The female pelvis inlet and outlet are wider, the pelvis is shallower, lighter, and rounder than that of the male, and the ischial tuberosities are farther apart. (p. 211)

13. The arches distribute the weight of the body. (p. 215)

14. Cleft palate: persistent opening in the palate interferes with sucking and can lead to aspiration of food into the lungs. Hip dysplasia: abnormality that allows the head of the femur to slip out of the acetabulum. (p. 218)

15. In a young adult skeleton, the bone mass is dense, water content is normal in discs, the vertebral column is strong. In old age, the discs decline in water content and become thinner and less elastic, the spine shortens and becomes an arc, and all the bones lose mass. (p. 218) The thorax becomes more rigid with increasing age, mainly due to the ossification of costal cartilage. The cranial bones lose less mass with age than most bones, but the facial contours of the aged change. (p. 219)

Critical Thinking and Application Questions

1. The entire shoulder region collapses medially when the clavicle is fractured. The medial two-thirds of the clavicle is convex anteriorly so it will collapse anteriorly rather than posteriorly toward the subclavian artery. (p. 203)

2. He experienced a dislocated jaw with the mandibular condyles popping out of the socket. (p. 190)

3. A lateral curvature is scoliosis due to an uneven pull of muscles. Since muscles on one side of the body were nonfunctional, those on the opposite side caused an uneven pull and forced the spine out of alignment. (p. 196)

4. The fracture of the neck of the femur is usually called a broken hip and is common in the elderly due to osteoporosis, which especially weakens the vertebrae and neck of the femur. (p. 219)

Key Figure Questions

Figure 7.1 Vertebra, rib, sternum. **Figure 7.4** All other bones of the cranium articulate with the sphenoid bone. **Figure 7.13** Because the lumbar region bears the greatest weight, its vertebrae are the sturdiest. **Figure 7.23** You would look for the styloid process of the ulna, which is on its medial aspect, and for two depressions on its lateral aspect, which receive the head and the distal end of the radius.

SUGGESTED READINGS

1. Basmajian, J.V. *Grant's Method of Anatomy.* 10th ed. Baltimore: Williams and Wilkins, 1980.

2. Clemente, C.D. *Anatomy: A Regional Atlas of the Human Body.* 2nd ed. Philadelphia: Lea and Febiger, 1981.

3. Gordon, K.R. "Adaptive Nature of Skeletal Design." *BioScience* 39 (Dec. 1989): 784-790.

4. Hamilton, W.J. *Textbook of Human Anatomy.* 2nd ed. St. Louis: C.V. Mosby, 1976.

5. Jaffe, C. "Medical Imaging." *American Scientist* 70 (Nov.-Dec. 1982): 576-585.

6. Keiffer, S.A., and E.R. Heitzman. *An Atlas of Cross-Sectional Anatomy.* New York: Harper & Row, 1979.

7. McMinn, R.M.H., and R.T. Hutchings. *Color Atlas of Human Anatomy.* Chicago: Year Book Medical Publishers, 1977.

8. Newschwander, G. E., et al. "Limb Lengthening with Llizaron External Fixator." *Orthopaedic Nursing* 8 (May-June 1989): 15-21.

9. Shipman, P. A. Walker, and D. Bichell. *The Human Skeleton.* Cambridge: Harvard University Press, 1985.

10. Woodburne, R.T. *Essentials of Human Anatomy.* 7th ed. New York: Oxford University Press, 1983.

11. Yokochi, C., and J.W. Rohen. *Photographic Anatomy of the Human Body.* 2nd ed. Tokyo, New York: Igaku-Shoin, Ltd., 1978.

8

Joints

Chapter Preview

This chapter describes the characteristics of the points of contact between two or more bones. These areas, called joints or articulations, help secure our bones together and allow our rigid skeleton to move with grace and strength. The three main types of joints, the types of movements allowed at these joints, homeostatic imbalances, and developmental aspects of the joints will also be presented.

INTEGRATING THE PACKAGE

Suggested Lecture Outline

I. Introduction to Articulations (p. 222)

II. Classification of Joints (p. 222; Table 8.1, pp. 223-225)

    A. Structural Classification (p. 222)

        1. Fibrous

        2. Cartilaginous

        3. Synovial

    B. Functional Classification (p. 222)

        1. Synarthroses

        2. Amphiarthroses

        3. Diarthroses

III. Fibrous Joints (pp. 223-225; Fig. 8.1, p. 223)

    A. General Characteristics (p. 223)

    B. Sutures (p. 223)

    C. Syndesmoses (p. 223)

    D. Gomphoses (p. 225)

IV. Cartilaginous Joints (p. 226; Fig. 8.2, p. 226)

    A. General Characteristics (p. 226)

    B. Synchondroses (p. 226)

    C. Symphyses (p. 226)

V. Synovial Joints (pp. 226-240)

    A. General Characteristics (p. 226; Fig. 8.3, p. 227)

    B. General Structure (pp. 226-227)

        1. Articular Cartilage

        2. Joint Cavity

        3. Articular Capsule

2. Hip (Coxal) Joint (Fig. 8.10, p. 236)

    a. Basic Features

    b. Ligaments

    c. Tendons

3. Elbow Joint (Fig. 8.11, p. 237)

    a. Basic Features

    b. Ligaments

    c. Tendons

4. Knee Joint (Fig. 8.12, p. 238; Fig. 8.13, p. 239)

    a. Basic Features

    b. Ligaments

    c. Tendons

    d. Analysis of Knee Movements

VI. Homeostatic Imbalances of Joints (pp. 240-244)

  A. Common Joint Injuries (p. 240)

    1. Sprains

    2. Cartilage Injuries

    3. Dislocations

  B. Inflammatory and Degenerative Conditions (pp. 241-244)

    1. Bursitis and Tendonitis

    2. Arthritis

      a. Basic Features

      b. Osteoarthritis (OA)

      c. Rheumatoid Arthritis (RA) (Fig. 8.14, p. 241)

      d. Gouty Arthritis

  C. Lyme Disease (p. 244)

VII. Developmental Aspects of Joints (p. 244)

  A. Embryonic and Fetal Development of the Joints

  B. Development of the Joints Through Adolescence

  C. Effects of Aging on the Joints

Review Items

1. Ligaments and tendons (dense connective tissue) (Chapter 4, p. 122)

2. Hyaline cartilage (Chapter 4, p. 123; Fig. 4.8G, p. 119)

3. Fibrocartilage (Chapter 4, p. 123; Fig. 4.8I, p. 120)

4. Epiphyseal plate (Chapter 6, p. 159)

5. Articular cartilage (Chapter 6, p. 159)

6. Periosteum (Chapter 6, p. 159)

7. Planes of the body (Chapter 1, pp. 14-17; Fig. 1.8, p. 16)

Cross References

1. The periodontal ligament is illustrated in Chapter 24.

2. Intervertebral discs are also discussed in Chapter 7, p. 195, and illustrated in Fig. 7.14, p. 195.

3. The stability/flexibility of the pectoral (shoulder) girdle is also discussed in Chapter 7, p. 203.

4. The role of synovial joints in the movement of the body is detailed in Chapter 10.

Laboratory Correlations

1. Marieb, E. N. *Human Anatomy and Physiology Laboratory Manual: Cat and Fetal Pig Versions.* 3rd. ed. Benjamin/Cummings, 1989.

   Exercise 13: Articulations and Body Movements

2. Marieb, E. N. *Human Anatomy and Physiology Laboratory Manual: Brief Version.* 3rd. ed. Benjamin/Cummings, 1992.

   Exercise 11: Articulations and Body Movements

Transparencies Index

8.1	Fibrous joints	8.3	Synovial joint
8.2	Cartilaginous	8.8	Types of synovial joints

Bassett Atlas Figures Index

Slide Number	Figure Number	Description
59	6.7A	Shoulder Joint
60	6.8	Shoulder joint
63	6.11	Elbow joint
71	7.3A	Hip Joint
78	7.10	Knee joint, patellar ligament
79	7.11	Knee joint, open
80	7.12	Knee joint, cruciates cut
86	7.18A	Ankle joint

INSTRUCTIONAL AIDS

Lecture Hints

1. Clearly distinguish between the two systems of joint classification (structural and functional).

2. Point out the difference between the joint of the first rib and sternum in contrast to ribs 2-10.

3. Emphasize that a muscle must cross a joint in order to cause movement.

4. Opposition of the thumb and great toe is possible in many primates (monkeys).

5. Compare and contrast the size and shape of the glenoid cavity and acetabulum.

6. If only one synovial joint will be studied in detail, the best choice is the knee.

Demonstrations/Activities

1. Audio-visual materials of choice.

2. Obtain an articulated skeleton to exhibit joints such as sutures, syndesmoses, gomphoses, and others.

3. Select a student to illustrate types of motion, such as abduction, flexion/extension, etc.

4. Call on students to demonstrate the various types of body movements: abduction, adduction, flexion, extension, etc., occurring at specific joints (e.g., flex your knee, rotate you hand).

5. Obtain a 3-D model of a joint, such as the knee, to illustrate the relationship of ligaments, cartilage, and muscle. A fresh beef knee joint could also be used.

6. Obtain X-rays of patients with gouty arthritis, osteo-arthritis, and rheumatoid arthritis.

7. Obtain a video or request that a local orthopedic surgeon visit the class and describe the techniques and advantages of arthroscopic knee surgery.

8. Obtain an X-ray showing a prosthetic joint.

Critical Thinking/Discussion Topics

1. Why are diarthroses found predominantly in the limbs while synarthroses and amphiarthroses are found largely in the axial skeleton?

2. What are the advantages of the shoulder joint being the most freely moving joint in the body?

3. Cortisone shots can readily reduce swelling that occurs in joints, such as the shoulder and knee, following athletic injuries. Why is it dangerous for athletes to continue getting these shots?

4. Physical therapists suggest various stretching exercises before proceeding with rigorous physical activity. Of what value are these exercises for the joint areas?

5. What does it mean to be "double-jointed"?

6. Most people can "crack" their knuckles. What does this term mean and what effect, if any, will this have on the knuckles in the future?

7. Bones appear to have numerous projections and protuberances. What do you suppose these are for?

Library Research Topics

1. Joints may often be injured during sports activities. What are the major joint injuries associated with football, basketball, baseball, and tennis?

2. Congenital dislocation of the hip is an orthopedic defect in which the acetabulum is too shallow and as a result the head of the femur has poor articulation. What is the current treatment for this defect?

3. The replacement of a damaged joint with an artificial one is becoming more common. Currently, which joints can be replaced?

4. Temporomandibular joint disorders are very painful. What methods of treatment are there and how successful are they? How do these joint disorders arise?

5. Much controversy surrounds the use of the drug dimethyl sulfoxide (DMSO). Why is the FDA so reluctant to provide full approval of this drug for use on humans when it's widely used for horses?

6. Contact an orthopedic surgeon in your area for information and/or videos on arthroscopic surgery.

7. Review the literature on the procedures and materials used for artificial joint replacements.

8. Rheumatoid arthritis appears to be an autoimmune disease. What are the current methods of treatment and what is the future prognosis for this disease and its cure?

9. What is the difference between the action of nonsteroid antiinflammatory drugs and steroid anti-inflammatory drugs? What are the advantages and disadvantages of each?

Audio-Visual Aids/Computer Software

Videotapes

1. Moving Parts (FHS, QB-830, VHS/BETA). Describes muscle coordination and activity. Describes the role of joints and shows a human knee joint. Also available as 16 mm film.

2. Gluteal Region and Hip Joint (TF, 19 min., C, VHS)

3. Knee Joint (TF, 16 min., C, VHS)

   See Guide to Audio-Visual Resources on page 339 for key to AV distributors.

LECTURE ENHANCEMENT MATERIAL

Clinical and Related Terms

1. Arthralgia — pain in a joint.

2. Arthroplasty — the artificial or plastic repair of a joint.

3. Arthrosis — a term referring to a joint or articulation, or to a disease of a joint.

4. Bursectomy — the surgical removal of a bursa.

5. Chondritis — an inflammation of cartilage.

6. Double-jointed — an unusually flexible joint that allows greater mobility than normal.

7. Kinesiology — the scientific study of the movement of the body.

8. Prosthesis — an artificial substitute for the replacement of a missing part.

9. Rheumatology — the field of medicine that deals with rheumatic disorders, their treatment, diagnosis, causes, and pathologic features.

10. Subluxation — a partial or incomplete dislocation of a joint.

11. TMJ Syndrome — a relatively common malady characterized by a dysfunction of the temporo-mandibular joint (TMJ). Symptoms include pain of the head and neck, a clicking or grinding sensation at the joint, tiredness and soreness of the jaw muscles, and stiffness of the jaw. TMJ syndrome may be triggered by stress, infection, arthritis, a blow to the head, or other factors. Treatment may be symptomatic, such as changing diet, use of tranquilizers, and stress management, or it may involve arthroscopic surgery.

Disorders/Homeostatic Imbalances

Arthritic Disorders

1. Arthrodesis — the surgical fusion or immobilization of a joint. Also known as surgical ankylosis.

2. Rheumatism — a generalized term used to denote any of a variety of disorders affecting the joints and related structures that are characterized by inflammation, pain, stiffness, limited motion, and degeneration of the tissue.

3. Tuberculosis Arthritis — an arthritis induced as a complication of disseminated tuberculosis. Usually affects the spine (Pott's disease or spondylitis) or the hip joint, knee, elbow, or wrist. The disease tends to become chronic and destructive, beginning with edema, congestion, and thickening of the synovial membranes. Bone destruction of the joint with a thick pannus formation usually occurs if treatment involving INH and/or rifampin is not begun immediately.

4. Supperative Arthritis — a pus and leukocyte infiltration of a joint due to a bacterial infection. Most commonly caused by gonococci, staphylococci, streptococci, and gram-negative bacilli such as *E. coli*. The infection usually results from direct bacterial invasion due to trauma or hematogenous spread. Symptoms include redness, swelling, pain, and tenderness. Lack of antibiotic treatment may lead to chronic involvement and destruction to the articular surfaces.

Neoplastic Disorders

1. Synovial Sarcoma — an uncommon malignant tumor that arises in the bursae, tendon sheaths, and in the articular surfaces of the joint, usually of the lower extremities. The lesions arise slowly to become painful, soft-to-gritty masses. The tumors tend to metastasize, but respond fairly well to aggressive therapy.

Miscellaneous Lesions

1. Ganglion — a small cystic lesion most commonly found in the collagenous connective tissue of a joint capsule or tendon such as the small joint of the wrist. It may arise from a degeneration and cystic softening of connective tissue. The lesions do not infiltrate and are easily removed surgically. The "Bible Treatment," or smashing the cyst with a heavy book, is not recommended.

2. Tenosynovitis — an inflammation of the sheaths surrounding a tendon, associated most often with athletes and those who put great stress on certain tendons during their activity. It may also arise from bacterial infections. Usually rest, heat, and other supportive measures appear most satisfactory for treatment.

ANSWERS TO END-OF-CHAPTER QUESTIONS

Multiple Choice/Matching

1. (1) c; (2) a; (3) a; (4) b; (5) c; (6) b; (7) b; (8) a; (9) c
2. b
3. d
4. d
5. b
6. d
7. d

Short Answer Essay Questions

8. Joints are defined as the site of contact between two or more bones. (p. 222)

9. Freely moveable joints provide mobility; slightly moveable joints provide strength with limited flexibility; immovable joints provide secure enclosures and protection. (pp. 222-223)

10. Bursae are synovial membrane-lined sacs that function to prevent friction, and are located where ligaments, muscles, skin, and/or muscle tendons overlie and rub against bone. In the latter case, the friction-reducing structures are called tendon sheaths. (p. 227)

11. Nonaxial movements mean slipping movements only, uniaxial movements mean movement in one plane, biaxial movements mean movement in two planes, and multiaxial movements mean movement in or around all three planes and axes. (p. 229)

12. Flexion and extension refer to decreasing or increasing the angle of a joint and bringing the two articulating bones together along the sagittal plane, while adduction and abduction refer to moving a limb closer to or away from the midline along the frontal plane. (p. 231)

13. Rotation refers to a bone turning around its own long axis, while circumduction means to move a limb so that it describes a cone in space, an action that involves a variety of movements. (pp. 231-232)

14. Uniaxial — hinge (elbow) and pivot (atlantoaxial); biaxial — condyloid (knuckle) and saddle (thumb); multiaxial — ball and socket (shoulder and hip). (pp. 232-233)

15. The knee menisci deepen the articulating surface of the tibia to prevent side-to-side rocking of the femur on the tibia and to absorb shock transmitted to the knee joint. The cruciate ligaments prevent anterior/posterior displacement of the articulating bone and help to secure the joint. (pp. 238-239)

16. The knees rely heavily on nonarticular factors for stability, since they must carry the total body weight. The knees can absorb an upward force of great intensity, but they must also absorb direct blows and blows from the side, though they are poorly designed to do so. (pp. 238-240)

17. Cartilages and ligaments are poorly vascularized and tend to heal very slowly. (p. 240)

Critical Thinking and Application Questions

1. Most likely bursitis of the subcutaneous prepatellar bursa. It is a good guess that Sophie spends a good deal of time on her knees (perhaps scrubbing the floors). (p. 241)

2. a. Yes and no.

   b. Ligaments.

   c. Returning bones back to position without an incision.

   d. Sprains heal slowly and need repair to stabilize joint.

   e. The examination of a joint by means of an endoscope.

   f. Using arthroscopic surgery, only small incisions are needed instead of an open surgical wound. There is less chance of infection and healing is considerably faster. (p. 240)

3. a. Probably gout, although it is more common in males.

   b. Caused by a deposition of uric acid crystals in soft tissues of joints. (p. 244)

Key Figure Questions

Figure 8.3 Mobility.

SUGGESTED READINGS

1. Allman, W.F. "The Knee." *Science* 83 (Nov. 1983)

2. Arehart-Treichel, J. "A Healing Scaffold." *Science News* 122 (Oct. 1982): 219.

3. Arehart-Treichel, J. "The Joint Destroyers." *Science News* (Sept. 1982): 156-157.

4. Aufranc, O.E., and R.H. Turner. "Total Replacement of the Arthritic Hip." *Hospital Practice* (Oct. 1971).

5. Barnett, C.H., et al. *Synovial Joints: Their Structure and Mechanics*. Springfield: Charles Thomas, 1961.

6. Beil, L. "Of Joints and Juveniles." *Science News* 134 (Oct. 1983): 190-191.

7. Edwards, D.D. "Arthritis: Looking for Immunotherapy." *Science News* 131 (Apr. 1987): 228.

8. Evans, F.G. *Studies On the Anatomy and Function of Bone and Joints.* New York: Springer-Verlag, 1966.

9. Fackelmann, K.A. "Chicken Cartilage Soothes Aching Joints." *Science News* 144 (Sept. 1993): 198.

10. Fackelmann, K.A. "The Nine-Month Arthritis 'Cure.'" *Science News* 144 (Oct. 1993): 144.

11. Frankel, V.H., and M. Nordine. *Basic Biomechanics of the Skeletal System.* Philadelphia: Lea and Febiger, 1980.

12. Higging, J.R. *Human Movement: An Integrated Approach.* St. Louis: C.V. Mosby, 1977.

13. Koerner, M.E., and G.R. Dickinson. "Adult Arthritis: A Look at Some of Its Forms." *American Journal of Nursing* 83 (Feb. 1983): 254-278.

14. Langone, J. "Back Surgery Without Stitches." *Time* (Sept. 1988): 55.

15. Simon. W.H. *The Human Joint in Health and Disease*. Philadelphia: University of Pennsylvania Press, 1978.

16. Sonstegard, D.A., et al. "The Surgical Replacement of the Human Knee Joint." *Scientific American* (1978).

17. Toufexis, A. "Treating an 'In' Malady." *Time* (Apr. 1988): 102-104.

18. Walker, P.S. "Joints to Spare." *Science* 85 (Nov. 1985): 57.

9

Muscles and Muscle Tissue

Chapter Preview

This chapter introduces the important functions and characteristics of the three types of muscle tissues that collectively make up about half of total body mass. The gross and microscopic features of skeletal and smooth muscle tissue will be described in detail. The physiology of contraction, including the sliding filament mechanism and the methods of energy production, will also be illustrated.

INTEGRATING THE PACKAGE

Suggested Outline

  2. Myofibrils (Fig. 9.2, p. 252)

      a. Characteristics

      b. Striations, Sarcomeres, and Myofilaments (Fig. 9.2, p. 252)

      c. Ultrastructure and Molecular Composition of the Myofilaments (Fig. 9.3, p. 253)

  3. Sarcoplasmic Reticulum and T-Tubules (Fig. 9.5, p. 255)

C. Contraction of a Skeletal Muscle Fiber (pp. 255-258)

  1. Sliding Filament Mechanism of Contraction (Fig. 9.6, p. 256)

  2. Regulation of Contraction

      a. The Neuromuscular Junction and the Nerve Stimulus

      b. Generation of an Action Potential Across the Sarcolemma

      c. Destruction of Acetylcholine

      d. Roles of Ionic Calcium in Muscle Contraction

      e. Excitation-Contraction Coupling (Fig. 9.11, p. 262)

D. Contraction of a Skeletal Muscle (pp. 263-267)

  1. Basic Characteristics

  2. The Motor Unit (Fig. 9.12, p. 263)

  3. The Muscle Twitch and Development of Muscle Tension (Fig. 9.13, p. 264)

  4. Graded Muscle Responses

      a. Basic Features

      b. Wave Summation and Tetanus (Fig. 9.14, p. 265)

      c. Multiple Motor Unit Summation

  5. Treppe: The Staircase Effect (Fig. 9.15, p. 266)

  6. Muscle Tone

  7. Isometric and Isotonic Contractions

E. Muscle Metabolism (pp. 267-269)

  1. Providing Energy for Contraction

      a. Stored ATP (Fig. 9.16, p. 267)

      b. Direct Phosphorylation of ADP by Creatine Phosphate

      c. Aerobic Respiration

      d. Anaerobic Glycolysis and Lactic Acid Formation

      e. Energy Systems Used During Sports Activities

  2. Muscle Fatigue

  3. Oxygen Debt

  4. Heat Production During Muscle Activity

F. Force, Velocity, and Duration of Muscle Contraction (pp. 270-274; Fig. 9.19, p. 271)

  1. Force of Contraction

      a. Number of Muscle Fibers Stimulated

      b. Relative Size of the Muscle

      c. Series-Elastic Elements

      d. Degree of Muscle Stretch

  2. Velocity and Duration of Contraction

      a. Load (Fig. 9.22, p. 272)

      b. Muscle Fiber Type

Review Items

1. Connective tissues (Chapter 4, pp. 112-124).
2. The structure of bone tissue (Chapter 6, pp. 158-162).
3. General cellular structural components (Chapter 3, pp. 75-86).
4. ATP (Chapter 2, p. 55).
5. Muscle tissue (Chapter 4, p. 124).
6. Ions (Chapter 2, p. 32).
7. Membrane transport (Chapter 3, p. 65).
8. Microfilaments (Chapter 3, p. 82).
9. Gap junctions (Chapter 3, p. 65).

Cross References

1. General structure and function of synapses, and neurotransmitters are described in detail in Chapter 11, pp. 360-371.
2. Skeletal muscles of the body are examined in detail in Chapter 10.

3. A detailed examination of the function of cardiac tissue is given in Chapter 19.

4. Smooth muscle utilization is illustrated throughout the text, and especially in Chapter 20 (Blood Vessels), Chapter 24 (Digestive System), Chapter 28 (Reproductive System).

5. Metabolic pathways of energy production (glycolysis, Kreb's cycle, and electron transport) are described in detail in Chapter 25.

6. Single-unit smooth muscle function is described in relation to peristalsis in Chapter 24 (Digestive System).

7. Joint stability as a function of skeletal muscle contraction is described in Chapter 8.

8. Membrane potentials are described in Chapter 3, p. 73.

9. Sensory receptors located in skeletal muscle are described in Chapter 13, p. 429.

10. Shivering as a heat production mechanism is discussed in Chapter 25, p. 888.

11. The interaction between muscle and bones is described in Chapter 10.

12. An example of the sliding filament mechanism of muscle contraction is presented in Chapter 19.

13. Motor neurons of the peripheral nervous system and the neuromuscular junction are further described in Chapter 13, p. 434.

Laboratory Correlations

1. Marieb, E. N. *Human Anatomy and Physiology Laboratory Manual: Cat and Fetal Pig Versions*. 3rd. ed. Benjamin/Cummings, 1989.

    Exercise 14: Microscopic Anatomy, Organization and Classification of Skeletal Muscles

    Exercise 16: Muscle Physiology

2. Marieb, E. N. *Human Anatomy and Physiology Laboratory Manual: Brief Version*. 3rd. ed. Benjamin/Cummings, 1992.

    Exercise 12: Microscopic Anatomy, Organization, and Classification of Skeletal Muscles

    Exercise 14: Muscle Physiology

Transparencies Index

9.1	Connective tissue of skeletal muscles	9.9	Sliding of actin filaments/contraction
9.3	Myofilament composition in skeletal muscle	9.11	Excitation-contraction coupling
9.8	Sequence of events involved in sliding of actin filaments	9.17	Skeletal muscle energy metabolism

---

INSTRUCTIONAL AIDS

Lecture Hints

1. The terminology used in the description of the functional characteristics of muscle is confusing to many students. Point out that "extend" is a root of the word *extensibility*, and it is easy to associate that stretching with extending. Elasticity (the ability to recoil) is the opposite of extensibility.

2. Since the prefixes endo-, epi-, peri- are often used in anatomical terminology, emphasize the meanings and indicate that students will see these again.

3. A nice way to pull together the overall structure of muscle tissue from myofilaments to the entire muscle is by a cross-sectional diagram. Students are often confused by the similarity in the names of the structures, e.g., myofilaments vs. myofibril vs. myofiber. Use analogies to help students remember: filaments are like those in light bulbs, very thin, forming the basic structure for causing an action (light); in contrast, fibers are larger structures and, like nerve fibers, are cells; the name *fibril* is intermediate between the two.

4. In the description of the sliding filament mechanism, be sure to use several drawings (or diagrams) of the relationships between the thick and thin filaments during contraction. Students are easily confused by the series of static diagrams presented. Be sure to indicate the dynamic nature of contraction.

5. Students often have difficulty with "all or none" as applied to individual muscle fibers. Clearly point out the distinction between muscle cells (and motor units) vs. whole muscle. Use the analogy of a light switch control of all the lamps attached to it—either on or off, no in-between.

6. Emphasize that graded muscle contraction is achieved by increasing the frequency of stimulation of motor units or increasing the number of motor units activated.

7. As a way of introducing isometric and isotonic contractions, ask the class if muscle contraction always results in movement. Whatever their answer, illustrate the point by attempting to lift a fixed object in the classroom. Point out that although force is increasing, the object is not being lifted; therefore the muscle must remain the same length. Ask what system of measuring length is used in science. Someone will answer "metric" at which time the definition of the prefix iso- should be given. Use of real-life analogies will help students remember these very similar terms.

8. To illustrate length-tension relationships, ask the class to comment on the amount of force generated if myosin and actin do not overlap at all, so that the myosin heads do not cross-bridge with actin. Conversely, ask what would happen if myosin and actin were completely overlapped.

9. When explaining the differences between slow and fast twitch fibers, it is helpful to give examples of different types of athletes and the types of muscle fibers that predominate as a result of specific exercises. Students remember these examples (and the principles) easily when they can relate to real-life experience.

10. Explain that all muscle types contain actin and myosin myofilaments, but that the arrangement (in part) accounts for the structural and functional differences.

11. Emphasize the relationship between structure and function in smooth muscle. In general, smooth muscle is controlled automatically so that we do not have to consciously think about controlling action. Single-unit muscle contracts sequentially in a wave and is found where something needs to be pushed, usually along a tubular structure. Due to the design of single-unit muscle, fine control of muscle movement would not occur. Multiunit smooth muscle is required where automatic but precise control is needed. An excellent example to relate is the multiunit control of pupil diameter in the eye. We have very precise control of pupillary diameter, and therefore the amount of light entering the eye. Ask the class what would happen if the internal eye muscles were formed of single-unit smooth muscle, or even skeletal muscle.

12. Be sure to inform students that the terms *striated* and *skeletal muscle* are interchangeable, and that although cardiac muscle is striated, the term *striated* should not be used as a name for cardiac muscle.

Demonstrations/Activities

1. Audio-visual materials of choice.

2. Demonstrate muscle contraction using a simple myograph or kymograph apparatus and the gastrocnemius muscle of a frog. A film loop showing these events might be used. It is important that students be able to visualize these events.

3. Use models that compare the three types of muscle tissue to point out the unique structural characteristics of each type.

4. Set up a microscope with a slide of a motor unit for class viewing.

5. Obtain a 3-D model of a sarcomere to exhibit tubules and myofibrils.

6. Obtain or construct a 3-D model of a sarcomere that illustrates the sliding filament mechanism of muscle contraction.

7. Use a microprojector and microslides of skeletal, smooth, and cardiac muscle to illustrate their microscopic similarities and differences.

8. Ask students to demonstrate examples of isometric and isotonic contractions and explain how individual muscle cells and motor units are behaving to cause muscle contraction.

9. Use an articulated skeleton to point out various origins and insertions, then ask students to specify the resulting movement.

10. Pick apart a piece of cooked chicken breast to demonstrate individual fascicles.

Critical Thinking/Discussion Topics

1. You are caught by a sudden snowstorm and you are without shelter or warm clothing. Why is it important for you to keep moving and exercising rather than sitting on the ground, motionless, waiting for rescue?

2. Muscles that are immobilized for long periods of time, as with a cast, frequently get smaller. Why? What is necessary to revitalize them?

3. Why do you suppose activities such as swimming and fast walking are so beneficial? Are there any negative attributes to activities such as racquetball and sprinting?

4. How can weightlifters have such enormous muscles, while long-distance runners have lean muscles?

5. What effect would there be on the body if intestinal peristaltic waves were stopped either by infection or injury?

6. Why do athletes "warm up" before a competitive event? Would you expect the warm-up period to make contraction more or less efficient? Why?

7. Visit a local gym frequented by body builders. Obtain information on the procedures used to build muscle mass and an explanation of how those procedures accomplish that goal.

8. If the number of myosin heads were doubled, what would be the effect on force production? ATP consumption?

9. What is a muscle spasm? How do you think it may be caused?

10. Are spasms and cramps related? Compare and contrast the different possible mechanisms of each.

11. Draw diagrams of different fascicle arrangements and describe what type of movement is charac-teristic of each (i.e., short range, powerful, etc.) Then have students apply these ideas to place the different fascicle types in logical locations on an articulated skeleton (e.g., the deltoid origin and insertion is a logical example of perfect use of a convergent pattern).

Library Research Topics

1. Why have muscle cells "lost" their ability to regenerate? What current research is being done in this area?

2. Investigate the long-term effect of anabolic steroid use on muscle tissue.

3. Why do you suppose the Olympic committees are so adamant against the use of "performance-enhancing" drugs such as anabolic steroids?

4. Trace the embryonic development of skeletal muscles, noting how they maintain constant contact with neural cells.

5. Explore the current theories for the etiology of muscular dystrophy.

6. What is the current status of the sliding filament model of muscle contraction? Do we know all there is to know?

7. Examine how biofeedback can reduce stress-induced muscle tension.

8. Describe several metabolic diseases of muscle (usually due to an enzyme or enzyme group deficiency).

9. Alcohol can induce a form of toxic myopathy. Describe the effects of alcohol on muscle tissue.

10. Define the term *myositis*. What causative agents could result in this form of muscle disease?

11. Are tumors (benign or malignant) associated with muscle tissue? Would cancer develop in the muscle cells themselves, or in the connective tissue coverings? Explore the possibilities.

Audio-Visual Aids/Computer Software

Videotapes

1. Muscle Power (FHS, QB-829, VHS/BETA). Illustrates microscopic view of muscles and compares all three types of muscle. Also available as 16-mm film.

2. Physical Fitness and Exercise: What Does This Mean to You? (PLP, CH-197013, VHS)

3. Your Body Series - Part 1, Your Muscular System (PLP, CH-140201, VHS)

4. What Can Go Wrong? (Muscular-Skeletal System) (PLP, CH-140503, VHS)

5. Steroids: Shortcut to Make-Believe Muscles (CBS, 49-3810-V, VHS)

Computer Software

1. Skeletal Muscle Anatomy and Physiology (PLP, R-513591, Apple). Examines three categories of muscles, sliding filament theory, and motor units.

2. Neuromuscular Concepts (PLP, R-510002, Apple). Surveys muscle action potentials, contraction, and movement.

3. Exercises in Muscle Contraction (EI, C-3032, 1985, Apple, IBM). Animated exercises in muscle stimulation.

4. Dynamics of the Human Muscular System (EI, C-3051, 1988, Apple or IBM)

5. Biochemistry of Muscle (EI, C-4557, 1987, Apple). Presents muscle contractions, the neuromuscular junction, metabolism, and all three muscle types.

6. The Body Electric (HRM, HRM577A, Apple). Monitor and measure brain waves, electrocardiograms, and the electrical activity of muscles.

7. Biofeedback Microlab (HRM, HRM546A, Apple). Collect and store data, analyze results, and solve problems relating to electrodermal activity, muscle tension, and skin temperature.

8. Mechanical Properties of Active Muscle (QUE, Com4203, IBM). Best Simulation, 1987. Simulates the lab experience with the frog gastrocnemius series of experiments.

9. Skeletal Muscle Anatomy and Physiology (QUE, INT4612A, Apple). Topics include the main categories of muscle tissue, sliding filament mechanism, motor units, and lever systems.

10. Neuromuscular Concepts (QUE, INT4609A, Apple). Tutorials covering muscle action potentials, using the electromyogram, and skeletal muscle disorders.

See Guide to Audio-Visual Resources on page 339 for key to AV distributors.

LECTURE ENHANCEMENT MATERIAL

Clinical and Related Terms

1. Convulsion—a rapid series of involuntary muscular contractions and relaxations, especially of the skeletal muscles. Usually symptomatic of some type of CNS disorder such as epilepsy, tetanus, poisonings, chemical disorders, etc. In children, the cause is most often high fever. Accurate observation and charting of the disorder is important to ascertain the cause.

2. Electromyography — the recording, study, preparation, and interpretation of graphic records of the contractions of muscles as a result of electrical stimulation. Similar to a kymograph. Electrodes are inserted into the muscles and can record the responses of muscles during contraction and at rest and may be used in the diagnosis of diseases such as polio and amyotrophic lateral sclerosis.

3. Fibrosis — the abnormal formation of, or degeneration into, fibrous tissue. May occur around the membranes of the lungs as the result of an inflammation or pneumonia. May also occur in and around the arteries, the uterus, the myocardium, and other areas due to inflammation or disease.

4. Fibrositis — an inflammation and hyperplasia of the white fibrous tissue, especially that which forms muscle sheaths and fasciae layers. Usually causes pain and stiffness. Commonly called muscular rheumatism.

5. Flaccid — a condition in which the muscles are uncommonly weak, flabby, or soft. May be indicative of a defect in muscular tone.

6. Kymograph — a recording instrument used in muscle physiology experiments. The device uses a writing pen that moves in response to the force applied, and marks that response on a rotating drum.

7. Myectomy —the surgical removal of a portion of a muscle.

8. Myoclonus — the spastic, shock-like contractions of a part of a muscle or the entire muscle. Usually a symptom of a convulsive disorder.

9. Myokymia — an involuntary twitching or quivering of a muscle.

10. Myoma — a tumor characterized by containing muscle tissue. May be benign or malignant.

11. Myomalacia —the abnormal softening of muscle tissue.

12. Myosclerosis — the abnormal hardening of muscle tissue.

13. Myositis — an inflammation of a muscle or muscles. May be caused by infection, infestation, injury, or other factors.

14. Myotonia — a condition involving irregular tonic spasms of a muscle or the temporary rigidity of a muscle after normal contraction.

15. Paralysis — a temporary or permanent loss of functional motion or sensation of a muscle or muscles due to an impairment of the neural or muscular mechanism. May result from any of a number of conditions or injuries, especially those that affect the CNS or PNS.

16. Tendonitis — an inflammation of a tendon and of a tendon-muscle attachment. Frequently affects the elbow (tennis elbow) and shoulder in athletes. One of the most common causes of pain around the joints. May involve calcium deposits in chronic cases.

Disorders/Homeostatic Imbalances

Microbiological Injuries or Effects on Muscles

1. Botulism — a severe bacterial food poisoning caused by the anaerobe *Clostridium botulinum*. The botulinum toxin produced in the food has a selective action on the synaptic end of myoneural junctions where it blocks the release of the neuromuscular transmitter, acetylcholine. The use of an antitoxin is the only effective treatment.

2. Gas gangrene — a form of gangrene or tissue wasting that results from the bacterial invasion of dirty, lacerated wounds in which the tissues and muscles become filled with gas. The bacterial agent is usually a member of the genus *Clostridium* that produces enzymes and toxins that dissolve the collagen fibers and produces the characteristic gas. Treatment requires debridement, antibiotic and oxygen therapy, and possibly amputation of the affected part.

3. Tetanus — an acute infectious disease caused by the toxins of the soil bacterium *Clostridium tetani*. The organisms gain access to an infected wound, usually through a puncture, which allows the tissues to become anaerobic. If the patient is not properly vaccinated, the toxin, which circulates in the blood, affects the spinal cord, causing mild-to-severe muscle spasms and contractions of skeletal muscles. Frequently the muscles of the jaw are affected, hence the term *lockjaw*. Antibiotics and/or tetanus antitoxin may be used in therapy, while the disease can be easily
prevented by means of the tetanus toxoid.

4. Trichinosis — a parasitic infection caused by the nematode *Trichinella spiralis* that encysts in muscle tissue following the consumption of infected, raw, or undercooked meat, especially pork. In the early stages of the disease, prior to encystment and when the worms are still in the stomach, there may be diarrhea, nausea, and abdominal pain. Later, the ingested worms release their young which travel to various parts of the body, especially to very active muscles, such as the eye and arms. Periorbital edema, pain of the eye, and soreness in arm movement frequently occur. The larvae eventually encyst, become dormant, and die. Symptomatic treatment is usually the only mode of therapy.

Muscular Dystrophies

1. Duchenne Muscular Dystrophy — see textbook p. 281 for a description of this disease.

2. Limb-Girdle Muscular Dystrophy — a rare autosomal recessive disease that begins in the third decade of life and involves either the shoulder or pelvic girdle. Other adjacent muscles may also be affected. Other types of autosomal muscular dystrophies, such as congenital myotonic and facioscapulohumeral dystrophy have been seen. Treatment, therapy, and prognosis for these dystrophies and Duchenne are generally the same.

Congenital Myopathies

1. Central Core Disease — a rare autosomal dominant disorder characterized by an amorphous central core of fibers with partial or complete absence of myofilaments. Generally characterized clinically as the floppy-infant syndrome. Since the muscle weakness is not progressive, the prognosis is generally fair.

2. Myotonia Congenita — a hereditary disease characterized by tonic spasms, stiffness, and rigidity of muscle groups during attempts at voluntary contractions, such as in walking. With continued exercise, the muscles tend to respond properly.

Tumors and Tumorlike Disorders

1. Rhabdomyosarcoma — a rare sarcoma of the muscle that is of unknown etiology and metastasizes early. Several variants have been seen with a diagnosis based on the observation of cross-striated myofibrils within the myoblastic cells. The prognosis has been poor; however, with surgical advances and radiation treatment, considerable improvement has been seen.

2. Traumatic Myositis Ossificans — the formation and deposition of fibrous tissue and bone following a traumatic injury to a muscle. Tumors of this type frequently arise from athletic injuries, especially to the quadriceps femoris and brachialis muscles. Hemorrhaging of the damaged muscle is followed by fibrous scarring, cartilage formation, and subsequent calcification. Symptoms include pain, tenderness, and swelling.

ANSWERS TO END-OF-CHAPTER QUESTIONS

Multiple Choice/Matching

1. c
2. b
3. c
4. (1)b; (2)a; (3)b; (4)a; (5)b; (6)a
5. c
6. a

7. a
8. d
9. a
10. (1)a; (2)a,c; (3)b; (4)c; (5)b; (6)b
11. a
12. c

13. (1)c,d,e; (2)a,b
14. c
15. c
16. b

Short Answer Essay Questions

17. The functions are: excitability — the ability to receive and respond to a stimulus; contractility — the ability to shorten; extensibility — the ability to be stretched; and, elasticity — the ability to resume normal length after contraction or having been stretched. (p. 249)

18. a. In direct attachment, the epimysium of the muscle is fused to the periosteum of a bone, and in indirect attachment, the muscle connective tissue sheaths extend beyond the muscle as a tendon; the tendon anchors to the periosteum of a bone. (p. 251)

    b. A tendon is a rope-like mass of fibrous tissue; an aponeurosis is a flat, broad sheet. (p. 251)

19. a. A sarcomere is the region of a myofibril between two successive Z-lines and is the smallest contractile unit of a muscle cell. The myofilaments are within the sarcomere. (p. 253)

    b. The theory proposes that the thin filaments are slid toward the center of the sarcomere through the rachet-like action of the myosin heads. The process is energized by ATP. (Fig. 9.7, p. 255)

20. AChE destroys the ACh after it is released. This prevents continued muscle fiber contraction in the absence of additional stimulation. (p. 260)

21. A slight contraction only involves a few motor units that may affect only a few muscle fibers of the muscle, whereas a strong contraction would involve many (or all) motor units. (p. 265)

22. Excitation-contraction coupling is the sequence of events by which an action potential traveling along the sarcolemma leads to the contraction of a muscle fiber. (p. 261)

23. A motor unit is the motor neuron and all the muscle fibers it controls. (p. 263)

24. Table 9.3, p. 273, illustrates the structural and functional characteristics of the three types of skeletal muscle fibers.

25. False. Most body muscles contain a mixture of fiber types that allows them to exhibit a range of contractile speeds and fatigue resistance. However, certain muscle fiber types may predominate in specific muscles, e.g., white fibers predominate in the occular muscles. (p. 274)

26. Muscle fatigue is the state of physiological inability to contract. It occurs due to ATP deficit, lactic acid buildup and ionic imbalance. (p. 270)

27. Oxygen debt is defined as the extra amount of oxygen that must be taken in by the body to provide for restorative processes, and it represents the difference between the amount of oxygen needed for totally aerobic respiration during muscle activity and the amount that is actually used. (p. 269)

28. Smooth muscle is located within the walls of hollow organs and around blood vessels. The tissue is under involuntary control. These characteristics are essential since the vessels and hollow organs must respond slowly, fill and expand slowly, and must avoid expulsive contractions. (pp. 275-280)

Critical Thinking and Application Questions

1. A strain is an excessively stretched or torn muscle, whereas a sprain is an injury to the ligaments reinforcing a joint. The symptoms are similar for both: pain and swelling. (p. 284)

2. Regular resistance exercise leads to increased muscle strength by causing muscle cells to hypertrophy or increase in size. The number of myofilaments increases in these muscles. (p. 276)

3. The reason for the tightness is rigor mortis. The myosin cross-bridges are "locked on" to the actin because of the lack of ATP necessary for release. No, peak rigidity occurs at 12 hours and then gradually dissipates over the next 48 to 60 hours as biological molecules begin to degrade. (p. 257)

Key Figure Questions

Figure 9.6 The H zone and the I zone. **Figure 9.8** The sarcomeres would remain in their partially contracted state. **Figure 9.11** 1. Because calcium binding to troponin frees the actin active sites for myosin cross bridge binding. 2. The initial trigger is neurotransmitter release which generates the action potential along the sarcolemma.

SUGGESTED READINGS

1. Amato, I. "Muscular Dystrophy Gene Cornered." *Science News* 130 (Oct. 1986).

2. Arehard-Treichel, J. "The Retreat of Multiple Sclerosis." *Science News* 123 (Feb. 1983): 142.

3. Carafoli, E., and J.T. Penniston. "The Calcium Signal." *Scientific American* (Nov. 1985).

4. Carlson, F.D., and D. R. Wilkie. *Muscle Physiology.* Englewood Cliffs, NJ: Prentice-Hall, 1974.

5. Cheung, W.Y. "Calmodulin." *Scientific American* (June 1982).

6. Close, R.I. "Dynamic Properties of Mammalian Skeletal Muscles." *Physiological Reviews* 52 (1972): 129.

7. Cohen, C. "The Protein Switch of Muscle Contraction." *Scientific American* (1975).

8. Cooke, R. "The Mechanism of Muscle Contraction." *Critical Review of Biochemistry* 21 (1986): 53-118.

9. DeRosier, D. "The Changing Shape of Actin." *Nature* 347 (Sept. 1990): 21.

10. Endo, M. "Calcium Release From the Sarcoplasmic Reticulum." *Physiological Reviews* 57 (1977): 71.

11. Fackelmann, K.A. "The Nine-Month Arthritis Cure." *Science News* 144 (Oct. 1993): 266.

12. Fackelman, K.A. "The Ups and Downs of Multiple Sclerosis." *Science News* 135 (Apr. 1989): 245.

13. Ferenczi, A., M. Irving, V. Lombardi, and G. Piazzesi. "Myosin Head Movements are Synchronous with the Elementary Force-Generating Process in Muscle." *Nature* 357 (May 1992): 156.

14. Goldman, Y., and H. Higuchi. "Sliding Distance Between Actin and Myosin Filaments per ATP Molecule Hydrolysed in Skin Muscle Fibres." *Nature* 352 (July 1991): 352.

15. Grady, D. "One Foot Forward." *Discover* 11 (Sept. 1990): 86-93.

16. Greenberg, J. "Exercise: A Matter of Life or Death." *Science News* 126 (Sept. 1984): 138-141.

17. Greenberg, J. "Psyching Out Reaches High-Tech Proportions." *Science News* 127 (May 1985).

18. Hall, S. "The Gene Bay." *Hippocrates* 3 (Nov.-Dec. 1989): 75-82.

19. Hartwig, J., S. Hvidt, P.A. Janmey, J. Lamb, G. F. Oster, and T.P. Stossel. "Effect of ATP on Actin Filament Stiffness." *Nature* 347 (Sept. 1990): 95.

20. Hoyle, G. "How a Muscle Is Turned On and Off." *Scientific American* (Apr. 1970).

21. Huddart, H., and S. Hunt. *Visceral Muscle*. New York: Halsted Press, 1975.

22. Huxley, H.E. "The Mechanism of Muscular Contraction," *Scientific American* 213 (Dec. 1985): 18-27.

23. Lester, H.A. "The Response to Acetylcholine." *Scientific American* (Feb. 1977).

24. Linari, M., V. Lombardi, and G. Piazzesi. "Rapid Regeneration of the Actin-Myosin Power Stroke in Contracting Muscle." *Nature* 355 (Feb. 1992): 638.

25. Lisak, R.P. "Myasthenia Gravis: Mechanisms and Management." *Hospital Practice* (Mar. 1983).

26. Margaria, R. "The Sources of Muscular Energy." *Scientific American* (Mar. 1972).

27. Marz, J.L. "Calmodulin: A Protein for All Seasons." *Science* 208 (1980): 728.

28. Merton, P.A. "How We Control the Contraction of Our Muscles." *Scientific American* (May 1972).

29. Murray, J.M., and A. Weber. "The Cooperative Action of Muscle Proteins." *Scientific American* (Feb. 1974).

30. Nadel, E.R. "Physiological Adaptation to Aerobic Training." *American Scientist* 73 (July-Aug. 1985): 334-343.

31. Oster, G. "Muscle Sounds." *Scientific American* (Mar. 1984).

32. Pennisi, E. "3-D Atomic View of Muscle Molecule." *Science News* 144 (July 1993): 4.

33. Pope, H.G., and D.L. Katz. "Affective and Psychotic Symptoms Associated With Anabolic Steroids." *American Journal of Psychiatry* 145 (1988): 487-490.

34. Rosse, C., and D.K. Clawson. *The Musculoskeletal System in Health and Disease*. New York: Harper and Row, 1980.

35. Rowland, L.P. "Duchenne Dystrophy: Behind the Discoveries." *MDA Newsmagazine* (Spring 1989): 16-20.

36. Sandow, A. "Skeletal Muscle." *Annual Review of Physiology* 32 (1970): 479.

37. Smith, K.K., and W.M. Kier. "Trunks, Tongues, and Tentacles: Moving with Skeletons of Muscle." *American Scientist* 77 (Jan.-Feb. 1989): 29-35.

38. Warrwick, H.M., and J.A. Spudick. "Myosin: Structure and Function in Cell Motility." *Annual Review of Cell Biology* 3 (1987): 379-422.

39. Weisburd, S. "Smooth Muscle Cells Twist and Clout." *Science News* 131 (June 1987): 389.

40. Weiss, R. "Muscular Dystrophy Protein Identified." *Science News* 133 (Jan. 1988): 4.

41. Wickelgren, I. "Muscles in Space Forfeit More Than Fibers." *Science News* 134 (Oct. 1988): 277.

42. Wilson, F. *The Musculoskeletal System: Basic Processes and Disorders*. 2nd ed. Philadelphia: Lippincott, 1983.

10

The Muscular System

Chapter Preview

This chapter begins by focusing on the principles of leverage, the interactions of the skeletal muscles in the body, and the methods used in naming the skeletal muscles. The remainder of the chapter describes the characteristics of the major skeletal muscles of the body, including their origin, insertion, action, and location.

INTEGRATING THE PACKAGE

Suggested Lecture Outline

I. Muscle Mechanics: Importance of Leverage and Fascicle Arrangement (pp. 287-290)
   A. Lever Systems: Bone-Muscle Relationships (pp. 287-289; Fig. 10.1, p. 288)
   B. Arrangement of Fascicles (pp. 289-290; Fig. 10.3, p. 290)

II. Interactions of Skeletal Muscles in the Body (pp. 290-291)
   A. General Characteristics (p. 289)
   B. Prime Movers (p. 290)
   C. Antagonists (p. 291)
   D. Synergists (p. 291)

III. Naming of Skeletal Muscles (pp. 290-291)
   A. Location of the Muscle
   B. Shape of the Muscle
   C. Relative Size of the Muscle
   D. Direction of Muscle Fibers
   E. Number of Origins
   F. Location of the Muscle's Origin and/or Insertion
   G. Action of the Muscle

IV. Major Skeletal Muscles of the Body (pp. 292-337; Figs. 10.4, 10.5, pp. 292-295)
   A. General Characteristics (pp. 292-294)
   B. Muscles of the Head, Part I: Facial Expression (pp. 296-297; Fig. 10.6, p. 297)
      1. Basic Features (Table 10.1, pp. 296-297)
      2. Muscles of the Scalp
      3. Muscles of the Face
   C. Muscles of the Head, Part II: Mastication and Tongue Movement (pp. 298-297; Fig. 10.7, p. 299)
      1. Basic Features (Table 10.2, p. 297)
      2. Muscles of Mastication
      3. Muscles Promoting Tongue Movements (Extrinsic Muscles)

D. Muscles of the Anterior Neck and Throat: Swallowing (pp. 300-301; Fig. 10.8, p. 301)

   1. Basic Features (Table 10.3, p. 300)

   2. Suprahyoid Muscles

   3. Infrahyoid Muscles

   4. Pharyngeal Muscles

E. Muscles of the Neck and Vertebral Column: Head and Trunk Movements (pp. 302-305; Fig. 10.9, pp. 303-305)

   1. Basic Features (Table 10.4, pp. 302-305)

   2. Anterolateral Neck Muscles

   3. Intrinsic Muscles of the Back

F. Muscles of the Thorax: Breathing (pp. 306-307; Fig. 10.10, p. 307)

   1. Basic Features (Table 10.5, p. 306)

   2. Muscles of the Thorax

G. Muscles of the Abdominal Wall: Trunk Movements and Compression of Abdominal Viscera (pp. 308-309; Fig. 10.11, p. 309)

   1. Basic Features (Table 10.6, p. 308)

   2. Muscles of the Anterolateral Abdominal Wall

H. Muscles of the Pelvic Floor and Perineum: Support of Abdominopelvic Organs (pp. 310-311; Fig. 10.12, p. 311)

   1. Basic Features (Table 10.7, p. 310)

   2. Muscles of the Pelvic Diaphragm

   3. Muscles of the Urogenital Diaphragm

   4. Muscles of the Superficial Space

I. Superficial Muscles of the Anterior and Posterior Thorax: Movements of the Scapula (pp. 32-313; Fig. 10-13, p. 313)

   1. Basic Features (Table 10.8, p. 312)

   2. Muscles of the Anterior Thorax

   3. Muscles of the Posterior Thorax

J. Muscles Crossing the Shoulder Joint: Movements of the Arm (Humerus) (pp. 314-316; Fig. 10.14, p. 316)

   1. Basic Features (Table 10.9, pp. 314-316)

   2. Muscles Moving the Arm

K. Muscles Crossing the Elbow Joint: Flexion and Extension of the Forearm (pp. 316-317; Fig. 10.14, p. 316)

   1. Basic Features (Table 10.10, p. 317)

   2. Posterior Muscles

   3. Anterior Muscles

L. Muscles of the Forearm: Movements of the Wrist, Hand, and Fingers (pp. 318-321; Fig. 10.15, p. 319; and Fig. 10.16, p. 321)

   1. Basic Features (Table 10.11, pp. 318-321)

   2. Anterior Superficial Muscles

   3. Anterior Deep Muscles

   4. Posterior Superficial Muscles

   5. Posterior Deep Muscles

Review Items

1. Skeletal muscle tissue (Chapter 9, pp. 249-275)

2. Bones of the skull (Chapter 7, pp. 179-193)

3. Facial bones (Chapter 7, pp. 190-191)

4. Bones of the vertebral column (Chapter 7, pp. 194-200)

5. The bony thorax (Chapter 7, pp. 200-202)

6. Pectoral bones (Chapter 7, pp. 203-204)

7. Upper extremity (Chapter 7, pp. 204-209)

8. Pelvic girdle (Chapter 7, pp. 209-212)

9. Lower extremity (Chapter 7, pp. 212-217)

10. Synovial joints (Chapter 8, pp. 226-240)

Cross References

1. The male and female perineum as related to reproductive anatomy is presented in Chapter 28.

2. The muscles of the pelvic floor are explained in further detail in Chapter 28.

3. Muscles involved in controlling micturition are mentioned in Chapter 26, p. 920.

4. The muscles of mastication and tongue movement are mentioned in Chapter 24, p. 803.

5. Abdominal muscles involved in respiration are discussed in detail in Chapter 23, p. 758.

Laboratory Correlations

1. Marieb, E. N. *Human Anatomy and Physiology Laboratory Manual: Cat and Fetal Pig Versions.* 3rd. ed. Benjamin/Cummings, 1989.

   Exercise 15: Gross Anatomy of the Muscular System

2. Marieb, E. N. *Human Anatomy and Physiology Laboratory Manual: Brief Version.* 3rd. ed. Benjamin/Cummings, 1992.

   Exercise 13: Identification of Human Muscles

Transparencies Index

10.1   Lever systems operating at a mechanical advantage and mechanical disadvantage
10.2   Lever systems
10.4   Anterior view of superficial muscles of the body
10.5   Posterior view of superficial muscles of the body
10.6   Lateral view of muscles of the scalp, face, and neck
10.7   Muscles promoting mastication and tongue movements
10.8   Muscles of the anterior neck and throat that promote swallowing
10.9   Muscles of the neck and vertebral column causing movements of the head and neck
10.10 Muscles of quiet respiration
10.11 Muscles of the abdominal wall
10.12 Muscles of the pelvic floor
10.13 Superficial muscles of the thorax and on the scapula and arm
10.14 Muscles crossing the shoulder and elbow joint, respectively causing movements of the arm and forearm
10.15 Muscles of the anterior fascial compartment of the forearm acting on the wrist and fingers
10.16 Muscles of the posterior fascial compartment of the forearm acting on the wrist and fingers
10.17 Summary of actions of muscles of the arm and forearm
10.18 Anterior and medial muscles promoting movements of the thigh and leg
10.19 Posterior muscles of the right hip and thigh
10.20 Muscles of the anterior compartment of the right leg
10.21 Muscles of lateral compartment of the right leg
10.22 Muscles of the posterior compartment of the right leg
10.23 Summary of actions of muscles of the thigh and leg

Bassett Atlas Figures Index

Slide Number	Figure Number	Description
9	1.9A,B	Spinal cord origin
10	1.10	Spinal cord, cauda equina
12	2.1A,B	Parotid gland and facial nerve
14	2.3A,B	Sagittal section, head and neck
15	2.4	Sagittal section, larynx
17	3.1A,B	Chest wall and breast
28	4.1	Abdominal wall, rectus uncovered
29	4.2A,B	Abdominal wall, rectus reflected
44	5.1A,B	Left inguinal area
45	5.2	Right inguinal area from within

INSTRUCTIONAL AIDS

Lecture Hints

1. It is easy for students to treat the muscular (or any other) system as an individual unit, without relating it to the rest of the body. Stress that students should associate any specific muscle (and its associated synergists, antagonists, etc.) with its fascicle arrangement, origin and insertion sites, and bones involved to "keep sight of the whole picture."

2. Be sure that students understand that motion is achieved by muscle shortening, never by muscle pushing.

3. Students often do not readily grasp the idea of the compromise between power and range of movement when discussing the relationship between origin and exact site of insertion (e.g., biceps brachii into radial tuberosity). Ask the class what would happen to power and range of movement if the biceps inserted several centimeters distal (or proximal) to the actual site.

4. When describing different muscles of the body, try coaxing names from the class by carefully indicating locations, properties, etc. For example, point to a diagram (or manikin) displaying

transverse abdominis and ask the class which way the fibers are running. They should answer "transversly." Then ask: "What general area of the body is this muscle located in?" Answer: abdominal. Finally ask: "What would be a logical name for this muscle?" One can apply this type of logic in many instances, all of which help students master the information rather than simply memorizing it.

Demonstrations/Activities

1. Audio-visual materials of choice.

2. Obtain a dissected, preserved animal such as a cat or a fetal pig and exhibit the major muscle groups.

3. Obtain a 3-D model or chart to illustrate the major human muscle groups.

4. Obtain implements such as scissors, a wheelbarrow, and forceps to illustrate the three types of lever systems.

5. Obtain a human cadaver to illustrate the major human muscle groups.

6. Select a volunteer and have him/her contract an arm or leg. Have students record: prime mover, synergists, and antagonists.

7. Have students work in pairs as follows: One should attempt to contract a particular muscle, while the second student provides resistance to prevent that movement. In this way the muscle will produce its maximal "bulge." Muscles being examined should be palpated in the relaxed and contracted states by both students. For example, the "demonstrator" can attempt to flex his/her elbow while the person providing the resistance holds the forearm to prevent its movement. The biceps brachii on the anterior arm will bulge and be easily palpated.

8. As muscles are being described, project 2 x 2 slides of cadaver dissection so that students can readily see "the real thing" as material is being presented in lecture. The microprojector can be running during the entire lecture, with cues placed in lecture notes on when to display specific slides.

Critical Thinking/Discussion Topics

1. Why is it necessary for pregnant women to strengthen their "pelvic floor"?

2. What do bones possess that allow them to act as effective levers?

3. What are the most appropriate modes of therapy for a pulled hamstring muscle or pulled groin muscle?

4. Injections are often made directly into the muscle tissue. What are the advantages and disadvantages?

5. If a prime mover muscle such as the pectoralis major is surgically removed, how will the actions provided by that muscle be replaced?

6. How would one design an upper appendage so that it would operate with a relatively higher degree of mechanical advantage than presently exists?

Library Research Topics

1. Since muscle cells do not regenerate, what methods of treatment are available if a major group of muscles are lost? What is the status of skeletal muscle transplants?

2. What affect does old age have on skeletal muscle? What type of research is under way concerning this topic?

3. Olympic athletes use scientific training methods to improve their physical performance. Find out what some of these methods are and try to employ them to better your performance.

Audio-Visual Aids/Computer Software

Slides

1. Muscular System and Its Functions (EI, SS-0355, Slides)

Videotapes

1. The Guides to Dissection Series (TF, C, VHS). A superb series of 42 programs demonstrating the dissection of all major areas of the human body. Also available as a 16mm film.

2. Muscles of the Anterior Forearm (TF, 14 min., C, VHS)

3. The Palmer Hand, Part II: Intrinsic Muscles (TF, 15 min., C, VHS)

4. Muscles of Mastication and Infratemporal Fossa (TF, 15 min., C, VHS)

Computer Software

1. Muscular System (PLP, CH-182005 Apple; CH-182006, IBM)

2. The Human Systems: Series 2 (PLP, CH-920009, Apple)

    See Guide to Audio-Visual Resources on page 339 for key to AV distributors.

LECTURE ENHANCEMENT MATERIAL

Clinical and Related Terms

1. Femoral hernia — a protrusion of an intestinal loop into the femoral canal.

2. Hernia — the abnormal projection or protrusion of an organ or part of an organ through the muscular wall of a vessel or cavity that normally contains it. Also known as a rupture; however no tissues are actually torn. Hernias can be congenital or acquired and usually require surgical repair.

3. Hiatal hernia — usually involves a protrusion of the stomach through the esophageal hiatus of the diaphragm.

4. Inguinal hernia — a hernia occurring in the groin.

5. Lockjaw — tetanic spasms of the muscles of the jaw. Caused by stress, strain, bacterial infection, or physiological imbalance.

6. Lumbago — a term used to denote lower back pain, usually affecting the lumbar region. Pain may result from arthritis, injury, poor posture, or other disorder.

7. Transcutaneous muscle stimulations — a battery-operated device is used by physical therapists to electrically stimulate muscles that may have atrophied due to immobilization by a cast, or to strengthen muscles weakened due to surgery.

8. Umbilical hernia — a protrusion of part of intestine through the umbilicus.

ANSWERS TO END-OF-CHAPTER QUESTIONS

Multiple Choice/Matching

1.	c	5.	d	9.	d
2.	(1)e; (2)c; (3)g; (4)f; (5)d	6.	c	10.	b
3.	a,c	7.	c	11.	a
4.	c	8.	b	12.	c

Short Answer Essay Questions

13. Student answers will vary. (pp. 291-292)

   a. Location of the muscle — frontalis, occipitalis, zygomaticus

   b. Shape of the muscle — rhomboids, serratus anterior, quadratus lumborum

   c. Relative size of the muscle — pectoralis major and minor, peroneus longus and brevis

   d. Direction of muscle fibers — rectus abdominus, external oblique, superficial transverse perineus

   e. Number of origins — triceps brachii, biceps femoris

   f. Location of the origin/insertion — stylohyoid, sternothyroid, coracobrachialis

   g. Action of the muscle — levator scapulae, pronator teres, carpi radialis, abductor longus

14. First class: effort — fulcrum — load

   Second class: fulcrum — load — effort

   Third class: fulcrum — effort — load (pp. 287-289)

15. When the load is far from the fulcrum and the effort is applied near the fulcrum, the effort applied must be greater than the load to be moved. This type of leverage can be advantageous because it allows the load to be moved rapidly through a large distance, with only minimal muscle shortening. (p. 288)

16. Pharyngeal constrictor muscles (p. 301)

17. To shake your head "no" — sternocleidomastoid. (p. 302) To nod "yes" — sternocleidomastoid and splenius. (pp. 302-303)

18. a. Rectus abdominus, external oblique, internal oblique, transversus abdominus (p. 308)

   b. Each pair is arranged at cross-directions to each other. (p. 308)

   c. External oblique and internal oblique (p. 308)

   d. Rectus abdominus (p. 308)

19. Flexion of the humerus: pectoralis major and deltoid

   Extension of the humerus: latissimus dorsi and deltoid

   Abduction of the humerus: deltoid.

   Circumduction of the humerus: combination of all above

   Rotation of the humerus laterally: infraspinatus and teres minor

   Rotation of the humerus medially: subscapularis (pp. 314-315)

20. a. Extensor carpi radialis longus and brevis (p. 320)

   b. Flexor digitorum profundus (p. 319)

21. Piriformis, obturator externus and internus, gemellus, and quadratus femoris (pp. 327-329)

22. Adductors, pectineus, and gracilus (p. 326)

23. a. Deltoid, vastus lateralis, gluteus maximus, gluteus medius (pp. 315, 324, 327)

    b. Vastus lateralis is used in infants because their hip and arm muscles are poorly developed. (p. 326)

Critical Thinking and Application Questions

1. When the forearm is pronated, the biceps brachii, a prime mover of forearm flexion, is unable to act. (p. 317)

2. Levator ani, coccygeus, and sphincter urethrae (p. 310)

3. There was a rupture of the Achilles or calcaneous tendon. (p. 333) The calf appears swollen because the gastrocnemius muscle is no longer anchored to the calcaneus.

Key Figure Questions

Figure 10.2 The third-class lever. **Figure 10.3** The sartorius and the biceps brachii could shorten the most. The pectoralis major at the rectus femoris would be the most powerful because they are the most fleshy. **Figure 10.5** The trapezius.

SUGGESTED READINGS

1. Allman, William F. "Steroids in Sports: Do They Work?" *Science* 83 (Nov. 1983): 14.

2. Basmajian, J.V. *Grant's Method of Anatomy*. 10th ed. Baltimore: Williams and Wilkins, 1980.

3. Clemente, C.D. *Anatomy: A Regional Atlas of the Human Body*. 2nd ed. Philadelphia: Lea and Febiger, 1981.

4. Franklin, D. "Steroids Heft Heart Risks in Iron Pumpers." *Science News* 126 (July 1984): 38.

5. Hamer, B. "Women Body Builders: Is Bigger Better?" *Science* 86 (Mar. 1986): 74-75.

6. Hamilton, W.J. *Textbook of Human Anatomy*. 2nd ed. St. Louis: C.V. Mosby, 1976.

7. Hildebrand, M. "The Mechanics of Horse Legs." *American Scientist* 75 (Nov.-Dec. 1987): 594-601.

8. Hinson, M.M. *Kinesiology*. 2nd ed. Dubuque, IA: Wm. C. Brown, 1981.

9. McMinn, R.M.H., and R.T. Hutchings. *Color Atlas of Human Anatomy*. Chicago: Year Book Medical Publishers, 1977.

10. Peachey, L.D. *Muscle and Motility*. New York: McGraw-Hill, 1974.

11. Woodburne, R.T. *Essentials of Human Anatomy*. 7th ed. New York: Oxford Univ. Press, 1983.

12. Yokochi, C., and J.W. Rohen. *Photographic Anatomy of the Human Body*. 2nd ed. Tokyo, New York: Igaku-Shoin, Ltd., 1978.

11

Fundamentals of the Nervous System and Nervous Tissue

Chapter Preview

This chapter begins with a brief overview of the organization of the nervous system. It then focuses on the functional anatomy of nervous tissue, especially that of the nerve cells, or neurons, that are the key to the subtle, efficient system of neural communication. The histology and neurophysiology of the tissue is presented and explained in detail. Concepts of neural integration, patterns of neural organization, the reflex arc, and the developmental aspects of the nervous tissue are also described.

INTEGRATING THE PACKAGE

Suggested Lecture Outline

I. Introduction (pp. 340-341)

    A. Basic Characteristics (p. 340)

    B. Functions (pp. 340-341)

II. Organization of the Nervous System (pp. 341-342; Fig. 11.1, p. 341)

    A. Central Nervous System (p. 341)

    B. Peripheral Nervous System (pp. 341-342)

        1. Afferent (Sensory) Division

        2. Efferent (Motor) Division

            a. Somatic Nervous System

            b. Automatic Nervous System

        1. Sympathetic Division

        2. Parasympathetic Division

III. Histology of Nervous Tissue (pp. 342-350)

    A. Supporting Cells (pp. 342-343; Fig. 11.2, p. 343)

        1. Supporting Cells in the CNS

            a. Astrocytes

            b. Microglia

            c. Ependymal

            d. Oligodendrocytes

        2. Supporting Cells in the PNS

            a. Schwann Cells

            b. Satellite Cells

B. Neurons (pp. 343-350)
1. Basic Characteristics (Fig. 11.3, p. 344)
2. Neuron Cell Body (p. 344)
    a. Perikaryon
    b. Nissl Bodies
    c. Neurofilaments
    d. Pigment Inclusions
    e. Nuclei
    f. Ganglia
3. Neuron Processes
    a. Dendrites
    b. The Axon
        1. Axon Hillock
        2. Axoplasm
        3. Axon Collaterals
        4. Telodendria
        5. Axonal Terminals
        6. Axonal Transport
        7. Axoplasmic Flow
        8. Myelin Sheath and Neurilemma (Table 11.1, p. 348)
            a. Myelinated Fibers
            b. Unmyelinated Fibers
            c. Schwann Cells
            d. Neurilemma
            e. Nodes of Ranvier
            f. White Matter
            g. Gray Matter
C. Classification of Neurons (pp. 346-350)
1. Structural Classification (Fig. 11.5, p. 346)
    a. Multipolar Neurons
    b. Bipolar Neurons
    c. Unipolar Neurons (Pseudounipolar Neurons)
2. Functional Classification
    a. Sensory (Afferent) Neurons
    b. Motor (Efferent) Neurons
    c. Association (Interneurons) Neurons
IV. Neurophysiology (pp. 350-371)
A. Basic Principles of Electricity (pp. 350-351)
1. Introduction to Electricity
2. Potential (Fig. 11.6, p. 351)
3. Current
4. Resistance

     5. Ohm's Law

     6. Membrane Channels (Fig. 11.5, p. 351)

     7. Electrochemical Gradients

B. The Resting Membrane Potential: The Polarized State (pp. 351-352)

     1. Resting Membrane Potential (Fig. 11.7, p. 352)

     2. Passive and Active Forces Regulating Potentials

C. Membrane Potentials that Act as Signals (pp. 352-359)

     1. General Characteristics (Fig. 11.8, p. 353)

       a. Depolarization

       b. Hyperpolarization

     2. Graded Potentials (Fig. 11.9, p. 353)

       a. Basic Characteristics

       b. Receptor Potentials

       c. Postsynaptic Potentials

     3. Action Potentials

       a. Basic Characteristics

       b. Generation of an Action Potential (Fig. 11.11, p. 355)

         1. Increase in Sodium Permeability

           a. Threshold Level

         2. Decrease in Sodium Permeability

         3. Increase in Potassium Permeability and Repolarization

       c. Propagation of an Action Potential (Fig. 11.12, p. 357)

       d. Threshold and the All-or-None Phenomenon

       e. Coding for Stimulus Intensity

       f. Absolute and Relative Refractory Periods

       g. Conduction Velocities of Axons

         1. Influence of Axon Diameter

         2. Influence of Myelin Sheath

           a. Saltatory Conduction (Fig. 11.15, p. 359)

D. The Synapse (pp. 360-362)

     1. General Characteristics

       a. Presynaptic Neuron

       b. Postsynaptic Neuron

     2. Electrical Synapses

     3. Chemical Synapses (Fig. 11.16, p. 360)

       a. Basic Characteristics

         1. Axonal Terminal

         2. Receptor Region

         3. Synaptic Cleft

       b. Information Transfer Across Chemical Synapses

       c. Termination of Neurotransmitter Effects

       d. Synaptic Delay

Review Items

4. Cytoskeletal elements (Chapter 3, p. 82)

5. Cell cycle (Chapter 3, p. 86)

6. Enzymes and enzyme function (Chapter 2, p. 51)

7. Synapse (neuromuscular junction) (Chapter 9, pp. 258-259)

Cross References

1. Cholinergic and adrenergic receptors and other neurotransmitter effects are described in Chapter 14, pp. 470-471.

2. Receptors for the special senses are presented in Chapter 16.

3. Nervous system modulation of endocrine function is examined in Chapter 17.

4. Membrane potential and the electrical activity of the heart is explained in Chapter 19.

5. Examples of receptors are given in Chapter 25, p. 887.

6. Sensory receptors and control of digestive processes are described in Chapter 24.

7. Chemoreceptors and stretch receptors related to respiratory function are covered in Chapter 23, pp. 773-777.

8. Baroreceptors and chemoreceptors in blood pressure and flow regulation are examined in Chapter 20, pp. 654-656.

9. Synapses involved in the special senses are examined in Chapter 16.

10. Neurotransmitters in the special senses are further described in Chapter 16.

11. Neural integration is presented in great detail in Chapters 14, pp. 473-476; and 15, pp. 483-491.

12. Membrane potentials are further described in Chapter 13, p. 429.

13. Autonomic synapses are presented in Chapter 14, p. 462.

Laboratory Correlations

1. Marieb, E. N. *Human Anatomy and Physiology Laboratory Manual: Cat and Fetal Pig Versions.* 3rd. ed. Benjamin/Cummings, 1989.

   Exercise 17: Histology of Nervous Tissue

   Exercise 20: Neurophysiology of Nerve Impulses

2. Marieb, E. N. *Human Anatomy and Physiology Laboratory Manual: Brief Version.* 3rd. ed. Benjamin/Cummings, 1992.

   Exercise 15: Neuron Anatomy and the Nerve Impulse

Transparencies Index

11.1 Levels of organization in the nervous system

11.2 Supporting cells of nervous tissue

11.3 Structure of a motor neuron

11.4 Relationship of Schwann cells to axons in PNS

11.5 Operation of gated channels

11.7 Passive & active forces/membrane potential

11.8 Depolarization and hyperpolarization of the membrane

11.11 The phases of the action potential

11.12 Propagation of an action potential

11.18 Events occuring at a chemical synapse

11.20 Neural integration of EPSPs and IPSPs

11.23 Neurotransmitter receptor mechanisms

11.26 A simple reflex arc

INSTRUCTIONAL AIDS

Lecture Hints

1. By this time the class has been exposed to only a few systems (integumentary, skeletal, and muscular), but enough information has been given so that students can understand the basics of nervous system function from the beginning of this section. Ask students questions, e.g.: (1) When you touch something hot, how do you react? (2) Do you have to consciously think about pulling your hand away? The idea for these basic, probing questions is to get students to come up with the idea that a neural pathway consists of a sensory structure, some means of conveying information to the brain, and some means of causing motor response. If you get students to come up with these "solutions," they will remember the logic used to derive the answers.

2. Emphasize strongly the three basic functions of the nervous system: sensory, integration, and motor. Students should "burn this into the brain," since it will be seen again and again in all systems.

3. Stress that although we discuss the nervous system in segments, it is actually tightly integrated.

4. Present a general introduction of the entire nervous system near the beginning of nervous system discussion so that students will be able to see the entire picture. In this way, they will better understand the relationships as material is covered.

5. Point out the similarities between skeletal muscle cells and neurons. It is also possible to introduce the electrical characteristics of cardiac pacemaker cells (modified muscle cells) and note the similarities to neuron function. It is worthwhile to point out that although function is totally different (muscle = contractile, nervous = impulse generation, propagation), the structural basis of each is a slight modification of a basic cellular blueprint.

6. Bring a model (or overheads, 2 x 2 slides) of a neuron to lecture to visually demonstrate the anatomy of a nerve cell.

7. Many students have difficulty understanding the difference between the myelin sheath and the neurilemma (sheath of Schwann). Use a diagram (blackline master) to point out that both are parts of the same cell.

8. Emphasize the difference in myelination between the CNS and PNS. Point out the regeneration capabilities of each.

9. Many students have trouble relating ion movements with electrical current. One way to approach neurophysiology is to (loosely) compare a 1.5V battery to the cell membrane (something students can relate to). The electrical potential between the positive and negative poles is analogous to the outside and inside of a cell. When a connection is made between positive and negative poles (ion gates opened), current is delivered.

10. Clearly distinguish the difference between graded potentials and action potentials. It helps to use a full-page acetate of a neuron to demonstrate the positive feedback nature of the action potential.

11. Most introductory physiology students will experience difficulty with the idea of saltatory conduction. Draw (or project) diagrams of myelinated vs. unmyelinated fibers and electrical propagation.

12. Do a diagram of a synapse, then use root word dissection to emphasize the distinction between pre- and postsynaptic neurons. This is a good introduction to the synapse, and establishes a reference point upon which students can build.

13. Use absolute numbers as an introductory example for summation, for example: If three presynaptic neurons each simultaneously deliver a one-third threshold stimulus, will the postsynaptic neuron fire? Use several examples to emphasize the difference between spatial and temporal summation.

14. Use diagrams when describing the different types of circuits.

Demonstrations/Activities

1. Audio-visual materials of choice.

2. Obtain a microprojector and microscope slides of neurons, neuroglia, and peripheral nerves to illustrate the histology of the tissue.

3. Obtain an oscilloscope and a neurophysiology kit to illustrate how an action potential can be registered.

4. Obtain 3-D models of motor and sensory neurons to illustrate their similarities and differences.

5. To illustrate proprioception, have students place their hands at their sides and then, without allowing them to look, have them visualize where the backs of their hands are.

6. Use a match to illustrate how an EPSP can work and how a graded potential will be intense at the receptor end and decrease thereafter; then use a fuse wire to illustrate how an action potential is carried down the wire.

Critical Thinking/Discussion Topics

1. How can drugs, such as novocaine, effectively block the transmission of pain impulses? Why don't they block motor impulses, or do they?

2. What effect does alcohol have on the transmission of electrical impulses?

3. How can rubbing one's nose decrease the possibility of a sneeze? Discuss in terms of EPSPs and IPSPs.

4. Acetylcholine has long been recognized as a neurotransmitter. Why has it been so difficult to identify other neurotransmitters?

5. How can some people eat extremely hot peppers without experiencing the same pain that others normally have?

6. What would happen at a synapse if one introduced an agent that blocked the activity of chemically gated Na^+ channels? K^+ channels?

Library Research Topics

1. Of what value is the development of recombinant DNA technology to our study of protein-based neurotransmitters?

2. What is the status of research on the repair and/or regeneration of nervous tissue of the CNS?

3. Why do most tumors of nervous tissue develop in neuroglia rather than neurons?

4. Could we use neurotransmitters to enhance our memory capacity?

5. How are experiments performed to test the anatomy and physiology of plasma membrane ion gates and channels?

Audio-Visual Aids/Computer Software

Slides

1. Histology of the Nervous System (EI, #613, Slides). Various stain techniques used to show cell bodies, myelin sheath, neurons, and glial cells.

2. Nervous System and Its Function (EI, SS-0350F, Slides). Reviews neural transmission, brain areas, and spinal cord.

3. Neurobiology I: Excitatory Membranes (EI, EP-2138, 1984, Slides). Describes resting membrane potential, action potential, and active transport mechanisms.

4. Neurobiology II: Neural Function (EI, EP-2139, 1984, Slides). Describes information processing, EPSPs and IPSPs.

Videotapes

1. Nerves at Work (FHS, QB-831, 26 min., C, VHS/BETA). Program explores nerve signals, impulse transmission, and reflex activities. Also available in 16mm.

2. Decision (FHS, QB-832, 26 min., C, VHS/BETA). Program shows how the brain organizes input and output, and how circuits of nerve cells operate. Also available in 16mm.

3. Brain Triggers, Part 1 - Neurotransmitters (PLP, CH-770234, VHS)

4. Your Body - Part 3, Your Nervous System (PLP, CH-140203, VHS)

Computer Software

1. The Nervous System: Our Information Network (CA, MLC 5131, Apple)

2. Dynamics of the Human Nervous System (EI, C-3050, Apple/IBM)

3. Nervous System (PLP, CH-381079, Apple 64K)

4. Body Language: Study of Human Anatomy, Nervous System (PLP, CH-182013, Apple; CH-182014, IBM)

5. The Human Systems: Series 2 (PLP, CH-920009, Apple)

6. The Human Brain (PLP, CH-510003, Apple)

See Guide to Audio-Visual Resources on page 339 for key to AV distributors.

LECTURE ENHANCEMENT MATERIAL

Clinical and Related Terms

1. Ependymitis — inflammation of the membrane lining the cerebral ventricles.

2. Ganglionitis —inflammation of a ganglion.

3. Synaptic plasticity — the ability of the synapse to change at both the molecular and the cellular level.

Disorders/Homeostatic Imbalances

Demyelinating Disorders

1. Tay-Sach's Disease — an inherited infantile form of amaurotic familial idiocy marked by degeneration of brain tissue due to an inborn error in metabolism. A deficiency of the enzyme hexosaminidase A results in a sphingolipodosis with an accumulation of GM2 gangliosides in the brain.

Motor Neuron Disorders

1. Amyotrophic Lateral Sclerosis (ALS) — a progressive neurologic disorder characterized by degeneration of lower motor neuron cell bodies in the gray matter of the anterior horns of the spinal cord, brain, and pyramidal tracts. It is also called motor neuron disease and "Lou Gehrig's Disease." There is no known cause or cure.

2. Wernig-Hoffman Disease — an inherited infantile motor neuron disease exhibiting type I spinal muscular atrophy.

Disorders on Neuroglial Cells

1. Astrocytoma —a usually slow-growing neoplasm derived from astrocytes. Classified in order of malignancy as: Grade I, consisting of fibrillary or protoplasmic astrocytes; Grade II, astroblastoma; and, Grades III and IV, glioblastoma multiforme.

2. Ependymoma — a slow-growing neoplasm derived from ependymal cells that line the ventricles of the brain and form the central canal of the spinal cord. Prognosis is extremely poor.

3. Glioblastoma Multiforme — an astrocytoma of Grade III or IV; usually rapid-growing and occurring in the cerebral hemispheres; composed of spongioblasts, astroblasts, and astrocytes. Prognosis is generally very poor.

4. Glioneuroma — a glioma combined with neuroma.

5. Oligodendroglioma — a neoplasm derived from and composed of oligodendroglia. They are among the most unpredictable gliomas and are capable of rapid progression.

6. Schwannoma — a neoplasm originating from Schwann cells that form the myelin sheath of neurons. These tumors include neurofibromas and neurilemomas.

ANSWERS TO END-OF-CHAPTER QUESTIONS

Multiple Choice/Matching

1. b

2. (1)d; (2)b; (3)f; (4)c; (5)a

3. b

4. c

5. a

6. c

7. b

8. d

9. c

10. c

11. a

12. (1)d; (2)b; (3)a; (4)c

Short Answer Essay Questions

13. Anatomical division includes the CNS (brain and spinal cord) and the PNS (nerves and ganglia). Functional division includes the somatic and autonomic motor divisions of the PNS. The autonomic division is divided into sympathetic and parasympathetic subdivisions. (pp. 341-342)

14. a. The cell body is the biosynthetic and metabolic center of a neuron. It contains the usual organelles, but lacks centrioles. (pp. 344-345)

    b. Dendrites and axons both function to carry electrical current. Dendrites differ in that they are short, transmit toward the cell body, and function as receptor sites. Axons are typically long, are myelinated, and transmit away from the cell body. (p. 345)

15. a. Myelin is a whitish, fatty, phospholipid insulating material (essentially the wrapped plasma membranes of oligodendrocytes or Schwann cells).

    b. CNS myelin sheaths are formed by flap-like extensions of oligodendrocytes and lack a neurilemma. Fibers cannot regenerate. PNS myelin is formed by Schwann cells. The sheaths have a neurilemma, and the fibers they protect are capable of regeneration. (p. 346)

16. Multipolar neurons have many dendrites, one axon, and are found in the CNS (and autonomic ganglia). Bipolar neurons have one axon and one dendrite, and are found in receptor end organs of the special senses such as the retina of the eye and olfactory mucosa. Unipolar neurons have one process that divides into an axon and a dendrite and is a sensory neuron with the cell body found in a dorsal root ganglion or cranial nerve ganglion. (p. 349)

17. A polarized membrane possesses a net positive charge outside, and a net negative charge inside, with the voltage across the membrane being at -70 mv. Diffusion of Na^+ and K^+ across the membrane establishes the resting potential because the membrane is slightly more permeable to K^+. The Na^+ - K^+ pump, an active transport mechanism, maintains this polarized state by maintaining the diffusion gradient for Na^+ and K^+. (pp. 351-352; see Fig. 11.7, p. 352)

18. a. The generation of an action potential involves: (1) an increase in sodium permeability and reversal of the membrane potential; (2) a decrease in sodium permeability; and (3) an increase in potassium permeability and repolarization. (Fig. 11.11, p. 355)

   b. The ionic gates are controlled by changes in the membrane potential and activated by local currents. (p. 355)

   c. The all-or-none phenomenon means that the local depolarizing current must reach a critical "firing" or threshold point before it will respond, and when it responds, it will respond completely by conducting the action potential along the entire length of its axon. (p. 357)

19. The CNS "knows" a stimulus is strong when the frequency or rate of action potential generation is high. (p. 358)

20. a. An EPSP is an excitatory (depolarizing) postsynaptic potential that increases the chance of a depolarization event. An IPSP is an inhibitory (hyperpolarizing) postsynaptic potential that decreases the chance of a depolarization event. (See Table 11.2, p. 364)

   b. It is determined by the type of neurotransmitter that binds at the postsynaptic neuron and the specific receptor subtype it binds to. (p. 362)

21. Each neuron's axon hillock keeps a "running account" of all signals it receives via temporal and spatial summation. (p. 364)

22. The neurotransmitter is quickly removed by enzymatic degradation or reuptake into the presynaptic axon. This insures discrete limited responses. (p. 362)

23. a. Absolute refractory period is when the neuron is incapable of responding to another stimulus because repolarization is still occurring. (p. 358)

   b. A-fibers have the largest diameter and thick myelin sheaths and conduct impulses quickly; B-fibers are lightly myelinated, have intermediate diameters, and are slower conductors. (p. 360)

   c. Node of Ranvier is an interruption of the myelin sheath due to wrapping of the individual Schwann cells. (p. 346)

24. In serial processing the pathway is constant and through a definite sequence of neurons. In parallel processing, impulses reach the final CNS target by multiple pathways. Parallel processing allows for a variety of responses. (p. 373)

25. First, they proliferate; second, they migrate to proper position; third, they differentiate. (p. 374)

26. Factors include "pathfinder" neurons; orienting glial fibers; and attracting substances such as nerve cell adhesion molecules and growth cones. (p. 373)

Critical Thinking and Application Questions

1. The resting potential would decrease, that is, become less negative, because the concentration gradient causing net diffusion of K^+ out of the cell would be smaller. Action potentials would be fired more easily, that is, in response to smaller stimuli, because the resting potential would be closer to threshold. Repolarization would occur more slowly because repolarization depends on net K^+ diffusion from the cell and the concentration gradient driving this diffusion is lower. Also, the after hyperpolarization would be smaller. (pp. 351-356)

2. Local anesthetics such as novocaine and sedatives affect the neural processes usually at the nodes of Ranvier, by reducing the membrane permeability to sodium ions. (p. 359)

3. The bacteria remain in the wound; however, the toxin produced travels via axonal transport to reach the cell body. (p. 346)

4. In MS, the myelin sheaths are destroyed. Loss of this insulating sheath results in shunting of current and eventual cessation of neurotransmission.

Key Figure Questions

Figure 11.3 The longer the neuron, the larger the cell body needed to service it. **Figure 11.8** The negativity of the internal membrane surface. **Figure 11.10** Those on the cell body which are closest to the axon hillock would have the greatest potential for producing a threshold stimulus. **Figure 11.15** In a myelinated axon, voltage-gated sodium channels are located only at the nodes of Ranvier, as opposed to being located along the entire length of an unmyelinated axon. **Figure 11.23** The term *first messenger* designates the original chemical messenger (the stimulus), which in this case is the neurotransmitter.

SUGGESTED READINGS

1. Adams, R.V. *Principles of Neurology*. 2nd ed. New York: McGraw-Hill, 1981.

2. Bainbridge, Jr., J.S. "Frogs That Sweat—Not Bullets, But a Poison for Darts." *Smithsonian* 19 (Jan. 1989): 70-76.

3. Bloom, F. "Brain Drugs." *Science* 85 (Nov. 1985): 100-101.

4. Bloom, F.E. "Neuropeptides." *Scientific American* 245 (Oct. 1981).

5. Carpenter, M.B. *Human Neuroanatomy*. 7th ed. Baltimore: Williams and Wilkins, 1976.

6. Catterall, W.A. "The Molecular Basis of Neuronal Excitability." *Science* 223 (1984): 653.

7. Dunant, Y., and M. Israel. "The Release of Acetylcholine." *Scientific American* 252 (Apr. 1985): 58-66.

8. Fischbach, G.D. "Mind and Brain." *Scientific American* 267 (Sept. 1992): 48.

9. Hodgkin, A.L. "The Ionic Basis of Nervous Conduction." *Science* 145 (Sept. 1964).

10. Gilman, A.G., and M.E. Linder. "G Proteins." *Scientific American* (1993): 70.

11. Goodman, C.S. and M.J. Bastiani. "How Embryonic Cells Recognize One Another." *Scientific American* (Dec. 1984): 58-66.

12. Gottlieb, D.I. "GABAergic Neurons." *Scientific American* 258 (Feb. 1988): 82-89.

13. Kalil, R.E. "Synapse Formation in the Developing Brain." *Scientific American* 261 (Dec. 1989): 76-85.

14. Kandel, E.R. "Small Systems of Neurons." *Scientific American* (Sept. 1979).

15. Keynes, R.D. "Ion Channels in the Nerve Cell Membrane." *Scientific American* (Sept. 1979).

16. Kimelberg, H.K., and M.D. Norenberg. "Astrocytes." *Scientific American* 260 (Apr. 1989): 66-76.

17. Kuffler, S.W., et al. *From Neuron to Brain: A Cellular Approach to Function of the Nervous System*. Sunderland, MA: Sinauer, 1984.

18. Lester, H.A. "The Response to Acetylcholine." *Scientific American* (Feb. 1977).

19. Levi-Montalcini, R., and P. Calissano. "The Nerve Growth Factor." *Scientific American* 240 (June 1979): 68-77.

20. Llinas, R.R. "Calcium in Synaptic Transmission." *Scientific American* 247 (Oct. 1982).

21. McGeer, P.L. "The Chemistry of the Mind." *American Scientist* 59 (Mar.-Apr. 1971).

22. Miller, J.A. "Grow, Nerves, Grow." *Science News* 129 (Mar. 1986): 204-206.

23. Morell, P., and W.T. Norton. "Myelin." *Scientific American* (May 1980).

24. Nauta, W.J.H., and M. Feirtag. *Fundamental Neuroanatomy.* New York: W.H. Freeman and Co., 1985.

25. Neher, E., and B. Sakmonn. "The Patch Clamp Technique." *Scientific American* 266 (March 1992): 44.

26. Patterson, P.H., et al. "The Chemical Differentiation of Nerve Cells." *Scientific American* 239 (July 1978): 50-59.

27. Pennisi, E. "Monitoring the Movements of Nerves." *Science News* 144 (July 1993): 68.

28. Schwartz, J.H. "The Transport of Substances in Nerve Cells." *Scientific American* (Apr. 1980).

29. Shatz, C.J. "The Developing Brain." *Scientific American* 267 (Sept. 1992): 60.

30. Shepherd, G.M. "Microcircuits in the Nervous System." *Scientific American* (Feb. 1978).

31. Snyder, S.H. "Opiate Receptors and Internal Opiates." *Scientific American* 236 (Mar. 1977): 44-56.

32. Stevens, C.F. "The Neuron." *Scientific American* 241 (Sept. 1979): 54-65.

33. Wurtman, R.J. "Nutrients That Modify Brain Function." *Scientific American* 246 (Apr. 1982): 50-59.

12

The Central Nervous System

Chapter Preview

This chapter focuses on the structures of the central nervous system and touches on the functions associated with its specific anatomical regions. The human brain and all its divisions, the meninges, and the cerebrospinal fluid circulation, along with the spinal cord and all its tracts, will be described in detail.

INTEGRATING THE PACKAGE

Suggested Lecture Outline

I.   Introduction (p. 379)

II. The Brain (pp. 380-408)

    A. Embryonic Development of the Brain (pp. 380-381)

        1. Ectodermal Derivations (Fig. 12.2, p. 380)

            a. Neural Plate

            b. Neural Folds

            c. Neural Tube

            d. Neural Crest

        2. Primary Brain Vesicles

            a. Prosencephalon (Forebrain)

            b. Mesencephalon (Midbrain)

            c. Rhombencephalon (Hindbrain)

        3. Secondary Brain Vesicles

            a. Telencephalon

            b. Diencephalon

            c. Metencephalon

            d. Myelencephalon

            e. Mesencephalon

    B. Regions and Organization of the Brain (p. 382; Fig. 12.5, p. 380)

    C. Ventricles of the Brain (p. 383; Fig. 12.6, p. 383)

        1. Lateral Ventricles

        2. Third Ventricle

        3. Fourth Ventricle

    D. The Cerebral Hemispheres (pp. 384-392)

        1. Surface Anatomy

            a. Gyri

E. The Diencephalon (pp. 392-395; Fig. 12.14, p. 394)

    1. The Thalamus

        a. Intermediate Mass

        b. Thalamic Nuclei

        c. General Function

    2. The Hypothalamus

        a. Mammillary Bodies

        b. Infundibulum

        c. Hypothalamic Nuclei

        d. General Functions

    3. The Epithalamus

        a. Pineal Gland

        b. Choroid Plexus

        c. General Functions

F. The Brain Stem (pp. 395-398)

    1. The Midbrain (Fig. 12.15, p. 396)

        a. Cerebral Peduncles

        b. Cerebral Aqueduct

        c. Corpora Quadrigemina

        d. Superior and Inferior Colliculi

        e. Substantia Nigra

        f. Red Nucleus

        g. General Functions

    2. The Pons (Fig. 12.16, p. 397)

        a. Pons Nuclei

        b. General Functions

    3. The Medulla Oblongata (Fig. 12.15, p. 396; Fig. 12.16, p. 397; Fig. 12.17, p. 398)

        a. Pyramids

        b. Decussation of the Pyramids

        c. Olives

        d. Vestibular Nuclear Complex

        e. General Functions

G. The Cerebellum (pp. 399-401; Fig. 12.18, p. 400)

    1. Anatomy of the Cerebellum

        a. Cerebellar Hemispheres

        b. Vermis

        c. Lobes

        d. Dentate Nuclei

        e. Arbor Vitae

        f. Cerebellar Peduncles

    2. Cerebellar Processing

b. Lateral Gray Masses
   1. Posterior Horns
   2. Anterior Horns
   3. Lateral Horns
c. Roots
   1. Ventral Roots
   2. Dorsal Roots
d. Dorsal Root Ganglia
e. Spinal Nerves
2. White Matter
   a. Funiculi
      1. Posterior
      2. Lateral
      3. Anterior
   b. Functional Generalizations
3. Ascending Pathways and Tracts (Table 12.2, p. 415; Fig. 12.29, p. 414)
   a. General Characteristics
   b. Fasciculus Cuneatus
   c. Fasciculus Gracilis
   d. Spinothalamic Tracts
   e. Spinocerebellar Tracts
4. Descending Pathways and Tracts
   a. General Characteristics
   b. Pyramidal (Corticospinal) Tracts
   c. Extrapyramidal Tracts
      1. Rubrospinal
      2. Vestibulospinal
      3. Reticulospinal
      4. Tectospinal
D. Spinal Cord Trauma (pp. 416-418)
   A. Paralysis
   B. Paresthesias
   C. Transections of the Cord
IV. Diagnostic Procedures for Assessing CNS Dysfunction (p. 418)
   A. Pneumoencephography
   B. Cerebral Angiograms
   C. Imaging Techniques
      1. CT Scans
      2. MRI Scans
      3. PET Scans

V. Developmental Aspects of the CNS (pp. 419-421)
   A. Embryonic and Fetal Development (p. 417)
   B. Congenital Malformations (pp. 417-419)
      1. Congenital Hydrocephalus
      2. Anencephaly
      3. Spina Bifida
   C. Effect of Aging on the Nervous System

Cross References

1. The role of the medulla in cardiac rate regulation is described in Chapter 19.

2. Testosterone and development of the brain is further explained in Chapter 28, pp. 967-968.

3. The role of the hypothalamus in regulation of fluid and electrolyte balance is detailed in Chapter 27.

4. The role of the hypothalamus in body temperature regulation is presented in Chapter 25, p. 887.

5. Central nervous system involvement in the reflex activity controlling digestive processes is mentioned in Chapter 24, p. 808.

6. The respiratory centers in the medulla and pons are covered in Chapter 23, pp. 772-773.

7. Cortical and hypothalamic involvement in respiration is explained in Chapter 23, p. 773.

8. The capillaries of the brain (blood-brain barrier) are further explained in Chapter 20, p. 646.

9. The medulla and regulation of blood vessel diameter (vasomotor center) is examined in Chapter 20, p. 656.

10. The hypothalamus and blood pressure regulation is mentioned in Chapter 20, p. 656.

11. The hypothalamus and hormone production is examined in great detail in Chapter 17, pp. 547, 553.

12. The role of the cerebral cortex and cerebellum in integration of sensory information is further examined in Chapter 16.

13. The role of the thalamus in the special senses is mentioned in Chapter 16.

14. The different brain areas and neural integration are examined in Chapter 15.

15. Spinal roots and peripheral nervous system function is presented in Chapters 13, p. 442; 14, p. 460.

16. The relationship between the peripheral nervous system and gray and white matter of the spinal cord are explained in Chapters 13 and 14.

Laboratory Correlations

1. Marieb, E. N. *Human Anatomy and Physiology Laboratory Manual: Cat and Fetal Pig Versions.* 3rd. ed. Benjamin/Cummings, 1989.

   Exercise 18: Gross Anatomy of the Brain and Cranial Nerves

   Exercise 19: Spinal Cord, Spinal Nerves, and Autonomic Nervous System

   Exercise 23: Electroencephalography

2. Marieb, E. N. *Human Anatomy and Physiology Laboratory Manual: Brief Version.* 3rd. ed. Benjamin/Cummings, 1992.

   Exercise 16: Gross Anatomy of the Brain and Cranial Nerves

   Exercise 17: Electroencephalography

   Exercise 18: Spinal Cord, Spinal Nerves, and Autonomic Nervous System

Transparencies Index

Bassett Atlas Figures Index

INSTRUCTIONAL AIDS

Lecture Hints

1. Study of the central nervous system is difficult for most students. The complexity of the material can overwhelm the best of individuals. Initially present the material from an overall conceptual perspective, then progress into greater levels of detail. In this way, students are not as likely to get lost.

2. When discussing the ventricles, do a rough diagram on the board (or an acetate) that shows a schematic representation of the chambers and connecting passageways. As students comprehend the serial nature of CSF flow, translate the sketches to actual cross-sectional photographs or accurate diagrams.

3. Students often have difficulty understanding how the cerebellum is involved in the control of motor activity. Try using a physical activity such as golf to illustrate cerebellar interaction, i.e., we all know how to swing a club but only well developed cerebellar coordination of muscle group action allows a "pro" to place the ball exactly where it should be.

4. Emphasize that the meningeal protection of the brain and spinal cord are continuous, but that the spinal cord has an epidural space, whereas the brain does not.

Demonstrations/Activities

1. Audio-visual materials of choice.

2. Obtain a 3-D model of a human brain and compare it to a real human brain and/or a dissected sheep brain.

3. Obtain a 3-D model of a spinal cord, both longitudinal section and cross section, to illustrate its features.

4. Obtain stained sections of brain tissue to illustrate the differences between gray and white matter and to show internal parts.

5. Obtain a 3-D model or cast of the ventricles of the brain.

6. Obtain a sheep brain with the cranium and/or meninges still intact.

7. Project or set up microslides to demonstrate cross-sectional anatomy of the spinal cord at several different levels to show how gray and white matter changes with level in the cord.

Critical Thinking/Discussion Topics

1. Discuss the difference between encephalitis and meningitis.

2. Prefrontal lobotomies have been used in psychotherapy along with electrical shock. How and why have these techniques been used?

3. Since a right-handed person's left hemisphere appears to dominate in cerebral functions, what could be done to increase the use of the left hemisphere?

4. Anencephalic children will always die soon after birth. There is currently a desire among some medical groups to use the organs of these children to help others. What are the pros and cons of this type of organ transplantation?

5. If a needle is used to deliver or remove fluids from the spaces surrounding the spinal cord, where is the best location (along the length of the cord) to perform the procedure? Why?

6. Trace the complete path of CSF from formation to reabsorption and examine the consequences if choroid plexus function were altered, or an obstruction developed in the path of CSF flow.

Library Research Topics

1. What techniques are currently used to localize and treat tumors of the brain?

2. How has the human brain changed in size and shape over millions of years of evolution? Explore the development of the human nervous system.

3. What drugs are being used to enhance memory? Where and how do they work?

4. The sensory and motor areas of the cerebral cortex around the pre- and post-central gyrus have been carefully mapped out. How was this done?

5. What methods of experimentation have been used to study the limbic system? What research has been done on determining whether some habitual criminals have defects in this system?

6. Describe the latest techniques used to examine structure/function of the CNS.

7. How can fetal tissues be used to repair adult CNS dysfunctions?

Audio-Visual Aids/Computer Software

Videotapes

1. The Addicted Brain (FHS, QB-1363, 26 min., C, VHS/BETA). Documentary explores drug use and effect on brain.

2. The Sexual Brain (FHS, QB-1416, 28 min., C, VHS/BETA). Study of the brain illustrating the differences between males and females.

3. Our Talented Brain (FHS, QB-833, 26 min., C, VHS/BETA). Program explores neural structure and physiology of the brain. Also available in 16mm.

4. The Anatomical Basis of the Brain Function Series (TF, C, 1988). Twenty titles, ranging from 14 to 23 minutes, presenting virtually every aspect of human brain neuroanatomy.

5. The Neuroanatomy Series (TF, C, 1988). Twenty-four titles, ranging from 15 to 51 minutes, presenting the dissection of the human brain.

6. Brain Triggers (PLP, CH-770234, VHS)

7. Your Body, Part 3 (PLP, CH-140203, VHS)

Computer Software

1. The Human Brain: Neurons (PLP, R-51003, Apple). Explores neuron structures, types of neurons, potentials, and neurotransmitters.

2. Body Language, Study of Human Anatomy, Nervous System (PLP, CH-182013, Apple; CH-182014, IBM)

3. The Human Brain: Neurons (PLP, CH-510003, Apple)

    See Guide to Audio-Visual Resources on page 339 for key to AV distributors.

LECTURE ENHANCEMENT MATERIAL

Clinical and Related Terms

1. Amyelencephalia — absence of both brain and spinal cord.

2. Amyotrophy — atrophy of the spinal cord.

3. Apraxia — the loss of ability to carry out normal, familiar movements in the absence of sensory or motor impairment.

4. Arachnitis — inflammation of the arachnoid membrane.

5. Bradykinesia — an abnormal slowness of voluntary movements.

6. Carotid angiography — the injection of dye into the carotid artery to see tumors.

7. Cephalalgia — headache.

8. Chordotomy — division of the anterolateral tracts of the spinal cord.

9. Craniocele — herniation of any part of the cranial contents through a defect in the skull.

10. Duraplasty — plastic repair of the dura mater.

11. Electrocorticography — electroencephalography where the electrodes are applied directly to the cerebral cortex.

12. Encephalomalacia — morbid softening of the brain tissue.

13. Endarterectomy — the surgical removal of plaques within an extracranial artery, usually the carotid artery. Done to prevent strokes.

14. Hemiballismus — violent motor restlessness, usually of the upper extremities.

15. Hemispherectomy — resection of a cerebral hemisphere.

16. Hydromyelia — accumulation of fluid within the spinal cord. Central canal tends to enlarge.

17. Hypokinesia — abnormal decrease in motor function.

18. Leptomeningitis — an inflammation or infection of the arachnoid and pia mater of the brain and spinal cord.

19. Metrizamide — a nonionic, water-soluble, iodinated, radiographic contrast medium used in myelography and other techniques.

20. Microgyrus — an abnormally small, malformed convolution of the brain.

21. Migraine — a severe headache, usually involving one side of the head, that is often accompanied by nausea and vomiting.

22. Myelodysplasia — defective development of the spinal cord.

23. Myelography — the X-ray visualization of the spinal cord by means of a contrast medium, such as metrizamide into the subarachnoid space.

24. Pachymeningitis — inflammation of the dura mater.

25. RINE — reversible ischemic neurological effect.

26. TORCH — a symbolism used for remembering the most common prenatal infections: toxoplasmosis, other (syphilis, etc.), rubella, cytomegalovirus, and herpes.

27. Tractotomy — a cross section of a nerve tract in the CNS used for the relief of pain.

Disorders/Homeostatic Imbalances

Neoplasms

1. Gliosarcoma — a glioma combined with fusiform cells of a sarcoma.

2. Hemangioblastoma — a capillary hemangioma of the brain, consisting of proliferated blood vessel cells or angioblasts.

3. Medulloblastoma — a soft, infiltrating, malignant tumor, developing on the roof of the fourth ventricle and cerebellum. Tends to infiltrate the meninges.

4. Meningioma — a slow-growing tumor that originates in the arachnoid tissue.

Infectious Diseases

1. Bacterial Meningitis — a meningitis commonly caused by *Hemophilus influenzae* that affects children under the age of five. May be a part of the normal flora of adults. A vaccine is available for all children after the age of two.

2. Cryptococcosis — a disseminating systemic fungal infection caused by *Cryptococcus neoformans* that has a predilection for the brain and meninges.

3. Echinococcosis — a tapeworm infection caused by *Echinococcus granulosis* that forms hydatid cysts through the body of the host, including the brain and spinal cord.

4. Kuru — a chronic, progressive, usually fatal disorder of the CNS, caused by a slow, unconventional virus. Seen only in the headhunting, cannibalistic Fore Indian tribes of New Guinea.

5. Rabies — an acute infectious disease of the CNS, caused by an RNA virus of the rhabdoviridae group. The virus is present in the host's saliva, with human infection occurring following a bite of a rabid animal. The virus appears to follow neural pathways, ultimately reaching the brain.

6. Tabes Dorsalis — sclerosis of the posterior columns of the spinal cord due to a syphilis infection of the CNS. Symptoms include postural instability and a staggering wide-base gait.

7. Toxoplasmosis — a protozoan disease caused by *Toxoplasma gondii* that may affect neonates and children under two years. The congenital form is marked by CNS lesions that may lead to blindness, brain defects, and death. Commonly transmitted in the saliva and excrement of infected cats and rodents.

8. Waterhouse-Friderichsen Syndrome — the fulminating or malignant form of epidemic cerebrospinal meningitis, usually caused by *Neisseria meningitidis*. Usually marked by sudden onset and short course, with high fever, coma, collapse, cyanosis, and severe hemorrhages of the skin.

Genetic and Developmental Disorders

1. Arnold-Chiari Malformation — an unusual congenital anomaly, usually associated with hydrocephalus, where the cerebellum and medulla protrude down the cervical spinal canal through the foramen magnum.

2. Friedreich's Ataxia — a hereditary sclerosis of the lateral and dorsal column of the spinal cord, usually appearing in early childhood. Characterized by ataxia, speech impairment, scoliosis, muscular paralysis, and other features.

3. Wilson's Disease — a rare, progressive, inherited disorder due to an inability to properly metabolize copper, resulting in accumulations in the brain, liver, cornea, and other tissues. Neurological symptoms most commonly occur in young adults.

Degenerative Disorders

1. Subacute Sclerosing Panencephalitis (SSP) — an encephalitis that usually occurs in young children and is characterized by a progressive intellectual deterioration, behavioral disorders, and involuntary muscular disorders. May be due to a viral infection.

2. Progressive Multifocal Leukoencephalopathy (PML) — a rare, demyelinating disorder that primarily affects middle-aged patients with some form of neoplastic or immunocompromising disorder. The disease appears to be caused by the JC virus or SV 40 virus, which affects the white matter of the brain, brain stem, and cerebellum.

ANSWERS TO END-OF-CHAPTER QUESTIONS

Multiple Choice/Matching

1. a
2. (1)c; (2)f; (3)e; (4)g; (5)b; (6)f; (7)i; (8)a
3. d

4. c
5. a
6. b
7. c

8. a
9. (1)a; (2)b; (3)a; (4)a; (5)b; (6)a; (7)b; (8)b; (9)a
10. d

Short Answer Essay Questions

11. See Fig. 12.3, p. 381, for a diagram of the embryonic brain vesicles.

12. a. Increases cortical surface area which allows more neurons to occupy the limited space. (p. 381)

    b. Sulci and fissures (p. 384); gyri (p. 384)

    c. Median longitudinal fissure (p. 384)

    d. Central sulcus; lateral sulcus (p. 384)

13. a. See Fig. 12.10, p. 386, for a drawing of the functional areas of the brain.

b. Primary motor cortex — All voluntary somatic motor responses arise from this region.

Premotor cortex — This region controls learned motor skills of a repetitious or patterned nature. Somatosensory association area — Acts to integrate and analyze different somatosensory inputs, such as temperature, touch, pressure, and pain.

Primary Somatosensory Cortex — Receives all somatosensory information from receptors located in the skin and from proprioceptors in muscles; identifies the body region being stimulated.

Visual area — Receives information that originates in the retinas of the eyes.

Auditory area — Receives information that originates in the hearing receptors of the inner ear.

Prefrontal cortex — Most involved with elaboration of thought, intelligence, motivation, and personality. It also associates experiences necessary for the production of abstract ideas, judgment, planning, and conscience, and is important in planning motor activity.

Wernicke's area — Speech area involved in the comprehension of language, especially when the word needs to be sounded out or related.

Broca's area — Previously called the motor speech area; now known to be active in many other activities as well.

14. a. Specialization of cortical functions. The "dominant" hemisphere excels at language and mathematical skills. The nondominant hemisphere is better at visual-spatial skills, intuition, emotion, and appreciation of art and music. (p. 390)

b. Both hemispheres have perfect and instant communication with each other so there is tremendous integration, therefore neither side is better at everything. However, each hemisphere does have unique abilities not shared by its partner. (p. 390)

15. a. Initiate slow and sustained movement; helps to coordinate and control motor activity (p. 392)

b. The putamen and globus pallidus (p. 392)

c. Caudate nucleus (p. 392)

16. a. Three paired fiber tracts (cerebellar peduncles) connect it to the brain stem. (pp. 399-400)

b. The cerebellum has a convoluted surface with gray matter on outside and white on inside; it has two hemispheres that have overlapping functions. (pp. 399-400)

17. The cerebellum acts like an automatic pilot by initiating and coordinating the activity of skeletal muscle groups. A step-by-step discussion is given on p. 401.

18. a. Medial aspect of each cerebral hemisphere.

b. Cingulate gyrus, parahippocampal gyrus, hippocampus, regions of the hypothalamus, mammillary bodies, septal nuclei, amygdaloid nucleus, anterior thalamic nuclei, and fornix.

c. Acts as our emotional or affective (feeling) brain. (pp. 401-402)

19. a. It extends through the central core of the medulla, pons, and midbrain.

b. RAS means reticular activating system, which is our cortical arousal mechanism. It helps to keep the cerebral cortex alert while filtering out unimportant inputs. (pp. 402-403)

20. CNS protected by: bony cranium, meninges, cerebrospinal fluid and blood-brain barrier. (pp. 403-406)

21. a. CSF is formed by the choroid plexus via a secretory process involving both active transport and diffusion and is drained by the arachnoid villi. See Fig. 12.23 for the circulatory pathway. (p. 404)

b. A condition called hydrocephalus can develop. In children, the fontanels allow expansion without brain damage, but in adults, the lack of expansion may cause severe damage due to brain compression. (p. 404)

22. The blood-brain barrier represents capillaries that are formed by endothelial cells joined by tight junctions. This characteristic makes them highly selective, ensuring that only certain substances can gain access to the neural tissue. (p. 405)

23. a. A concussion occurs when brain injury is slight and the symptoms are mild and transient. Contusions occur when marked tissue destruction takes place. (p. 406)

    b. Due to injury of the RAS. (p. 402)

24. The spinal cord is somewhat flattened from front to back and is marked by two grooves: the anterior median fissure and the posterior median sulcus. The gray matter of the spinal cord is shaped like the letter H, consisting of lateral gray masses connected by the gray commissure. The two posterior (dorsal) projections are the posterior horns and the anterior (ventral) projections are the anterior horns. Somatic motor neurons send their axons out via the ventral roots, and sensory axons enter the cord via the dorsal roots. The white matter of the cord is composed of myelinated and unmyelinated fibers that run in three possible directions: (1) up to higher CNS centers, (2) down to the cord from the brain, and (3) across from one side of the cord to the other. The white matter on each side of the cord is divided into three white columns, posterior, lateral, and anterior funiculi. (pp. 408-418)

25. Touch and pressure: fasciculus cuneatus, fasciculus gracilis, and anterior spinothalamic. Proprioception: anterior and posterior spinocerebellar. Pain and temperature: lateral spinothalamic. (Table 12.2, p. 415)

26. a. Voluntary skeletal movements (pyramidal tracts) — lateral and anterior corticospinal

    b. Tectospinal, vestibulospinal, rubrospinal (Table 12.3, p. 418)

27. Spastic paralysis — due to damage to upper motor neurons of the primary motor cortex. Muscles can respond to reflex arcs.

    Flaccid paralysis — damage to ventral root or anterior horn cells. Muscles cannot respond. (p. 416)

28. Paraplegia — damage to cord (lower motor neurons) between T1 and L1 that causes paralysis of both lower limbs.

    Hemiplegia — damage, usually in the brain, that causes paralysis of one side of the body.

    Quadriplegia — damage to cord in cervical area affecting all four limbs. (pp. 416-418)

29. a. CVA, also known as stroke, occurs when blood circulation to a brain area is blocked and vital brain tissue dies. A new hypothesis targets the release of glutamate by oxygen-starved neurons (and subsequent entry of excess Ca^{++}) as the culprit.

    b. Any event that kills brain tissue due to a lack of oxygen; includes blockage of a cerebral artery by a blood clot, compression of brain tissue by hemorrhage or edema, and arteriosclerosis. Consequences include paralysis, sensory deficits, language difficulties, and speech problems. (p. 407)

30. a. Continued myelination of neural tissue accounts for growth and maturation of the nervous system.

    b. There is a decline in brain weight and volume in aging. (p. 421)

Critical Thinking and Application Questions

1. a. Only likely diagnosis is hydrocephalus.

   b. CT or sonograms, but most importantly pneumoencephalography.

   c. Lateral and third ventricles enlarge; fourth ventricle, central canal and subarachnoid space are not affected. If arachnoid villi are obstructed, all CSF areas will be enlarged. (p. 405)

2. Alzheimer's disease (pp. 407-408)

3. Probably the frontal lobes, specifically the prefrontal cortex, which mediates personality and moral behavior. (p. 389)

4. In myelomeningocele, a cyst containing parts of the spinal cord, nerve roots, and meninges protrudes from the spine. Pressure during vaginal delivery could cause the cyst to rupture, leading to infection and further damage. A C-section is preferable.

Key Figure Questions

Figure 12.4 Telencephelon, particularly the cerebral hemispheres. **Figure 12.7** The growth of the cerebral hemisphere is restricted by the presence of the forming embryonic skull such that the cerebral hemispheres are forced to grow posteriorly and inferiorly. Hence their cavities (ventricles) become horn-shaped. **Figure 12.10** Central sulcus. **Figure 12.11** Because the anatomical representations indicate the relative amount of the cerebral cortex dedicated to that particular body area. **Figure 12.13** Because the fibers of the corona radiata running through them make them appear striped. **Figure 12.23** Arachnoid villi.

SUGGESTED READINGS

---

1. Angier, N. "Storming the Wall." *Discover* 11 (May 1990): 60-67.

2. Begley, S., J. Carey, and R. Sawhill. "How the Brain Works." *Newsweek* 7 (Feb. 1983).

3. Bower, B. "Inside the Autistic Brain." *Science News* 130 (Sept. 1986): 154-155.

4. Bower, B. "The Language of the Brain." *Science News* 132 (July 1987): 40-41.

5. Bredt, D.S., and S.H. Snyder. "Biological Roles of Nitric Oxide." *Scientific American* (1993): 22.

6. Cowan, W.M. "The Development of the Brain." *Scientific American* 241 (Sept. 1979): 112-133.

7. Fine, A. "Transplantation in the CNS." *Scientific American* 255 (Aug. 1986): 52-58.

8. Galaburda, A.M., et al. "Right-Left Asymmetries in the Brain." Science 199 (1978): 852.

9. Geschwind, N. "Specializations of the Human Brain." *Scientific American* 241 (Sept. 1979): 180-199.

10. Gluhbegovic, N., and T.H. Williams. *The Human Brain: A Photographic Guide*. New York: Harper & Row, 1980.

11. Goldberger, A.L., D.R. Rigney, and B.J. West. "Chaos and Fractals in Human Physiology." *Scientific American* 262 (Feb. 1990): 43-49.

12. Goldstein, G.W., and A.L. Betz. "The Blood-Brain Barrier." *Scientific American* 255 (Sept. 1986): 74-83.

13. Guillemin, R. "Peptides in the Brain." *Science* 202 (1978): 390.

14. Holloway, M. Profile: Vive la Difference." *Scientific American* 263 (Oct. 1990): 40-42.

15. Hubel, D.H. "The Brain." *Scientific American* (Sept. 1979).

16. Iversen, L.L. "The Chemistry of the Brain." *Scientific American* 241 (Sept. 1979): 134-149.

17. Johanson, C.E., and R. Spector. "The Mammalian Choroid Plexus." *Scientific American* 261 (Nov. 1989): 68.

18. Johnson, G.T. "Is Spinal Anesthesia Best for You?" *Mayo Clinic Health Letter* (Oct. 1984).

19. Kety, S.S. "Disorders of the Human Brain." *Scientific American* 241 (Sept. 1979): 202-214.

20. Krieger, D.T. "Brain Peptides: What, Where, and Why." *Science* 222 (1983): 975.

21. Lancaster, J.R. "Nitric Oxide in Cells." *American Scientist* 80 (May-June 1992): 248.

22. Loscalzo, J., D.J. Singel, and J.S. Stamler. "Biochemistry of Nitric Oxide and Its Redox-Activated Forms." *Science* (Dec. 1992): 189.

23. Miller, J.A. "Sex Differences Found in Human Brains." *Science News* 126 (July 1984).

24. Miller, J.A. "Sex in the Spinal Cord." *Science News* 118 (Nov. 1980): 329.

25. Nathanson, J.A., and P. Greengard. "Second Messenger in the Brain." *Scientific American* 237 (Aug. 1977): 108-119.

26. Nauta, J.H., and M. Feirtag."The Organization of the Brain." *Scientific American* (Sept. 1979).

27. Pajk, M., et al. "Alzheimer's Disease." *American Journal of Nursing* 84 (Feb. 1984): 215-232.

28. Romero, J.H. "The Critical Minutes After Spinal Cord Injury." *RN* (Apr. 1988): 61-67.

29. Routenberg, A. "The Reward System of the Brain." *Scientific American* 239 (Nov. 1978): 154-164.

30. Schmidt, K.F. "Method Probes Chemistry of Stroke, Aging." *Science News* 143 (May 1993): 343.

31. Silberner, J. "Alzheimer's Disease: Source Searching." *Science News* 128 (July 1985): 24.

32. Snyder, S. "Nitric Oxide: First in a New Class of Neurotransmitters." *Science* 257 (July 1992): 494.

33. Spector, R. and C.E. Johnson. "The Mammalian Choroid Plexus." *Scientific American* 261 (Nov. 1989): 68-74.

34. Springer, S.P., and G. Deutsch. *Left Brain, Right Brain*. San Francisco: W.H. Freeman, 1981.

35. Tangley, L . "Female Brain Anatomy May Differ." *Science News* 136 (Nov. 1989).

36. Thompson, R.F. *The Brain: An Introduction to Neuroscience*. New York: W.H. Freeman and Co., 1985.

37. Thomanen, E. "Breaking the Blood-Brain Barrier." *Scientific American* 268 (Feb. 1993): 80.

38. Vellutino, F.R. "Dyslexia." *Scientific American* 256 (Mar. 1987): 34-41.

39. Weiss, R. "Neurons Regenerate Into Spinal Cord." *Science News* 132 (Nov. 1987).

40. Weiss, R. "Women's Skills Linked to Estrogen Levels." *Science News* 134 (Nov. 1988): 341.

41. Weiss, R. "New Therapies Brighten Stroke Horizon." *Science News* 136 (Nov. 1989): 292.

42. Wurtman, R.J. "Alzheimer's Disease." *Scientific American* (Jan. 1985).

13

The Peripheral Nervous System and Reflex Activity

Chapter Preview

This chapter will focus on the cranial and spinal nerves of the peripheral nervous system that communicate with the brain and spinal cord of the central nervous system. It will then describe some of the important spinal reflexes, especially as they involve somatic function.

INTEGRATING THE PACKAGE

Suggested Lecture Outline

I. Overview of the Peripheral Nervous System (pp. 425-434)

    A. Sensory Receptors (pp. 426-432)

        1. Classification by Location

        2. Classification by Stimulus Type Detected

        3. Classification by Structural Complexity

        4. Free Dendritic Endings

        5. Encapsulated Dendritic Endings

            a. Meissner's Corpuscles

            b. Krause's End Bulbs

            c. Pacinian Corpuscles

            d. Ruffini's Corpuscles

            e. Muscle Spindles

            f. Golgi Tendon Organs

            g. Joint Kinesthetic Receptors

        6. Sensory Receptor Potentials

    B. Nerves and Associated Ganglia (pp. 432-434)

        1. Structure and Classification

        2. Regeneration of Nerve Fibers

    C. Motor Endings (p. 434)

        1. Axonal Terminals (Boutons) of Somatic Fibers

        2. Varicosities of Autonomic Fibers

II. Cranial Nerves (pp. 434-436; Table 13.2, pp. 436-442)

    A. Olfactory Nerves (I) (p. 434)

    B. Optic Nerves (II) (p. 435)

    C. Oculomotor Nerves (III) (p. 435)

Review Items

8. Synapses (Chapter 11, p. 360)

9. Neurotransmitters (Chapter 11, pp. 366-370)

10. Ascending tracts of the spinal cord (Chapter 12, p. 413)

11. Descending tracts of the spinal cord (Chapter 12, p. 416)

12. Spinal roots (Chapter 12, p. 411)

13. Gray and white matter of the spinal cord (Chapter 12, pp. 411-413)

Cross References

1. Cutaneous sensory receptors are presented in Chapter 5, p. 146.

2. Sensory receptors for the special senses and generator potentials are further described in detail in Chapter 16.

3. Spinal reflexes and the physiology of the sexual response is described in Chapter 28, p. 965.

4. Reflex activity and the control of digestive secretions is described in detail in Chapter 24, p. 809.

5. The nerve plexuses involved in digestion are mentioned in Chapter 24, p. 792.

6. The function of the vagus nerve in parasympathetic control is examined in Chapter 24, p. 808.

7. The cranial nerves associated with their special senses are covered in Chapter 16.

8. Refex activity of the special senses is examined in Chapter 16.

9. Receptor and generator potentials as related to neural integration are described in Chapter 15, p. 483.

10. Spinal reflex control of micturition is examined in Chapter 26, p. 920.

Laboratory Correlations

1. Marieb, E. N. *Human Anatomy and Physiology Laboratory Manual: Cat and Fetal Pig Versions.* 3rd. ed. Benjamin/Cummings, 1989.

    Exercise 22: Human Reflex Physiology

2. Marieb, E. N. *Human Anatomy and Physiology Laboratory Manual: Brief Version.* 3rd. ed. Benjamin/Cummings, 1992.

    Exercise 19: Human Reflex Physiology

Transparencies Index

13.2	Structure of a nerve	13.9	The lumbar plexus
13.4	Location and function of cranial nerves	13.10	The sacral plexus
13.5	Distribution of spinal nerves, posterior view	13.12	Components of all human reflex arcs
13.6	Formation and branches of spinal nerves	13.13	Anatomy of muscle spindle and Golgi tendon organ
13.8	The brachial plexus	13.15	Events of the stretch reflex

Bassett Atlas Figures Index

Slide Number	Figure Number	Description
9	1.9A,B	Spinal cord, origin
10	1.10	Spinal cord, cauda equina
11	1.11A,B	Spinal cord, detail
12	2.1A,B	Parotid gland and facial nerve
19	3.3A,B	Chest wall removed
22	3.6A,B	Lateral view of mediastinum
29	4.2A,B	Abdominal wall, rectus reflected
58	6.6A,B	Brachial plexus and axilla
64	6.12A,B	Palmar forearm
72	7.4A,B	Inguinal area
73	7.5	Inguinal area
76	7.8	Popliteal fossa
84	7.16A,B	Foot, medial view

INSTRUCTIONAL AIDS

Lecture Hints

1. Emphasize the distinction between central and peripheral nervous system, although stressing that the nervous system functions as a continuous unit, even though we like to study its anatomy in bits and pieces. Students often treat each section as if it operates autonomously, without regard to what may be happening in other parts of the nervous system.

2. Many students will have a difficult time with the difference between receptor potentials, generator potentials, and action potentials. It is worth some time to be sure the distinction is clear.

3. As the anatomy of the nerve is discussed, point out the similarity between the basic structure of muscle tissue and nervous tissue. Also bring to the students' attention the similarity in nomenclature. Point out that by knowing the structure of muscle, they already know nerve anatomy (with slight changes in names).

4. Students often have problems with neuron regeneration and myelination (i.e., CNS and PNS neurons are both myelinated and why regeneration occurs in the PNS and not in the CNS). Spend time explaining the difference or refer the class to Chapter 11 to review myelination, the sheath of Schwann, and oligodendrocytes.

5. Try a diagram (cross-section) of the spinal cord indicating the dorsal and ventral roots and an extension into a short section of the spinal nerve. Draw arrows in these pathways indicating the direction of information flow. Remind the class that the brain must always receive information from an area in order to effect a change (the reason for two-way traffic in each level of the cord). Students are more likely to remember the anatomical relationship between these structures since they can logically relate material from a previous chapter to the material presented in this chapter.

6. Try asking specific questions of the class in order to promote student involvement. This technique holds student attention and, more importantly, enforces the logical thought processes necessary in order to thoroughly comprehend physiological concepts. The reflex arc is an excellent tool to employ this strategy, since by this time the class has a general knowledge of all the components necessary to construct a generalized arc. After a brief introduction to the reflex arc and what its general function is, ask questions such as: "If we wanted to construct a reflex arc, what could we use to convert a stimulus to a nervous impulse?" Lead the class by a series of questions to the complete construction of the basic reflex arc, then go into the modifications of the basic blueprint to describe specific arc types and their functions. Students will not forget the reflex arc since they have constructed it themselves.

Demonstrations/Activities

1. Audio-visual materials of choice.

2. Select a student to help in the illustration of reflexes, such as patellar, plantar, abdominal, etc.

3. Obtain a skull to illustrate the locations, exit, and entrance of several cranial nerves, such as the olfactory, optic, and trigeminal.

4. Obtain a sheep brain with the cranial nerves intact to illustrate their locations.

5. Obtain a 3-D model of the peripheral nervous system to illustrate the distribution of the spinal nerves.

6. Obtain a 3-D model of a spinal cord cross section to illustrate the five components of a reflex arc and to illustrate terms such as ipsilateral, contralateral, monosynaptic, etc.

Critical Thinking/Discussion Topics

1. How can the injection of novocaine into one area of the lower jaw anesthetize one entire side of the jaw and tongue?

2. How can seat belts for both the front and back seat passengers of a car prevent serious neurological damage? How can using only lap belts cause severe damage?

3. Some overly eager parents swing their newborn infants around by the hands. What damage could this cause?

4. Pregnant women often experience numbness in their fingers and toes. Why?

5. Animals have considerably more reflexive actions than humans. Why?

Library Research Topics

1. How does acupuncture relate to the distribution of spinal nerves?

2. Will all victims of polio be rendered paralyzed? What different forms are there?

3. How has microsurgery been used to reconnect severed peripheral nerves?

4. What techniques can be employed to increase our reflexive actions?

LECTURE ENHANCEMENT MATERIAL

Clinical and Related Terms

1. Achilles reflex — a reflex involving the plantar extension of the foot, following a tap on the Achilles tendon.

2. Anal reflex — a reflex involving the contraction of the anal sphincter or irritation of the anal skin.

3. Areflexia — absence of reflexes.

4. Carpal Tunnel Syndrome — a syndrome resulting from the compression of the median nerve in the carpal tunnel, resulting in pain, and burning and tingling of the fingers and hands.

5. Corneal reflex — a reflex involving the closure of the eyelids due to irritation of the cornea.

6. Cremasteric reflex — a reflex involving the contraction of the ipsilateral cremaster muscle that draws the testes upward following stroking of the inner aspect of the thigh.

7. Crutch palsy — paralysis to the brachial plexus due to pressure to the axilla from prolonged use of a crutch.

8. Delayed Nerve Grafting — a surgical technique involving the removal of damaged sections of a spinal cord and replacement with nerve segments from an arm or leg.

9. Gangliectomy — the surgical removal of a ganglion.

10. Nerve block — regional anesthesia obtained by an injection of anesthetics in close proximity to the appropriate nerve.

11. Neurectomy — excision of a portion of a nerve.

12. Neurectopia — displacement of a nerve.

13. Neuroanastomosis — the surgical anastomosis of one nerve to another.

14. Neurolysis — surgical breaking up of perineural adhesions.

15. Neurorrhaphy — the suturing of a nerve.

16. Neurotomy — dissection of a nerve.

17. Phrenicotomy — excision or resection of the phrenic nerve.

18. Polyneuritis — inflammation of multiple nerves.

19. Radiculitis — inflammation of a spinal nerve root.

20. Radiculoneuritis (Guillain-Barre Syndrome) — a syndrome characterized by absence of fever, pain or tenderness of muscle, motor weakness, and absence of tendon reflexes. A type of encephalitis.

21. Rhizotomy — interruption of the nerve roots in the spinal canal.

22. Thoracic Outlet Syndrome — a compression of the brachial plexus resulting in pain to the arms, numbness of the fingers, and wasting of the muscles of the hand.

Disorders/Homeostatic Imbalances

Tumors of Peripheral Nerves

1. Acoustic Neuroma — a benign tumor within the auditory canal arising from the 8th cranial (acoustic) nerve.

2. Neuroblastoma — a sarcoma derived from nervous tissue, chiefly neuroblasts. Primarily affects young children, usually arising in the autonomic nervous system.

3. Neurofibroma — a benign tumor of peripheral nerves due to an abnormal proliferation of Schwann cells. Also known as a fibroneuroma.

4. Neurilemoma — a tumor of a peripheral nerve sheath (neurilemma). Also known as a Schwannoma.

5. Neurosarcoma — a sarcoma with neuromatous elements.

Infections and Inflammations

1. Guillain-Barre Syndrome — idiopathic polyneuritis. Characterized by a widespread patchy demyelination of spinal nerves and nerve roots, with some inflammatory changes. Symptoms include muscular weakness, beginning in the legs and spreading to the trunk. Recovery is slow.

2. Peripheral Neuritis (Polyneuritis) — an inflammation of spinal nerves, characterized by muscular weakness, numbness, and tingling, tenderness, and pain. Usually caused by systemic chronic disorders such as diabetes, alcoholism, etc.

3. Shingles — an acute viral infection, manifested along sensory root ganglia and caused by the herpes simplex virus. The disease exhibits vesicular eruptions along the area of distribution of the sensory nerve.

ANSWERS TO END-OF-CHAPTER QUESTIONS

Multiple Choice/Matching

1. b

2. c

3. c

4. (1)f; (2)i; (3)b; (4) g,h,l; (5)e; (6)i; (7)c; (8)k; (9)l; (10)c, d, f, k

5. (1)b 6; (2)d 1,8; (3)c 2; (4)c 5; (5)a 4; (6)a 3,9; (7) a 7; (8) a 7; (9)d 1

6. b, 1; a, 3 and 5; a, 4; a, 2; c, 2

7. b

Short Answer Essay Questions

8. The PNS enables the CNS to receive information and carry out its decisions. (p. 425)

9. The PNS includes all nervous tissue outside the CNS, that is, the sensory receptors, the peripheral nerves (cranial or spinal), the ganglia, and motor nerve endings. The peripheral nerves transmit sensory and motor impulses, the ganglia contain cell bodies of sensory or autonomic nerve fibers, the sensory receptors receive stimuli, and the motor end plates release neurotransmitters that regulate the activity of the effectors. (p. 425)

10. A receptor potential is like an EPSP in that it is a graded potential that changes membrane permeability and can help to generate an action potential. It differs from an EPSP in its stimulus, i.e., a type of energy (light pressure, etc.) versus a chemical neurotransmitter in the case of the EPSP. (p. 429)

11. Schwann cells aid the regeneration process physically and chemically. Oligodendrocytes die and thus do not aid fiber regeneration. (p. 433)

12. a. Spinal nerves form from dorsal and ventral roots that unite distal to the dorsal root ganglion. Spinal nerves are mixed. (See Fig. 13.6, p. 443)

    b. The ventral rami contribute large plexi that supply the anterior and lateral parts of the body, trunk, and the limbs. The dorsal rami supply the muscles and skin of the back (posterior trunk). (p. 443)

13. a. A plexus is a branching nerve network that ensures that any damage to one nerve root will not result in total loss of innervation to that part of the body. (p. 444)

    b. See Figs. 13.7 to 13.10, and Tables 13.3 to 13.6, pp. 444-445, for detailed information about each of the four plexuses.

14. Ipsilateral reflexes involve a reflex affecting the same side of the body (p. 454); contralateral reflexes involve a reflex that is initiated on one side of the body and affects the other side. (p. 455)

15. The flexor or withdrawal reflex is a protective mechanism to withdraw from a painful stimulus. (p. 455)

16. Flexor reflexes are protective ipsilateral, polysynaptic, and prepotent reflexes, whereas crossed extensor reflexes consist of an ipsilateral withdrawal reflex and a contralateral extensor reflex that aids usually in balance. (pp. 455-456)

17. Reflex tests assess the condition of the nervous system. Exaggerated, distorted, or absent reflexes indicate degeneration or pathology of specific regions of the nervous system before other signs are apparent. (p. 454)

18. Dermatomes are related to the sensory innervation regions of the spinal nerves. The spinal nerves correlate with the segmented body plan, as do the muscles (at least embryologically). (p. 456)

Critical Thinking and Application Questions

1. Precise realignment of cut, regenerated axons with their former effector targets is highly unlikely. Coordination between nerve and muscle will have to be relearned. Additionally, not all fibers regenerate. (pp. 432-434)

2. He would have problems dorsiflexing his right foot and his knee joint would be unstable (more rocking of the femur from side-to-side on the tibia). (p. 448)

3. Damage to the brachial plexus occurred when he suddenly stopped his fall by grabbing the branch. (p. 446)

4. The left trochlear nerve (IV) which innervates the superior oblique muscle responsible for this action. (p. 437)

5. The region of motor and sensory loss follows the course of the sciatic nerves (and their divisions); hence they must have been severely damaged by the shooting accident. (pp. 447-448)

Key Figure Questions

Figure 13.3 They act as phagocytes and reform the channel that guides axonal growth. **Figure 13.5** The peripheral nerves serving the limbs arise in the lumbar and cervical regions. **Figure 13.12** The fiber of the motor neuron.

SUGGESTED READINGS

1. Barr, M.L., and J.A. Kiernan. *The Human Nervous System*. 4th ed. New York: Harper & Row, 1983.

2. Cowan, R. "Antibodies Enhance Spinal Nerve Regrowth." *Science News* 137 (Jan. 20, 1990): 38.

3. Easton, T.A. "On the Normal Use of Reflexes." *American Scientist* 60 (Sept.-Oct. 1972).

4. Guyton, A. *Basic Neuroscience: Anatomy and Physiology*. Philadelphia: Saunders, 1987.

5. Mathers, L.H. *The Peripheral Nervous System*. Stoneham, MA: Butterworth Pub., 1984.

6. Noback, C.T., and R. Demarest. *The Human Nervous System*. 3rd ed. New York: McGraw-Hill, 1980.

7. Schmidt, R.F. *Fundamentals of Sensory Physiology*. 3rd. ed. NewYork: Springer-Verlag, 1986.

8. Shepherd, G.M. *Neurobiology*. New York: Oxford Univ. Press, 1983.

9. Terry, G. "The Nervous System: Repairs to the Network." *New Scientist* 10 (June 1989): 1-4.

10. Weiss, R. "Neurons Regenerate Into Spinal Cord." *Science News* 132 (Nov. 21, 1987): 324.

11. Weiss, R. "Regenerated Nerves Send First Messages." *Science News* 136 (Oct. 14, 1989): 244.

12. Willis, W.D., and R.G. Grossman. *Medical Neurobiology*. 3rd ed. St. Louis: C.V. Mosby, 1981.

14

The Autonomic Nervous System

Chapter Preview

This chapter describes the complex internal regulating system, the autonomic nervous system. The anatomy and physiology of its two divisions, parasympathetic and sympathetic, will be illustrated in detail. How the two divisions interact to control most of our visceral organs and how they regulate homeostasis will then be presented.

INTEGRATING THE PACKAGE

Suggested Lecture Outline

I.   Introduction (p. 460)

II. Overview of the Autonomic Nervous System (pp. 461-463)

    A. Comparison of the Somatic and Autonomic Nervous System (pp. 461-462)

        1. Effectors

        2. Efferent Pathways and Ganglia

            a. Preganglionic Neuron

            b. Ganglion

            c. Postganglionic Neuron

        3. Neurotransmitter Effects

        4. Overlap of Somatic and Autonomic Function

    B. The Divisions of the Autonomic Nervous System (pp. 462-463)

        1. Role of the Parasympathetic Division

        2. Role of the Sympathetic Division

III. Anatomy of the Autonomic Nervous System (pp. 463-469)

    A. Parasympathetic (Craniosacral) Division (pp. 463-466)

        1. General Characteristics

            a. Craniosacral Fibers

            b. Terminal Ganglia

        2. Cranial Outflow

            a. Oculomotor Nerves, Ciliary Ganglia

            b. Facial Nerves, Sphenopalatine Ganglia, Submandibular Ganglia

            c. Glossopharyngeal Nerves, Otic Ganglia

            d. Vagus Nerve

                 1. Intramurine Ganglia

                 2. Thoracic Plexi

                 3. Sacral Outflow

B. Sympathetic (Thoracolumbar) Division (pp. 466-469)

   1. General Characteristics

      a. White Ramus Communicans

      b. Paravertebral Ganglia

      c. Sympathetic Trunk

      d. Collateral Ganglia

   2. Pathways with Synapses in a Paravertebral (Sympathetic Chain) Ganglion

      a. Gray Communicantes

      b. Superior Cervical Ganglia

      c. Middle and Inferior Cervical Ganglia

   3. Pathways with Synapses in a Prevertebral Ganglion

      a. Splanchnic Nerves

      b. Abdominal Aortic Plexi

         1. Celiac

         2. Mesenteric (Superior and Inferior)

         3. Hypogastric

   4. Pathways with Synapses in the Adrenal Medulla

      a. Norepinephrine

      b. Epinephrine

C. Visceral Sensory Neurons (p. 469)

IV. Physiology of the Autonomic Nervous System (pp. 469-478)

A. Neurotransmitters and Receptors (pp. 469-471)

   1. Cholinergic Receptors

      a. Nicotinic Receptors

      b. Muscarinic Receptors

   2. Adrenergic Receptors

      a. Alpha Receptors

      b. Beta Receptors

B. The Effects of Drugs (p. 471)

C. Interactions of the Autonomic Divisions (pp. 471-473)

   1. Basic Features (Table 14.4, p. 472)

   2. Antagonistic Interactions

   3. Sympathetic and Parasympathetic Tone

   4. Cooperative Effects

   5. Unique Roles of the Sympathetic Division

      a. Thermoregulatory Responses to Heat

      b. Release of Renin from the Kidneys

      c. Metabolic Effects

   6. Localized Versus Diffuse Effects

D. Control of Autonomic Functioning (pp. 473-478)

   1. Brain Stem and Spinal Cord Controls

   2. Hypothalamic Controls

3. Cortical Controls
    a. Influence of Meditation on Autonomic Function
    b. Influence of Biofeedback on Autonomic Function

V. Homeostatic Imbalances of the Autonomic Nervous System (p. 478)

  A. Hypertension
  B. Raynaud's Disease
  C. Mass Reflex Reaction

VI. Developmental Aspects of the Autonomic Nervous System (pp. 478-479)

  A. Embryonic and Fetal Development of the Autonomic Nervous System

    1. Neural Tube Derivation
    2. Neural Crest Derivation
    3. Nerve Growth Factor
    4. Hirschsprung's Disease

  B. Development of the Autonomic Nervous System Through Adolescence
  C. Effect of Aging on the Autonomic Nervous System

Review Items

1. Membrane functions (Chapter 3, pp. 65-75)
2. Membrane receptors (Chapter 3, p. 74)
3. Nervous tissue (Chapter 4, p. 126)
4. Membrane potentials (Chapter 11, p. 351)
5. Neuronal integration (Chapter 11, pp. 371-373)
6. Serial and parallel processing (Chapter 11, p. 373)
7. Synapses (Chapter 11, p. 360)
8. Neurotransmitters (Chapter 11, pp. 366-370)
9. Ascending tracts of the spinal cord (Chapter 12, p. 413)
10. Descending tracts of the spinal cord (Chapter 12, p. 416)
11. Spinal roots (Chapter 12, p. 411)
12. Gray and white matter of the spinal cord (Chapter 12, pp. 411-413)

Cross References

1. The role of the sympathetic and parasympathetic pathways (and epinephrine, norepinephrine) in medullary control of cardiac rate is explained in Chapter 19.
2. Sympathetic and parasympathetic effects in human sexual response are detailed in Chapter 28, p. 965.
3. Sympathetic control of blood vessels to the kidney is described in Chapter 26, p. 900.
4. Parasympathetic pelvic splanchnic nerves and the urinary system are further described in Chapter 26, p. 920.
5. Sympathetic and parasympathetic control of digestive processes is explained in Chapter 24.
6. Sympathetic control of blood vessel diameter is further examined in Chapter 20, p. 654.

Laboratory Correlations

1. Marieb, E. N. *Human Anatomy and Physiology Laboratory Manual: Cat and Fetal Pig Versions.* 3rd. ed. Benjamin/Cummings, 1989.

   Exercise 19: Spinal Cord, Spinal Nerves, and Autonomic Nervous System

2. Marieb, E. N. *Human Anatomy and Physiology Laboratory Manual: Brief Version.* 3rd. ed. Benjamin/Cummings, 1992.

   Exercise 18: Spinal Cord, Spinal Nerves, and Autonomic Nervous System

Transparencies Index

14.2   Somatic/Autonomic nervous systems
14.5   Sympathetic pathways

Bassett Atlas Figures Index

Slide Number	Figure Number	Description
18	3.2	Anterior mediastinum
19	3.3A,B	Chest wall removed
22	3.6A,B	Lateral view of mediastinum

INSTRUCTIONAL AIDS

Lecture Hints

1. Since the autonomic nervous system is more complex than the somatic nervous system, it is worthwhile to spend some time comparing and contrasting the anatomy of each.

2. Figure 14.4 is a good 3-D representation of sympathetic pathways; however, as an initial introduction during lecture, it is of benefit for the instructor to draw a 2-D schematic diagram of sympathetic and parasympathetic pathways so that the class can follow the construction of the circuit logically and understand how it is "wired." Then refer students to the overall construction presented in Figure 14.4, p. 464.

3. Emphasize that somatic efferent pathways consist of a motor neuron cell body in the CNS whose axon extends out through the PNS to directly innervate the skeletal muscle effector. In contrast, autonomic efferent pathways follow the same general plan, but consist of two motor neurons in series.

4. Point out that in many cases sympathetic and parasympathetic synapses use different neurotransmitters, an essential characteristic in the dual nature of autonomic function. This will be illustrated when discussing fight/flight and rest/digest responses.

5. Many students have difficulty with the idea of neurotransmitter/receptor function. Point out that many substances similar in chemical construction to the actual neurotransmitter are capable of generating the same response. Emphasize that it is the binding of a substance to a receptor that generates the cellular response.

6. To illustrate sympathetic tone, use the example of vasomotor control. Point out that dilation (to decrease blood pressure) is not a muscle contraction response, but that relaxation of the smooth muscle in the wall of the blood vessel is the actual cause. To vasoconstrict, increase sympathetic stimulation. Therefore in order for dilation to be possible, there must be a certain amount of constant sympathetic stimulation (tone) even during a relaxed state.

7. Emphasize that there is a constant level of parasympathetic stimulation (tone) to many visceral organs and that there is just enough sympathetic stimulation to keep systems in homeostasis. To probe the class for understanding, ask "What would happen to resting heart rate if parasympathetic stimulation were cut?"

Demonstrations/Activities

1. Audio-visual materials of choice.

2. Set up a live, exposed frog or turtle heart demonstration to illustrate the effects of acetylcholine and epinephrine.

3. Without announcing what you will be doing, walk quietly into the lecture room, set your notes down, and yell very loudly (to startle the students). Then, have each student prepare a list of all those organs that were affected and what the effect was.

4. Obtain a preserved cat and dissect it to illustrate the sympathetic nerve trunk, coeliac ganglia, splanchnic nerves, and other portions of the ANS.

5. Obtain a 3-D model of a spinal cord cross section and longitudinal section that illustrates the parts of the ANS and especially the sympathetic and gray and white rami.

6. Use diagrams followed by Bassett Atlas slides to emphasize the anatomy of the autonomic pathways.

Critical Thinking/Discussion Topics

1. Describe the role of beta blockers in treating certain types of visceral disorders.

2. At certain times when people are very excited or are shocked suddenly, their bowels and/or urinary sphincters lose control. In terms of the role of the ANS, why does this happen?

3. Some individuals, following a very stressful event such as final exams, frequently come down with colds. Is there any relationship between the ANS, stress, and the onset of an illness? Discuss.

4. Most people feel very tired after they eat a big meal. Why?

5. How can biofeedback be used to reduce effects of constant pain and stress?

6. Why is sympathetic action diffuse and long-lasting while parasympathetic is local and short-lived? What would happen to body systems during a stressful situation if these characteristics were reversed? How would anatomy have to be changed?

Library Research Topics

1. Do all animals have an autonomic nervous system? If so, is it more or less advanced than ours?

2. The ANS regulates peristaltic waves of the GI tract. If the ganglia and/or fibers controlling this activity were damaged, what would happen? What bacterial agents or type of trauma could cause this?

3. Ulcers seem to occur in hypertensive individuals. What are the causes of this problem and what treatment is available?

4. Nicotine and muscarine are substances that bind at specific receptors. What exactly do these receptors look like? Draw out a cell membrane and illustrate how the receptors might look.

LECTURE ENHANCEMENT MATERIAL

Clinical and Related Terms

1. Achalasia — failure to relax the smooth muscle fibers of the gastrointestinal tract, especially of the lower esophagus, due to degeneration of the ganglion cells in the walls of the organ.

2. Belladonna — a deadly nightshade plant, Atropa belladonna. Source of various alkaloids, such as atropine and scopolamine. Used as a sedative and for the management of gastrointestinal disorders.

3. Cardiospasm — a failure to relax the esophagus and an absence of esophageal motility.

4. Gangliosympathectomy — excision of a sympathetic ganglion.

5. Sympathicotripsy — surgical crushing of a sympathetic nerve or ganglion.

6. Sympathectomy — a resection or transection of a sympathetic nerve.

7. Transcendental Meditation — a technique used to attain a state of complete physical and psychological relaxation.

8. Vagotomy — transection of the vagus nerve.

Disorders/Homeostatic Imbalances

Tumors

1. Paraganglioma — a tumor of the autonomic nervous system tissue, composed of a collection of chromaffin cells occurring outside the adrenal medulla, usually near the sympathetic ganglia. Most tend to secrete epinephrine or norepinephrine.

2. Sympathicoblastoma — a malignant tumor containing embryonic cells that normally develop into sympathetic nerve cells.

3. Sympathogonioma — a tumor composed of undifferentiated embryonic cells that normally would develop into sympathetic cells.

Hereditary Disorders

1. Familial Dysautonomia — a hereditary condition marked by defective lacrimation, skin blotching, emotional instability, motor incoordination, total absence of pain sensation, and hyporeflexia.

Degenerative and Other Disorders

1. Autonomic Dysreflexia — an exaggerated reflex of the autonomic nervous system to stimulation, mainly as a result of spinal cord injury. Condition requires immediate care, especially to correct the high blood pressure as a result of the dysreflexia.

2. Hyperreflexic Bladder — an exaggerated reflex of the bladder due to spinal cord injury or to unknown causes.

3. Olivopontocerebellar Atrophy — a degenerative disorder of the neurons around the olive, pons, and cerebellum, as well as the basal ganglia and spinal cord. May also be inherited.

4. Striatonigral Degeneration — atrophy of the putamen and caudate nuclei. Similar to paralysis agitans.

ANSWERS TO END-OF-CHAPTER QUESTIONS

Multiple Choice/Matching

1. d

2. (1)S; (2)P; (3)P; (4)S; (5)S; (6)P; (7)P; (8)S; (9)P; (10)S; (11)P; (12)S

Short Answer Essay Questions

3. Involuntary nervous system is used to reflect its subconscious control; emotional-visceral system reflects the fact that the hypothalamus is the major regulatory center for both the emotional (limbic) response and visceral controls. The term *visceral* also indicates the location of most of its effectors. (p. 460)

4. White rami contain myelinated and preganglionic fibers that leave the spinal nerve to enter the paravertebral ganglion; gray rami represent axons of postganglionic neurons, are unmyelinated, and enter the spinal nerve. (pp. 466, 468)

5. Sweat glands — increase the production of sweat; eye — pupils enlarge; adrenal medulla — releases norepinephrine and epinephrine; heart — increase in rate and force of contraction; lungs — bronchodilation; liver — glycogenolysis and the release of glucose to the blood; blood vessels to the skeletal muscles — dilation; blood vessels to digestive viscera — constriction; salivary glands — constriction of blood vessels supplying the gland causing a decrease in saliva production. (p. 472)

6. All except the effects on the adrenal medulla, liver, and blood vessels. (p. 472)

7. All preganglionic fibers and postganglionic fibers of the parasympathetic division secrete acetylcholine. Some postganglionic sympathetic fibers secrete acetylcholine. Only postganglionic fibers of sympathetic division release norepinephrine. (p. 469)

8. Sympathetic tone means that the vascular system is under a partial state of contraction. Parasympathetic tone maintains the tone of the digestive organs and keeps heart rate at the lowest level that maintains homeostasis. (pp. 471-472)

9. Acetylcholine — nicotinic and muscarinic; norepinephrine — a1, a2, b1, b2. (pp. 469-471)

10. The reticular formation nuclei in the brain stem, particularly in the medulla. (p. 473)

11. The hypothalamus is the main integration center that coordinates heart rate, blood pressure, body temperature, etc. (p. 474)

12. The premise of biofeedback training is that we do not routinely exert voluntary controls over our visceral activities because we have little conscious awareness of our internal environment. The training allows subjects to become aware of the body's signals and subsequently make subtle internal changes to help them control such things as migraine headaches, stress, etc. (p. 478)

13. Elderly people often complain of constipation and dry eyes, and faintness when they change position, e.g., stand up abruptly after sitting. (p. 472)

Critical Thinking and Application Questions

1. Parasympathetic stimulation of the bladder via the release of acetylcholine increases bladder tone and releases the urinary sphincters, a result which will be reproduced by bethanechol. He will probably experience dizziness due to low blood pressure (decreased heart rate), deficient tear formation, wheezing, diarrhea, cramping, and undesirable erection of the penis — all parasympathetic effects. (pp. 469-472)

2. Referred pain is the sensation of pain that appears to originate from a site other than that of the painful stimulus. Damage to the heart gives rise to pain impulses that enter the spinal cord in the thoracic region that also receives impulses from the left chest and arm. (p. 469)

3. Raynaud's disease. Smoking causes vasoconstriction, i.e., the nicotine mimics the affects of acetylcholine on sympathetic nicotinic receptors of the skin blood vessels. (p. 478)

4. Hirschsprung's disease (congenital megacolon); the parasympathetic plexus supplying the distal part of the large intestine fails to develop normally, thus allowing feces to accumulate in the bowel. (p. 478)

Key Figure Questions

Figure 14.2 The axon of the preganglionic neuron spans the distance between the CNS and the autonomic ganglion, whereas the postganglionic neuron's axon lies distal to the ganglion.

SUGGESTED READINGS

---

1. Bower, B. "Monkeying with Stress." *Science News* 125 (Apr. 1984): 234-235.

2. Bower, B. "Shaping Up Your Mind." *Science News* 130 (Aug. 1986).

3. Decara, L.V. "Learning in the Autonomic Nervous System." *Scientific American* (Jan. 1970).

4. Goldberg, L.I., and S.I. Rajfer. "The Role of Adrenergic and Dopamine Receptors." *Hospital Practice* (June 1985).

5. Greenberg, J. "Psyching Out Reaches High-Tech Proportions." *Science News* 127 (May 1985).

6. Hammond, R.E. "Taming the Wild Emotions." *Carolina Tips* 46 (Oct. 1983).

7. Herbert, W. "Sources of Temperament: Bashful at Birth?" *Science News* 121 (Jan. 1982).

8. Herbert, W. "Punching the Biological Timeclock." *Science News* 122 (July 1982).

9. House, M.A. "Cocaine." *American Journal of Nursing* 90 (Apr. 1990): 40-45.

10. Kalin, N.H. "The Neurobiology of Fear." *Scientific American* 268 (May 1993): 94.

11. Long, M.E. "What Is This Thing Called Sleep?" *National Geographic* 172(6) (Dec. 1987): 786-782.

12. Mandell, A.J. "Neurobiological Barriers to Euphoria." *American Scientist* 61 (Oct. 1973).

13. Revkin, A. "Hunting Down Huntington's." *Discover* (Dec. 1993): 100.

14. Silberner, J. "Hypnotism Under the Knife." *Science News* 129 (Mar. 1986).

15. Treichel, J.A. "How Emotions Affect Involuntary Nerves." *Science News* 124 (Sept. 1983).

16. Wacker, R. "The Good Die Younger." *Science 85* (Dec. 1985).

17. Waid, W.M., and M.T. Orne. "The Physiological Detection of Deception." *American Scientist* 70 (1981).

18. Wallace R.K., and H. Benson. "The Physiology of Meditation." *Scientific American* 226 (Feb. 1972): 85-90.

19. Wilbur, R. "A Drug to Fight Cocaine." *Science 86* (Mar. 1986).

15

Neural Integration

Chapter Preview

This chapter explores the sensory and motor pathways, how they integrate with our higher learning centers, and how input is processed. Activities such as walking, running, and posture will be examined in detail to illustrate how they are accomplished. The last portion of this chapter looks at how we think, remember, and relate to others. The concepts of sleep, wakefulness, and memory will be explained.

INTEGRATING THE PACKAGE

Suggested Lecture Outline

I. Sensory Integration: From Reception to Perception (pp. 483-485)
    A. General Organization of the Somatosensory System (pp. 483-485)
        1. General Characteristics
        2. Processing at the Receptor Level
        3. Processing at the Circuit Level
            a. Basic Features
            b. Nonspecific Ascending Pathways
            c. Specific Ascending Pathways
        4. Processing at the Perceptual Level
            a. Basic Features
            b. Perceptual Detection
            c. Magnitude Estimation
            d. Spatial Discrimination
            e. Feature Abstraction
            f. Quality Discrimination
            g. Pattern Recognition
II. Motor Integration: From Intention to Effect (pp. 486-489)
    A. Levels of Motor Control (pp. 486-488)
        1. General Characteristics
        2. The Segmental Level
            a. Segmental Circuits
            b. Central Pattern Generators
        3. The Projection Level
            a. Command Neurons
            b. The Direct (Pyramidal) System
            c. The Indirect (Multineuronal) System

4. The Programs/Instructions Level
   a. Precommand Areas
   b. Role of Cerebellum
   c. Role of Basal Nuclei
B. Homeostatic Imbalances of Motor Integration (pp. 488-489)
   1. Disorders of the Cerebellum
      a. Ataxia
      b. Lack of Check
      c. Scanning Speech
   2. Disorders of Basal Nuclei
      a. Dyskinesia
      b. Parkinson's Disease
      c. Huntington's Disease
III. Higher Mental Functions (pp. 489-498)
   A. Brain Wave Patterns and the EEG (pp. 489-491)
      1. Electroencephalogram
      2. Brain Waves
         a. Alpha Waves
         b. Beta Waves
         c. Theta Waves
         d. Delta Waves
      3. Abnormal Electrical Activity of the Brain: Epilepsy
         a. Epileptic Seizures
         b. Petit Mal
         c. Psychomotor Epilepsy
         d. Grand Mal
   B. Sleep and Sleep-Awake Cycles (pp. 491-493)
      1. General Features
         a. Circadian Rhythms
         b. Reticular Activating System
      2. Types of Sleep
         a. NREM Sleep
         b. REM Sleep
      3. Sleep Patterns
      4. Importance of Sleep
      5. Homeostatic Imbalances of Sleep
         a. Narcolepsy
         b. Insomnia
   C. Consciousness (p. 493)
      1. Levels of Consciousness
      2. Holistic Information Processing
   D. Memory (pp. 494-497)

   1. Stages of Memory
      a. Short-Term Memory
      b. Long-Term Memory
   2. Categories of Memory
      a. Fact Memory
      b. Skill Memory
   3. Brain Structures Involved in Memory
      a. Hippocampus
      b. Amygdala
      c. Diencephalon
      d. Ventromedial Prefrontal Cortex
      e. Basal Forebrain
   4. Mechanisms of Memory
E. Language (pp. 497-498)
   1. Wernicke's Area
   2. Broca's Area

Review Items

1. Neuronal integration (Chapter 11, pp. 371-373)

2. Diencephalon (Chapter 12, pp. 392-395)

3. Limbic system (Chapter 12, p. 401)

4. Thalamus (Chapter 12, p. 393)

5. Hypothalamus (Chapter 12, p. 395)

6. Ascending tracts of the spinal cord (Chapter 12, p. 413)

7. Reticular activating system (Chapter 12, p. 402)

8. Descending tracts of the spinal cord (Chapter 12, p. 416)

9. Spinal roots (Chapter 12, p. 411)

10. Gray and white matter of the spinal cord (Chapter 12, pp. 411-413)

11. Cerebellum (Chapter 12, pp. 399-401)

12. Basal nuclei (Chapter 12 p. 391)

13. Sensory receptor potentials (Chapter 13, p. 429)

14. General reflexes (Chapter 13, pp. 450-456)

15. Sensory receptors (Chapter 13, pp. 427-429)

Laboratory Correlations

1. Marieb, E. N. *Human Anatomy and Physiology Laboratory Manual: Cat and Fetal Pig Versions*. 3rd. ed. Benjamin/Cummings, 1989.

   Exercise 21: General Sensations

2. Marieb, E. N. *Human Anatomy and Physiology Laboratory Manual: Brief Version*. 3rd. ed. Benjamin/Cummings, 1992.

   Exercise 20: General Sensations

Transparencies Index

INSTRUCTIONAL AIDS

Lecture Hints

1. Clearly distinguish between sensation and perception.

2. Since Chapter 15 relates to several previous chapters, refer students to specific areas for review.

3. Students are sometimes confused with the process of transduction and the relationship between receptor potentials, generator potentials, and action potentials.

4. Emphasize the distinction between processing at the circuit level and processing at the perceptual level.

5. This section of the course is an ideal opportunity to integrate the information from Chapters 11-14 so that students can pull together many concepts and get an overall sense of how the nervous system functions.

6. Fixed-action patterns are almost like the subroutines of a computer program; once initiated, they will procede to conclusion. Relate information many students have some foundation in (computers) with new information (processing).

7. Use examples of homeostatis imbalances to reinforce the information about normal function.

8. Thoroughly explain how electrical activity described by brain waves is measured. Students often do not realize that it is the total electrical activity of the brain being measured at the surface of the body and not that of individual neurons, and therefore the activity should not look like an action potential.

9. Memory is fascinating to most students, and it is worthwhile to spend some time on the proposed mechanisms.

Demonstrations/Activities

1. Audio-visual materials of choice.

2. Select a male volunteer to perform a two-point threshold test. Compare his sensitivity with that of a female volunteer.

3. Connect an EEG to a volunteer to exhibit normal brain wave patterns.

4. There are various tests for long- and short-term memory. Present your students with a series of numbers or words and, after a period of time, see how many have recalled the sequence.

5. Conscious thought, memory, and language are inextricably interwoven. Have the students repeat a sequence of words or actions, then have them write down those processes they felt were necessary to listen, interpret, formulate, and repeat the sequence.

Critical Thinking/Discussion Topics

1. Why can we say that pain is "merely a figment of one's imagination"?

2. How can the study of brain waves be used to diagnose disorders of the brain?

3. Sensory deprivation seems to have a great effect on individuals, physiologically and psychologically. What seems to be the basis for this? When we developed in the womb, were we not deprived of sensory input? Or were we?

4. How can some individuals have a higher pain tolerance than others?

5. Lie detector machines have been used for decades. In view of sensory and motor integrative functions, how and why do they work?

6. Thomas Edison never slept for an extended period of time, but always took "catnaps." Is this healthy and more productive? How much sleep do we really need?

Library Research Topics

1. Some individuals that have lost a limb experience a phenomenon called "phantom limb" pain. What methods of treatment are available for these individuals?

2. Bradykinin has been found to be a potent pain inducer. What chemicals have been found to counteract this chemical?

3. What are the current theories for the mechanism of action of acupuncture?

4. Some individuals with Parkinson's disease, when receiving treatment with L-dopa, experience schizophrenia. What seems to be the relationship between Parkinson's disease and behavior changes?

5. Some researchers indicate that we experience "circadian rhythms." How are these rhythms coordinated and do we really have them? Do they occur in other animals?

Audio-Visual Aids/Computer Software

Videotapes

1. Memory: Fabric of the Mind (FHS, QB-1738, 28 min., C, VHS/BETA). Program describes recent research on memory, brain chemistry, etc.

2. Dreams: Theater of the Mind (FHS, QB-1737, 28 min., C, VHS/BETA). Program examines theories about dreams.

3. Dream Voyage (FHS, QB-824, 26 min., C, VHS/BETA). Programs study recent experimentation on sleep.

4. As If by a Stroke of Lightning — A Program About Epilepsy (TF, 36 min., C). The video helps to educate the patient, family, employer, and general public about epilepsy.

See Guide to Audio-Visual Resources on page 339 for key to AV distributors.

LECTURE ENHANCEMENT MATERIAL

Clinical and Related Terms

1. Acataphasia — an inability to express thoughts in a connected manner because of a central lesion.

2. Acrasia — lack of self-control.

3. Amnesia — loss of memory.

4. Causalgia — a burning pain, often associated with skin changes in the hand or foot, caused by peripheral nerve damage.

5. Chorea — St. Vitus's dance.

6. Echolalia — meaningless repetition of words.

7. Hyperalgesia — excessive sensitivity to pain.

8. Megalomania — delusions of grandeur.

9. Misanthropia — hatred of mankind.

10. POSTS — positive occipital sharp transients of sleep.

11. Spindles — a particular wave form in the electroencephalogram during sleep exhibited by short bursts of 12-14 hertz activity.

12. St. Vitus's Dance (Sydenham's Chorea) — a disorder of the CNS that is a manifestation of rheumatic fever, marked by purposeless, irregular movements of voluntary muscles.

13. Stereognosis — the sense by which the form of objects is perceived.

14. Transcutaneous Electrical Nerve Stimulation (TENS) — the placement of electrodes on the body surface directly over nerve fibers that transmit pain. Stimulation of the area appears to relieve pain transmission.

Disorders/Homeostatic Imbalances

1. Korsakoff's Psychosis — a psychosis generally associated with chronic alcoholism and a vitamin-B deficiency. The psychosis exhibits retrograde and anterograde amnesia, disorientation, lack of insight into the memory deficit, and polyneuritis.

ANSWERS TO END-OF-CHAPTER QUESTIONS

Multiple Choice/Matching

1. c

2. d

3. c

4. e

5. b

6. (1)d; (2)e; (3)d; (4)a

Short Answer Essay Questions

7. *Sensation* means awareness of changes in the internal and external environment. Perception means conscious interpretation of those sensations. (p. 483)

8. *Analytic discrimination* means that each quality retains its individual nature. Synthetic discrimination means that the qualities merge together, creating a new sensation with new properties. (p. 485)

9. a. CPGs are local circuits that control locomotion; may involve networks of spinal cord neurons arranged in reverberating circuits.

   b. CPGs are controlled by command neurons, interneurons located at the projection level of the motor hierarchy, expecially the brain stem. (p. 486)

10. See Fig. 15.3, p. 487, for a diagram of the hierarchy of motor control.

11. The direct system involves neurons of the large pyramidal tracts that synapse in the cord and mainly produce fine or skilled voluntary muscle contractions of the distal muscles of the limbs. The indirect system involves neurons of the reticular, vestibular, and red nuclei that are mainly involved in involuntary movements of the proximal muscles of the limbs and trunk muscles that serve, for example, in posture and arm-swinging. (p. 486)

12. They control the outputs of the cortex and brainstem motor centers and stand at the highest level of motor hierarchy. (p. 488)

13. An EEG is a record of the electrical activity of cortical neurons, which is made by using electrodes placed at different locations on the scalp. (p. 490)

14. REM sleep occupies about 50% of the total sleeping time in infants, but then declines with age to stabilize at 25%. Stage 4 sleep declines steadily from birth and disappears completely in those over 60 years. The elderly often remain in a perpetual state of light sleep. (p. 493)

15. a. Narcolepsy means lapsing involuntarily into sleep during waking hours. Insomnia is the inability to obtain the amount or quality of sleep needed.

    b. Narcoleptics seem to have a problem controlling circuits involved with REM sleep. In their evening sleep they spend little time in REM cycle, hence they may not be acquiring enough REM sleep at night. Their narcoleptic episodes indicate REM sleep.

    c. Sleep episodes can occur at any time without warning, which would be dangerous if the person were driving or using a power saw, for example. (p. 493)

16. a. Epilepsy is a disorder characterized by abnormal electrical discharges of groups of brain neurons.

    b. Petit mal, seen in children, is the least severe involving only temporary lapses of attentiveness. Grand mal is most severe, involving loss of consciousness and intense convulsions. (p. 490)

17. Flashes of light, sensory hallucinations of taste or smell that sometimes occur prior to an epileptic seizure. (p. 491)

18. Holistic processing supposes that consciousness involves simultaneous activity of large areas of the cerebral cortex, that consciousness is superimposed on other types of neural activity, and that consciousness is totally interconnected. (p. 493)

19. STM is fleeting memory that serves as a sort of temporary holding bin for data and is limited to seven or eight chunks of data. LTM seems to have unlimited capacity for storage and is very long-lasting unless altered. (pp. 494-495)

20. a. Emotional state, rehearsal, association of new information with information already stored, automatic memory.

    b. Memory consolidation involves fitting new facts into the network of preexisting consolidated knowledge stored in the cerebral cortex. (p. 495)

21. Fact memory is the ability to learn explicit information and is related to our conscious thoughts and our ability to manipulate symbols and language. Skill memory is concerned with motor activities acquired through practice. (p. 495)

22. One function cannot exist without the other. To even speak a word, one must choose words that fit, use grammar to organize the words, and activate the muscles for speech, all in a split second. Therefore, the process cannot be entirely under conscious control. Nor can it be entirely reflexive, because it is often so original. Diseases such as aphasia help to illustrate this interrelationship. (p. 497)

Critical Thinking and Application Questions

1. a. Direct system; inferior precentral gyrus on the left, Broca's area, and the adjacent premotor area, also other frontal lobe language areas. (pp. 486, 497-498)

    b. Abdominal reflexes absent; Babinski's sign present. (p. 487)

    c. Speech apraxia or motor aphasia. (p. 493)

2. In spinal cord trauma leading to paralysis, all bladder and bowel reflexes stop, at least temporarily (see spinal shock, p. 418). Urinary retention encourages renal infection. Lack of movement leads to pressure on soft tissues underlying bony prominences and local deprivation of O_2 and nutrients, which encourages bedsores (p. 152) to form. Skeletal muscle paralysis, accompanied by functional lower motor neurons, results in spastic paralysis evidenced by uncontrollable muscle movements. (p. 416)

3. a. Parkinson's disease.

    b. Basal nuclei, substantia nigra; inadequate dopamine synthesis.

    c. Treat with the drug L-dopa. (p. 489)

Key Figure Questions

Figure 15.2 The smaller the two-point threshold, the more receptors there are in a given area, and the more sensitive the area. **Figure 15.3** The middle level.

SUGGESTED READINGS

1. Altman, J. "The Intricate Wiring That Lets Us Move." *New Scientist* 10 (Mar. 1990): 60-63.

2. Alkon, D.L. "Memory Storage and Neural Systems." *Scientific American* 261 (July 1989): 42-50.

3. Binkley, S.A., et al. "The Pineal Gland: A Biological Clock in Vitro." *Science* 202 (1978): 1198.

4. Bliss, T.V.P. , and G.L. Collingridge. "A Synaptic Model of Memory: Long-Term Potentiation in the Hippocampus." *Nature* 361 (Jan. 1993): 31.

5. Bower, B. "Neuroleptic Backlash." *Science News* 128 (July 1985): 45-46.

6. Bower, B. "Million-Cell Memories." *Science News* 130 (Nov. 1986): 313-315.

7. Bredt, D.S., and S.H. Snyder. "Biological Roles of Nitric Oxide." *Scientific American* (1993): 22.

8. Carney, R.M. "Clinical Applications of Relaxation Training." *Hospital Practice* (July 1983).

9. Damasio, A.R., and H. Damasio. "Brain and Language." *Scientific American* (Sept. 1992): 89.

10. Erickson, R.P. "On the Neural Bases of Behavior." *American Scientist* 72 (May-June 1984).

11. Ericsson, K.A., and W.G. Chase. "Exceptional Memory." *American Scientist* 70 (Nov.-Dec. 1982).

12. Evarts, E.V. "Brain Mechanisms of Movement." *Scientific American* 241 (Sept. 1979): 164-179.

13. Ezzell, C. "Memories Might Be Made of This." *Science News* 139 (May 1991): 328.

14. Geschwind, N. "Language and the Brain." *Scientific American* (Apr. 1972): 76-83.

15. Goldman-Rakic, P.S. "Working Memory and the Mind." *Scientific American* (Sept. 1992): 111.

16. Gould, J.L., and P. Marler. "Learning by Instinct." *Scientific American* 256 (Jan. 1987): 74-85.

17. Hawkins, R.D., and E.R. Kandel. "The Biological Basis of Learning and Individuality." *Scientific American* (Sept. 1992): 79.

18. Herbert, W. "Forgotten Dreams." *Science News* 124 (1983): 188.

19. Hinton, G.E. "How Neural Networks Learn From Experience." *Scientific American* (Sept. 1992): 144.

20. Hinton, G.E., D.C. Plant, and T. Shallice. "Simulating Brain Damage." *Scientific American* 269 (Oct. 1993): 76.

21. Kiester, W. "Spare Parts for Damaged Brains." *Science* 86, 7 (Mar. 1986): 33-38.

22. Kindel, S. "Give Me a Bowl of Texas Chili." *Forbes* (Sept. 1985): 184-186.

23. Lancaster, J.R. "Nitric Oxide in Cells." *American Scientist* 80 (May-June 1992): 248.

24. Loftus, E.F. "The Malleability of Human Memory." *American Scientist* 67 (May-June 1979): 312-320.

25. Long, M.E. "What Is This Thing Called Sleep?" *National Geographic* 172 (Dec. 1987): 787-821.

26. Loscalzo, J., D.J. Singel, and J.S. Stamler. "Biochemistry of Nitric Oxide and Its Redox-Activated Forms." *Science* (Dec. 1992): 189.

27. McKean, K. "New Parts for Damaged Brains." *Discover* 5 (Feb. 1984): 68-72.

28. Miller, G.A., and P.M. Gildea. "How Children Learn Words." *Scientific American* 257 (Sept. 1987): 94-99.

29. Mischkin, M., and T. Appenzeller. "The Anatomy of Memory." *Scientific American* 256 (June 1987): 80-89.

30. Mischkin, M., and H.L. Petri. "Behaviorism, Cognitivism, and the Neuropsychology of Memory." *American Scientist* 82 (Jan.-Feb. 1994): 30.

31. Montgomery, G. "Molecules of Memory." *Discover* 10 (Dec. 1989): 46-55.

32. Montgomery, G. "The Mind in Motion." *Discover* (Mar 1989): 58-68.

33. Morrison, A.R. "A Window in the Sleeping Brain." *Scientific American* 248 (Apr. 1983): 94-102.

34. Motley, M.T. "Slips of the Tongue." *Scientific American* 253 (Sept. 1985): 116-127.

35. Pearson, K. "The Control of Walking." *Scientific American* (Dec. 1976): 72-86.

36. Regan, D. "Electrical Responses Evoked From the Human Brain." *Scientific American* 241 (Dec. 1979): 134-145.

37. Revkin, A. "Hunting Down Huntington's." *Discover* (Dec. 1993): 100.

38. Seyle, H. *Stress in Health and Disease.* London: Butterworth, 1976.

39. Snyder, S. "Nitric Oxide: First in a New Class of Neurotransmitters." *Science* 257 (July 1992): 494.

40. Sperry, R. "Some Effects of Disconnecting the Cerebral Hemispheres." *Science* 217 (1982): 1223.

41. Tulving, E. "Remembering and Knowing the Past." *American Scientist* 77 (July-Aug. 1989): 361-367.

42. Weiss, R. "Bypassing the Brain." *Science News* 136 (Dec. 1989): 136, 379.

43. West, B.A. "Understanding Endorphins, Our Natural Pain Relief System." *Nursing* 81, 11 (Feb. 1981): 50-53.

44. Winter, P.M., and J.N. Miller. "Anesthesiology." *Scientific American* 252 (Apr. 1985): 124-131.

45. Wolff, B.B. "Perceptions of Pain." *The Sciences* (July-Aug. 1980): 10-13, 28.

16

The Special Senses

Chapter Preview

In this chapter, the functional anatomy of each of the special sense organs—smell, taste, sight, hearing, and equilibrium—will be considered. How they communicate and integrate with the central nervous system will be explored along with the homeostatic imbalances.

INTEGRATING THE PACKAGE

Suggested Lecture Outline

I. The Chemical Senses: Taste and Smell (pp. 502-506)

    A. Taste Buds and the Sense of Taste (pp. 502-504)

        1. Localization and Structure of Taste Buds (Fig. 16.1, p. 502)

           a. Papillae

           b. Taste Buds

        2. Basic Taste Sensations (Fig. 16.2, p. 503)

           a. Basic Qualities

           b. Sensitivity

        3. Physiology of Taste

           a. Activation of Taste Receptors

           b. Mechanism of Taste Transduction

        4. The Gustatory Pathway

           a. Facial Nerve

           b. Glossopharyngeal Nerve

           c. Vagus Nerve

        5. Influence of Other Sensations on Taste

    B. The Olfactory Epithelium and the Sense of Smell (pp. 504-506)

        1. Localization and Structure of Olfactory Receptors (Fig. 16.3, p. 505)

           a. Olfactory Epithelium

           b. Olfactory Nerve

        2. Physiology of Smell

           a. Activation of Olfactory Receptors

           b. Mechanism of Smell Transduction

        3. The Olfactory Pathway

           a. Olfactory Bulbs

           b. Olfactory Tracts

C. Homeostatic Imbalances of the Chemical Senses (p. 506)
   1. Anosmias
   2. Uncinate Fits

II. The Eye and Vision (pp. 506-526)
   A. Basic Characteristics (p. 506)
   B. Accessory Structures of the Eye (pp. 507-510; Fig. 16.5, p. 507)
      1. Eyebrows
      2. Eyelids
         a. Canthi
         b. Caruncle
         c. Tarsal Plates
         d. Eyelashes
         e. Glands
      3. Conjunctiva
         a. Basic Features
         b. Conjunctival Sacs
      4. Lacrimal Apparatus
         a. Lacrimal Glands
         b. Lacrimal Canals (Ducts)
         c. Lacrimal Sacs
         d. Nasolacrimal Ducts
         e. Lacrimal Secretions
      5. Extrinsic Eye Muscles (Fig. 16.6, p. 509)
         a. Annular Ring
         b. Rectus Muscles
         c. Oblique Muscles
         d. Neural Control
   C. Structure of the Eyeball (pp. 505-509; Fig. 16.7, p. 505)
      1. Tunics Forming the Wall of the Eyeball
         a. The Fibrous Tunic
            1. Sclera
            2. Cornea
         b. The Vascular Tunic (Uvea)
            1. Choroid
            2. Ciliary body
            3. Iris
            4. Pupil
         c. The Sensory Tunic (Retina)
            1. Pigmented Layer
            2. Neural Layer (Fig. 16.9, p. 512)
            3. Optic Disc
            4. Photoreceptors
            5. Macula Lutea
            6. Fovea Centralis

2. Internal Chambers and Fluids (Fig. 16.12, p. 514)
    a. Posterior Cavity
        1. Vitreous Humor
        2. Functions of Humor
    b. Anterior Cavity
        1. Anterior Chamber
        2. Posterior Chamber
        3. Aqueous Humor
        4. Functions of Humor
        5. Canal of Schlemm
        6. Glaucoma
3. Lens
    a. Basic Features
    b. Lens Epithelium
    c. Lens Fibers
    d. Cataracts

D. Physiology of Vision (pp. 515-526)
1. Overview: Light and Optics
    a. Wavelength and Color (Fig. 16.13, p. 515)
    b. Refraction and Lenses (Figs. 16.14 and 16.15, p. 516)
2. Focusing of Light on the Retina
    a. Basic Features
    b. Focusing for Distance Vision (Fig. 16.16, p. 517)
    c. Focusing for Close Vision
        1. Accommodation of the Lenses
        2. Constriction of the Pupils
        3. Convergence of the Eyeballs
    d. Homeostatic Imbalances of Refraction (Fig. 16.17, p. 518)
3. Photoreception
    a. Functional Anatomy of the Photoreceptors (Fig. 16.18, p. 519)
        1. Outer Segment
        2. Inner Segment
    b. The Chemistry of Visual Pigments
        1. Retinal
        2. Isomers
4. Stimulation of the Photoreceptors
    a. Excitation of Rods
        1. Rhodopsin
        2. Scotopsin
        3. Bleaching
    b. Excitation of Cones
        1. Photopsins
    c. Color Blindness

5. Light Transduction in Photoreceptors
   a. Cyclic GMP
   b. Transducin
6. Light and Dark Adaptation
7. Night Blindness
8. The Visual Pathway to the Brain (Fig. 16.21, p. 519)
   a. Neural Pathways
   b. Optic Chiasm
   c. Optic Nerves
   d. Neural Centers
9. Stereoscopic Vision and Depth Perception
   a. Panoramic Vision
   b. Depth Perception
   c. Three-Dimensional Vision
10. Visual Processing
    a. Retinal Processing
       1. On-Line Pathway
       2. Off-Line Pathway
       3. Center-Surround Antagonism
    b. Thalamic Processing
    c. Cortical Processing
       1. Simple Cortical Neurons
       2. Complex Cortical Neurons

III. The Ear: Hearing and Balance (pp. 526-539)
   A. Structure of the Ear (pp. 527-530; Fig. 16.25, p. 527)
      1. The Outer (External) Ear
         a. Auricle (Pinna)
         b. External Auditory Canal
            1. Ceruminous Glands
            2. Tympanic Membrane
      2. The Middle Ear
         a. Oval Window
         b. Round Window
         c. Epitympanic Recess
         d. Mastoid Antrum
         e. Auditory (Eustachian) Tube
         f. Ear Ossicles (Fig. 16.26, p. 528)
            1. Malleus (Hammer)
            2. Incus (Anvil)
            3. Stapes (Stirrup)
         g. Tensor Tympani
         h. Stapedius Muscle
         i. Tympanic Reflex

   3. The Inner (Internal) Ear
     a. Labyrinth (Fig. 16.27, p. 529)
       1. Bony Labyrinth
       2. Membranous Labyrinth
       3. Perilymph
       4. Endolymph
     b. The Vestibule
       1. Saccule
       2. Utricle
     c. The Semicircular Canals
       1. Semicircular Ducts
       2. Ampulla
       3. Crista Ampullaris
     d. The Cochlea (Fig. 16.28, p. 529)
       1. Modiolus
       2. Cochlear Duct
       3. Organ of Corti and Spiral Lamina
       4. Scala Vestibuli
       5. Scala Media
       6. Scala Tympani
       7. Helicotrema
       8. Vestibular Membrane
       9. Basilar Membrane
B. Sound and Mechanisms of Hearing (pp. 530-535)
   1. Properties of Sound
     a. Basic Features
     b. Frequency (Fig. 16.30, p. 532)
     c. Amplitude
   2. Transmission of Sound to the Inner Ear
   3. Resonance of the Basilar Membrane (Fig. 16.31, p. 533)
   4. Excitation of Hair Cells in the Organ of Corti
     a. Organ of Corti
     b. Cochlear Hair Cells
     c. Tectorial Membrane
   5. The Auditory Pathway to the Brain (Fig. 16.32, p. 534)
   6. Auditory Processing
     a. Perception of Pitch
     b. Detection of Loudness
     c. Localization of Sound
C. Homeostatic Imbalances of Hearing (p. 535)
   1. Deafness
     a. Conduction Deafness
     b. Sensorineural Deafness

    2. Tinnitus

    3. Meniere's Syndrome

  D. Mechanisms of Equilibrium and Orientation (pp. 535-539)

    1. Basic Features

    2. The Maculae and Static Equilibrium (Fig. 16.33, p. 536)

      a. Anatomy of the Maculae

        1. Supporting Cells

        2. Hair Cells

        3. Otolithic Membrane

        4. Vestibular Nerve

      b. Transducing Gravity and Linear Acceleration Stimuli

    3. The Crista Ampullaris and Dynamic Equilibrium

      a. Anatomy of the Crista Ampullaris

        1. Cupula

      b. Transducing Rotational Stimuli

    4. The Equilibrium Pathway to the Brain (Fig. 16.36, p. 539)

      a. Vestibular Nuclear Complex

      b. Cerebellum

      c. Motion Sickness

IV. Developmental Aspects of the Special Senses (pp. 539-542)

  A. Embryonic and Fetal Development of the Senses

  B. Effect of Aging on the Senses (p. 534)

Review Items

1. Epithelia (Chapter 4, pp. 103-112)

2. Exocrine glands (Chapter 4, pp. 110-112)

3. Connective tissues (Chapter 4, pp. 112-114)

4. Sebaceous glands (Chapter 5, p. 145)

5. Sudoriferous glands (Chapter 5, p. 144)

6. Synovial joints (Chapter 8, pp. 226-240)

7. Skeletel muscle naming (Chapter 10, pp. 291-292)

8. Chemoreceptors (Chapter 13, p. 426)

9. Synapses (Chapter 11, pp. 360-362)

10. Neurotransmitters (Chapter 11, pp. 366-371)

11. Cerebral cortex (Chapter 12, pp. 384-392)

12. Thalamus (Chapter 12, p. 393)

13. Receptor and generator potentials (Chapter 13, p. 429)

14. Cranial nerves (Chapter 13, pp. 434-436)

15. Reflex activity (Chapter 13, pp. 450-456)

Cross References

1. Formation of the aqueous humor is similar to CSF formation described in Chapter 12, p. 404.

2. Inflammation is described in detail in Chapter 22, pp. 710-712.

3. Secretion of saliva (p. 798) and gastric juice (p. 808) is described in detail in Chapter 24.

4. The salivary reflex is presented in Chapter 24, pp. 798-799.

5. Papillae and taste buds are further explained in Chapter 24, pp. 796-797.

6. The relationship between the auditory tube and the respiratory system is described in Chapter 23, p. 747.

Laboratory Correlations

1. Marieb, E. N. *Human Anatomy and Physiology Laboratory Manual: Cat and Fetal Pig Versions.* 3rd. ed. Benjamin/Cummings, 1989.

   Exercise 24: Special Senses: Vision

   Exercise 25: Special Senses: Hearing and Equilibrium

   Exercise 26: Special Senses: Taste and Olfaction

2. Marieb, E. N. *Human Anatomy and Physiology Laboratory Manual: Brief Version.* 3rd. ed. Benjamin/Cummings, 1992.

   Exercise 21: Special Senses: Vision

   Exercise 22: Special Senses: Hearing and Equilibrium

   Exercise 23: Special Senses: Taste and Olfaction

Transparencies Index

Bassett Atlas Figures Index

INSTRUCTIONAL AIDS

Lecture Hints

1. Emphasize that each taste sensation is not localized to a specific area, but that there is significant overlap of the different sensation areas. Students often assume that a particular point on the tongue responds to a single type of substance.

2. Point out the importance of other sensations (especially smell) on the perception of taste.

3. During the lecture on olfactory anatomy, ask the class what would happen to olfaction if mucus glands below the olfactory epithelium were absent.

4. Emphasize that olfactory receptors are the only renewable neurons in the body and are therefore the one exception to the rule that neurons do not replicate.

5. Point out that special sense receptor cells develop potential just like neurons (sodium and potassium gradients) do, and that even with all the apparent complexity of the body, often slight modifications to a single basic blueprint is all that is necessary to achieve a totally new function.

6. There is often confusion in the terminology of the chambers of the eye. Point out that the anterior segment is divided into anterior and posterior chambers by the iris.

7. Initially, it is difficult for even the best students to grasp the concept of ciliary muscle contraction leading to lens thickening (for close focus). Intuitively, most think of the process of stretching the lens as a consequence of muscle contraction, not relaxation. Spend some time reinforcing this concept.

8. Have students try out focusing on objects at night. Explain that they should not look directly at the object, but slightly to one side, and the object should appear brighter. Relate this exercise to the distribution of rods and cones in the eye.

9. As a point of interest, mention that the ossicles are joined by the smallest synovial joints in the body.

10. Emphasize the difference between static and dynamic equilibrium by comparing and contrasting anatomy of each type of equilibrium.

Demonstrations/Activities

1. Audio-visual materials of choice.

2. Obtain a 3-D model of an eye and ear to illustrate the various anatomical parts of each.

3. Use an onion, orange, and apple to test the interaction of taste, smell, and sight. Have two volunteers first taste and smell each item normally. Then cover their eyes and have them taste each. Then pinch shut their noses and taste each.

4. Select four volunteers and spray a different strong cologne on their wrists. Then determine how long it takes for each to "adapt" to the cologne.

5. Dissect a fresh (or preserved if fresh is not available) beef eye to illustrate the anatomical structure and nature of the tissues and fluids.

6. Obtain a skull to illustrate the locations of any bony structures associated with the senses.

7. Obtain a set of ear ossicles to illustrate how tiny they are.

8. Bring a convex lens to class and have students hold the glass up and focus on a distant object. They will notice that it is upside down and reversed. Then explain that the human eye is also a single-lens system. The question should arise: "Why don't we see things upside down?"

Critical Thinking/Discussion Topics

1. Most people with sinus infections can't smell or taste. Why?

2. Wine tasting can be a real art. Why are some people more adept at tasting than others? What effect does smoking, alcohol, and/or sweets have on wine tasting? Why is it useful to swirl a glass of wine and then sniff it?

3. Amplified rock music has long been implicated in auditory deafness. What suggestions could be made to alleviate this problem in teenagers?

4. Fish and other animals have eyes on the sides of their heads. Why is this useful? Why do birds of prey, cats, and other animals have both eyes focused on the same object (stereoscopic vision)? Why is this useful? In a few cases, genetically deformed children are born with eyes on the lateral sides of their heads. How could this happen and what could this suggest embryologically?

5. Certain types of sunglasses can cause more harm than good. What could be wrong with inexpensive sunglasses?

6. What would happen to gustatory and olfactory sensations if the receptors for taste and smell were specific to a single substance?

7. Since the sclera is avascular, why do we see blood vessels in the white of the eye?

8. How is it possible that the cornea is transparent and the sclera is opaque when they are both constructed of the same material and continuous with each other?

9. Examine the consequences to the anatomy of the eye and vision if aqueous humor drainage exceeded production.

10. Explain why depth perception is lost if one eye is not functioning.

11. If the number of cones feeding into a single ganglion cell was increased ten-fold, what would be the consequence to color visual acuity?

12. Examine the consequences to sound perception if the tympanic membrane increased two-fold in surface area. What would happen if the oval window had increased surface area? Would sounds be perceived if the round window became rigid?

Library Research Topics

1. How successful are cochlear implants? What surgical techniques are employed?

2. Some permanently deaf individuals have been helped by means of computers and electrical probes connected to certain areas of the brain. How is this possible and what is the current research in this area?

3. Contact lenses have long been used to correct vision problems. What is the status of contact lens implants and why is there a hesitance by ophthalmologists to do them?

4. What substances are found in wines such as cabernet, chardonnay, chenin blanc, and others that provide the tremendous variety of tastes and smells?

5. If hearts, lungs, and livers can be transplanted, why not eyes? What would be some of the technical difficulties?

Audio-Visual Aids/Computer Software

Slides

1. Histology of the Sensory System (EI, 614, Slides). Illustrates structural and functional correlations of a variety of sensory tissue, including the eye and ear.

2. Eyes and Their Function (EI, SS-0870F, Slides). Functional description of the eyeball and accessory structures.

3. Ears and Their Function (EI, SS-0330F, Slides). Studies middle and inner ear.

Videotapes

1. Eyes and Ears (FHS, QB-823, 26 min., C, VHS/BETA). Fascinating camera action used to show functions of the eye and ear. Also available in 16mm.

2. Anatomy of the Human Eye Series (TF, C, 1987). A series of seven videotapes, ranging from 13 to 19 minutes, explaining the gross anatomy of the human eye.

3. Optics of the Human Eye Series (TF, C, 1987). A series of four videotapes, each ten minutes long, which illustrate basic optics relating to the eye.

4. Dissection and Anatomy of the Beef Eye (CBS, 49-2300, VHS)

Computer Software

1. Dynamics of the Human Eye (EI, C-3060, Apple or IBM)

2. Dynamics of the Human Ear (EI, C-3061, Apple or IBM)

3. Dynamics of the Human Senses of Touch, Taste, and Smell (EI, C-3063, Apple or IBM)

4. Senses: Physiology of Human Perception (PLP, CH184006, Apple II, CH184007 IBM, CH184008 MAC)

5. The Eye (QUE, COM4210A, Apple)

   See Guide to Audio-Visual Resources on page 339 for key to AV distributors.

LECTURE ENHANCEMENT MATERIAL

Clinical and Related Terms

1. Achromatopsia — total color blindness.

2. Ageusia — lack or impairment of the sense of taste.

3. Amblyopia — dimness of vision.

4. Ametropia — a defect in the refractive powers of the eye where images cannot be brought into proper focus on the retina.

5. Anisopia — an inequality of vision in the two eyes.

6. Astereognosis — an inability to recognize familiar objects by feeling their shape.

7. Audiometer — a device used to determine the degree of hearing impairment.

8. Blepharectomy — the surgical excision of a lesion of the eyelids.

9. Cochlear implant — a surgical technique involving the implantation of an artificial device into the cochlea that receives impulses from an external receiver that in turn transforms the signals to the vestibulocochlear nerve, resulting in rudimentary sound perception.

10. Esotropia — strabismus in which there is a deviation of the visual axis of one eye toward the other, resulting in diplopia.

11. Eustachitis — infection or inflammation of the eustachian tube.

12. Iridectomy — excision of the iris.

13. Iritis — infection or inflammation of the iris of the eye.

14. Keratitis — infection or inflammation of the cornea of the eye.

15. Kinesthesis — the sense by which weight, position, and movement are perceived.

16. Labyrinthectomy — the excision of the labyrinth of the ear.

17. Myringitis — inflammation or infection of the eardrum.

18. Ophthalmoscope — an instrument used for the examination of the interior of the eye.

19. Osmesis — the act of smelling.

20. Otoscope — an instrument used for the examination of the external auditory meatus and middle ear.

21. Ptosis — the paralytic drooping of the upper eyelid.

22. Radial keratotomy — a surgical technique designed to improve myopia by making small incisions in the cornea, allowing it to stretch and become flatter.

23. Rhinne tests — the use of a vibrating tuning fork held against the bone behind the ear to diagnose conductive deafness.

24. Sclerostomy — the creation of a fistula through the sclera for the relief of glaucoma.

25. Snellen chart — a chart printed with varying size letters used in testing visual acuity.

26. Tonometry — the measurement of tension or pressure, especially intraocular pressure of the eye.

27. Uveitis — inflammation or infection of the uvea that includes the iris, ciliary body, and choroid.

28. Vertigo — dizziness, especially due to heights.

29. Weber test — the use of a vibrating tuning fork pressed against the forehead to diagnose conductive deafness.

Disorders/Homeostatic Imbalances

Infectious Disorders

1. Congenital Rubella Syndrome — a complication occurring in the unborn fetus due to a rubella infection in the mother, usually during the first two to three months of gestation. Common fetal anomalies include cataracts, heart defects, and others.

2. Mastoiditis — an inflammation of the air cells of the mastoid process. Condition is characterized by fever, chills, tenderness, and leukocytosis. May cause perisinus abscesses, periphlebitis, and lateral sinus thrombosis.

3. Ophthalmia Neonatorum — a severe purulent conjunctivitis of the eyes of newborns, usually caused by infection from *Neisseria gonorrhoea* as the infant passes through an infected birth canal. May be prevented by applying a 1% silver nitrate solution or antibiotic ointment to the eyes shortly after birth.

4. Trachoma — a chronic, highly contagious form of conjunctivitis caused by Chlamydia trachomatis. The disease is characterized by hypertrophy of the conjunctiva, formation of granulation tissue, and subsequent scar formation. If untreated, could cause blindness.

Disorders of the Eye

1. Chorioretinitis — an inflammation of the choroid layer and the retina.

2. Cyclopia — a developmental anomaly characterized by a single eye.

3. Nyctalopia — night blindness, usually caused by a deficiency of vitamin A.

4. Retinitis Pigmentosa — a chronic progressive disease that begins in early childhood, characterized by degeneration of the retinal epithelium, especially the rods; atrophy of the optic nerve; and extensive pigment changes in the retina.

5. Retinoblastoma — a tumor that arises from retinal blast cells. Also known as a retinal glioma. Usually occurs in young children and tends to show a hereditary pattern.

6. Sty — an inflammation of the sebaceous glands of the eyelids. Also called hordeolum.

7. Xerophthalmus — abnormally dry eyes, usually caused by a deficiency of vitamin A.

Disorders of the Ear

1. Eustachitis — infection or inflammation of the eustachian tube.

2. Macrotia — abnormally large ears.

3. Otopyorrhea — discharge of pus from the ear.

4. Tympanosclerosis — hardening of the tympanic membrane.

ANSWERS TO END-OF-CHAPTER QUESTIONS

Multiple Choice/Matching

1.	d	10.	b	19.	b
2.	a	11.	c	20.	c
3.	d	12.	d	21.	d
4.	c	13.	b	22.	b
5.	d	14.	a	23.	b
6.	b	15.	b	24.	e
7.	c	16.	b	25.	b
8.	d	17.	b	26.	c
9.	b	18.	d		

Short Answer Essay Questions

27. Sweet — anterior portion of the tongue; salty — anterior portion and lateral edges of the tongue; sour — lateral edges of the tongue; bitter — posterior-most portion of the tongue. The sensation for sweet substances overlaps sour and salty areas. (p. 503)

28. The receptors are located in the roof of each nasal cavity. The site is poorly suited because air entering the nasal cavities must make a hairpin turn to stimulate the receptors. (p. 504)

29. The nasolacrimal duct empties into the nasal cavity. (p. 508)

30. Rods are dim-light visual receptors, while cones are for bright-light and high-acuity color vision. (pp. 518-519)

31. The fovea contains only cones and provides detailed color vision for critical vision. (p. 520)

32. Retinal changes to the all-trans form; the retinal-scotopsin combination breaks down, separating retinal and scotopsin (bleaching) in the sequence: prelumirhodopsin $\rightarrow$ lumirhodopsin $\rightarrow$ metarhodopsin I $\rightarrow$ metarhodopsin II $\rightarrow$ pararhodopsin. The net effect is to "turn off" sodium entry into the cell, effectively hyperpolarizing the rod. (p. 521)

33. Each cone responds maximally to one of these colors of light, but there is overlap in their absorption spectra that accounts for the other hues. (p. 522)

34. With age, the lens loses its crystal clarity and becomes discolored, and the dilator muscles become less efficient. Atrophy of the organ of Corti reduces hearing acuity, especially for high-pitch sounds. The sense of smell and taste diminish due to a gradual loss of receptors, thus appetite is diminished. (pp. 539-542)

Critical Thinking and Application Questions

1. Papilledema — a nipplelike protrusion of the optic disc into the eyeball, which is caused by conditions that increase intracranial pressure. A rise in cerebrospinal fluid pressure caused by an intracranial tumor will compress the walls of the central vein resulting in its congestion and bulging of the optic disc. (p. 542)

2. Pathogenic microogranisms spread from the nasopharynx through the auditory tube into the tympanic cavity. They may then spread posteriorly into the mastoid air cells via the mastoid antrum resulting in mastoiditis. They may then extend medially to the inner ear causing secondary labyrinthitis. If unchecked, the infection may spread to the meninges causing meningitis and possibly an abscess in the temporal lobe of the brain or in the cerebellum. They may also enter into the blood causing septicemia. The cause of her dizziness and loss of balance is a disruption of the equilibrium apparatus due to the infectious process. (pp. 527-528)

3. Conjunctivitis. The foreign object probably would be found in the conjunctival sac near the orifice of the lacrimal canals. (p. 507)

4. This is known as a detached retina. The condition is serious, but the retina can be reattached surgically using lasers before permanent damage occurs. (p. 512)

5. The inability to hear high-pitched sounds is called presbycusis, a type of sensineural deafness. It is caused by the gradual loss of hearing receptors throughout life, but is accelerated if one is exposed to loud rock music for extended periods. (p. 541)

Key Figure Questions

Figure 16.2 Bitter. **Figure 16.3** It serves as a dissolving medium for odor molecules. **Figure 16.9** Ganglion. **Figure 16.18** By having it in the discs, surface area is increased. **Figure 16.21** The whole process will come to a halt because excitation depends upon having the proper ionic concentrations on the two sides of the membrane. **Figure 16.25** The tympanic membrane. The round membrane.

SUGGESTED READINGS

---

1. Abu-Mostafa, Y.S., and D. Psaltis. "Optical Neural Computers." *Scientific American* 256 (Mar. 1987): 88-96.

2. Barlow, Jr., R.B. "What the Brain Tells the Eye." *Scientific American* 262 (Apr. 1990): 90-95.

3. Brou, P., et al. "The Color of Things." *Scientific American* 255 (Sept. 1986).

4. Cain, W.S. "To Know with the Nose: Keys to Odor Identification." *Science* 203 (1979): 467.

5. Daw, N.W. "Neurophysiology of Color Vision." *Physiological Reviews* 53 (1973): 571.

6. Durrant, J.D., and J.H. Lovrinic. *Basics of Hearing Science.* (Baltimore: Williams & Wilkins, 1988).

7. Engen, T. "Remembering Odors and Their Names." *American Scientist* 75 (Sept.-Oct. 1987).

8. Fackelmann, K.A. "Smokers Suffer from Impaired Smell." *Science News* 137 (Mar. 1990).

9. Franklin D. "Crafting Sound from Silence." *Science News* 126 (Oct. 1984).

10. Freeman, W.J. "The Physiology of Perception." *Scientific American* 264 (Feb. 1991): 78.

11. Gibbons, B. "The Intimate Sense." *National Geographic* 170 (Sept. 1986): 324-360.

12. Gilman, A.G., and M.E. Linder. "G Proteins." *Scientific American* (Sept. 1992): 70.

13. Glickstein, M. "The Discovery of the Visual Cortex." *Scientific American* 295 (Sept. 1988): 118-127.

14. Gourse, L. "Patchwork Medicine." *Science* 85 (Oct. 1985): 79, 81.

15. Grady, D. "Sounds Instead of Silence." *Discover* 4 (Oct. 1983).

16. Greenberg, J. "Early Hearing Loss and Brain Development." *Science News* 131 (Mar. 1987): 149.

17. Hubbard, A. "A Traveling-Wave Amplifier Model of the Cochlea." *Science* 259 (Jan. 1993): 68.

18. Hubel, D.H. and T.N. Wiesel. "Brain Mechanisms of Vision." *Scientific American* 241 (Sept. 1979): 150-162.

19. Hudspeth, A.J. "The Hair Cells of the Inner Ear." *Scientific American* (Jan. 1983).

20. Hurley, J.B. "Sensing Calcium in Rod Cells." *Nature* 361 (Jan. 1993): 20.

21. Koretz, J.F., and G.H. Handelman. "How the Human Eye Focuses." *Scientific American* 259 (July 1988): 92-99.

22. Loeb, G.E. "The Functional Replacement of the Ear." *Scientific American* (Feb. 1985).

23. Largolskee, R.F., P.J. McKinnon, and S.K. McLaughlin. "Gustducin is a Taste-Cell-Specific G Protein Closely Related to the Transducins." *Nature* 357 (June 1992): 563.

24. Melamed, M.A. "Cataracts: Recognition and Assessment." *Hospital Medicine* (July 1982).

25. O'Brien, D.F. "The Chemistry of Vision." *Science* 218 (1982).

26. Parker, D.E. "The Vestibular Apparatus." *Scientific American* 243 (Nov. 1980).

27. Ramachandran, V.S. "Blind Spots." *Scientific American* 266 (May 1992).

28. Ramachandran, V.S., and S.M. Anstis. "The Perception of Apparent Motion." *Scientific American* 254 (June 1986).

29. Ross, P.E. "Smelling Better." *Scientific American* 262 (Mar. 1990): 32.

30. Rushton, W.A. "Visual Pigments and Color Blindness." *Scientific American* (Mar. 1975).

31. Schnapf, J.L., and D.A. Baylor. "How Photoreceptor Cells Respond to Light." *Scientific American* 256 (Apr. 1987): 40-47.

32. Shreeve, J. "Touching the Phantom." *Discover* 14 (June 1993): 34.

33. Stone, J. "Scents and Sensibility." *Discover* 10 (Dec. 1989): 26-31.

34. Weiss, R. "Transplanting the Light Fantastic." *Science News* 136 (Nov. 1989): 297, 300.

35. Weiss, R.L. "Eye Diving." *Science News* 138 (Sept. 1990): 170-172.

36. Wickelgren, I. "Vitamins C and E May Prevent Cataracts." *Science News* 135 (May 1989): 308.

37. Wolfe, J.M. "Hidden Visual Processes." *Scientific American* (Feb. 1983).

38. Wurtz, R.H., et al. "Brain Mechanisms of Visual Attention." *Scientific American* (June 1982).

39. Zeki, S. "The Visual Image in Mind and Brain." *Scientific American* (Sept. 1992): 68.

17

The Endocrine System

Chapter Preview

This chapter deals with one of the two controlling systems of the body, the endocrine system. This system acts with the nervous system to coordinate and integrate the activity of body cells. The endocrine system uses hormones carried in the blood to achieve its function; therefore, endocrine response is slower than that of the nervous system but does take place over a longer period of time. Hormonal targets include most cells of the body. This chapter describes the various endocrine glands, the hormones they produce, the functions of those hormones, and hypo- and hypersecretion disorders.

INTEGRATING THE PACKAGE

Suggested Lecture Outline

I. The Endocrine System and Hormone Function: An Overview (pp. 540-541)
    A. Endocrine Glands (Fig. 17.1, p. 541)
        1. Functions
        2. Locations
II. Hormones (pp. 548-553)
    A. The Chemistry of Hormones (pp. 549-551)
        1. Definition
        2. Groups
            a. Amino Acid–Based
            b. Steroids
            c. Prostaglandins
    B. Hormone-Target Cell Specificity (p. 548)
    C. Mechanism of Hormone Action (pp. 548-551)
        1. Second-Messenger System (Fig. 17.2, p. 549)
        2. Direct Gene Activation (Fig. 17.3, p. 551)
    D. Half-life, Onset and Duration of Hormone Activity (p. 551)
    E. Control of Hormone Release (pp. 551-553)
        1. Endocrine Gland Stimuli
            a. Humoral Stimuli
            b. Neural Stimuli
            c. Hormonal Stimuli
        2. Nervous System Modulation
III. Major Endocrine Organs
    A. The Pituitary Gland Hypophysis (pp. 553-576)

1. Introduction
    a. Posterior Lobe (Neurohypophysis)
    b. Anterior Lobe (Adenohypophysis)
2. Pituitary-Hypothalamic Relationships
    a. Hypophyseal Hypothalamic Tract
    b. Vascular Connection
3. Adenohypophyseal Hormones
    a. Pro-opiomelanocortin
    b. Growth Hormone (GH)
        1. Function
        2. Regulation
        3. Structural Abnormalities
            a. Gigantism
            b. Acromegaly
            c. Pituitary Dwarfism
    c. Thyroid-Stimulating Hormone (TSH)
    d. Adrenocorticotropic Hormone (ACTH)
    e. Gonadotropins
    f. Prolactin (PRL)
4. The Neurohypophysis and Hypothalamic Hormones
    a. Oxytocin
    b. Antidiuretic Hormone (ADH)
B. The Thyroid Gland (pp. 554-558)
    1. Location and Structure
    2. Thyroid Hormone (TH)
        a. Function
        b. Synthesis
        c. Transport and Regulation
        d. Metabolic Disturbances
            1. Myxedema
            2. Endemic Goiter
            3. Cretinism
            4. Graves's Disease
    3. Calcitonin
C. The Parathyroid Glands (p. 565)
    1. Location (Fig. 17.10, p. 565)
    2. Parathyroid Hormone
        a. Action
        b. Hyperparathyroidism
        c. Hypoparathyroidism

V. Developmental Aspects of the Endocrine System (pp. 577-580)

   A. Origin of Hormone-Producing Glands

   B. Structural Changes with Age

Review Items

1. Negative feedback (Chapter 1, pp. 10-12)

2. Steroids (Chapter 2, p. 46)

3. Amino acids (Chapter 2, pp. 47-48)

4. General cellular function (Chapter 3)

5. Endocrine glands (Chapter 4, p. 110)

6. Bone homeostasis (Chapter 6, pp. 166-172)

7. Epiphyseal plate (Chapter 6, pp. 158-159)

8. Enkephalin and beta-endorphin (Chapter 11, pp. 367-368)

9. Hypothalamus (Chapter 12, p. 395)

10. Norepinephrine and epinephrine (Chapters 11, p. 367, and 14, p. 469)

Cross References

1. The hepatic portal system as an example of portal systems is described in Chapter 20, p. 688.

2. Function of the gonadotropins is explained in Chapter 28, p. 967.

3. Stimulation of milk production by the mammary gland (due to prolactin secretion) is detailed in Chapter 29, p. 1022.

4. Antidiuretic hormone function is further explained in Chapters 26, p. 914, and 27, p. 937.

5. The results of oxytocin and prolactin release are described in Chapter 29, pp. 1019, 1023.

6. Aldosterone effects on renal tissue is explained in Chapters 26, p. 911, and 27, pp. 934-936.

7. The renin-angiotensin mechanism of blood pressure regulation is detailed in Chapters 26, p. 908, and 27, p. 936.

8. Blood pressure control is detailed in Chapter 20, pp. 653-658.

9. Chapter 25, pp. 875-876, has references to insulin and glucagon effects.

10. The effect of thymic hormones is examined in Chapter 22, p. 718.

11. The role of parathyroid hormone and calcium balance related to development is presented in Chapters 27, p. 940; 29, p. 1018.

12. Functions of human placental lactogen and human chorionic thyrotropin are described in Chapter 29, pp. 1005, 1018.

13. Prostaglandins and reproductive physiology are described in Chapter 29, p. 1019.

14. Testosterone production is further described in Chapter 28, p. 969.

15. The role of FSH and LH related to reproduction is covered in Chapter 28, pp. 967, 980-982.

16. Relaxin and inhibin and reproduction are mentioned in Chapter 28, p. 967; relaxin — Chapter 29, p. 1017.

17. Ovarian physiology is explained in great detail in Chapter 28, pp. 969-972.

18. The brain-testicular axis is mentioned in Chapter 28, p. 969.

19. The role of atrial natriuretic factor and fluid-electrolyte balance is presented in Chapters 26, p. 911, and 27, p. 937.

20. Estrogen and glucocorticoid function in fluid and electrolyte balance are detailed in Chapter 27, p. 937.

21. Hormone function related to general body metabolism is described in detail in Chapter 25.

22. Gastrin and secretin (hormones of the digestive system) are detailed in Chapter 24, pp. 809-810.

23. Atrial natriuretic factor and blood pressure regulation is examined in Chapter 20, p. 656.

24. The effects of calcitonin and calcium balance is examined in Chapter 27, p. 940.

25. Estrogen and progesterone function is described in detail in Chapter 28, Table 28.1, p. 968.

26. Human chorionic gonadotropin is detailed in Chapter 29, p. 1005.

Laboratory Correlations

1. Marieb, E. N. *Human Anatomy and Physiology Laboratory Manual: Cat and Fetal Pig Versions.* 3rd. ed. Benjamin/Cummings, 1989.

   Exercise 27: Anatomy and Basic Function of the Endocrine Gland

   Exercise 28: Experiments on Hormonal Action

2. Marieb, E. N. *Human Anatomy and Physiology Laboratory Manual: Brief Version.* 3rd. ed. Benjamin/Cummings, 1992.

   Exercise 24: Anatomy and Basic Functions of the Endocrine Glands

   Exercise 25: Experiments on Hormonal Action

Transparencies Index

17.1   Location of the major endocrine organs
17.2   Second messenger mechanisms of protein-peptide hormones
17.3   Direct gene activation mechanism of steroid hormones
17.5   Structural and functional relationships of the pituitary and hypothalamus
17.6   Scheme to classify metabolic actions of growth hormone
17.8   Gross and microscopic anatomy of the thyroid gland

17.10   The parathyroid glands
17.11   Parathyroid hormones effects on bone, intestines, kidneys
17.13   Major mechanisms controlling aldosterone release from the adrenal cortex
17.14   Role of the hypothalamus, adrenal medulla, and adrenal cortex in the stress response
17.16   Regulation of blood sugar levels by glucagon
17.18   Symptomatic results of insulin deficit

Bassett Atlas Figures Index

Slide Number	Figure Number	Description
18	3.2	Anterior mediastinum
42	4.15A,B	Retroperitoneum, kidneys dissected
43	4.16	Kidney, dissected, detail

INSTRUCTIONAL AIDS

Lecture Hints

1. A flowchart structure is an ideal method to condense a large volume of complex material into a compact package. Suggest to the class that on a single, large sheet of butcher paper, they map the entire endocrine system, flowchart style. Have them start at the top with the hypophysis and trace the path of each hormone to its target tissue. If students are able to see the entire picture on a single sheet, they will easily master the concepts of endocrine function.

2. Emphasize that minute quantities of hormone are all that is necessary to have rather large effects in the body.

3. Students are often confused regarding the actual site of neurohypophyseal hormone production. Point out that the hypophysis is the actual production site and that the axons from the hormone-producing neurons terminate in the neurohypophysis (where the neurohormones are released).

4. Point out the importance of receptor regulation in non-insulin-dependent diabetes.

5. The mechanism of hormone action is an ideal way to introduce some critical thought questions for the class. Ask the class: "Knowing the properties of steroids and proteins, how should these hormones be carried in the blood, and which mechanism (second messenger or intracellular receptor) demands what class of hormone?"

6. Students in an introductory level anatomy and physiology course often do not make the connection between the neurotransmitter norepinephrine and norepinephrine from other sources. Point out that it does not matter what the source of epinephrine and norepinephrine is, the effects on the sympathetic nervous system will be the same. In other words, if epinephrine is injected into the body or is released from the adrenal medulla, sympathetic activity will increase.

7. Use root word definitions to emphasize function of the parts of the pituitary: adeno = gland; neuro = nervous.

8. Point out the advantage of a portal system (like that in the digestive system) for the direct delivery of releasing and inhibiting hormones from hypothalamus to hypophysis.

9. Wherever possible, point out antagonistic hormone pairs (glucagon-insulin, calcitonin-parathyroid hormone) and indicate direct control vs. control by regulating factors (hormones).

Demonstrations/Activities

1. Audio-visual materials of choice.

2. Use a torso model and/or dissected animal model to exhibit endocrine glands.

3. Use photographs to demonstrate various endocrine disorders such as goiter, giantism, cretinism, acromegaly, etc.

4. Use the thermostat in the classroom (or one found in a home) to illustrate how a negative feedback system works.

Critical Thinking/Discussion Topics

1. Discuss how the negative feedback mechanism controls hormonal activity and yet allows hypo- and hypersecretion disorders to occur.

2. Discuss why the pancreas, ovaries, testes, thymus gland, digestive organs, placenta, kidney, and skin are considered to have endocrine function. Relate the endocrine functions to their non-endocrine functions.

3. Discuss the role of the endocrine system in stress and stress responses.

4. Explain the basis of the fact that nervous control is rapid but of short duration while hormonal control takes time to start, but the effects last a long time. How would body function change if the rate of hormone degradation increased? Decreased?

5. On the basis of their chemical properties, why do protein-based and steroid-based hormones utilize, respectively, second messenger and intracellular receptor mechanisms of action?

6. Examine the consequences of increasing receptor number, decreasing receptor number, and increasing or decreasing rates of hormone release.

7. Give students the following scenario: You have just finished a large meal and are relaxing when suddenly threatened by a mugger. Have students explain autonomically and hormonally what occurs in the body. Encourage them to think in logical terms and to be as complete as possible.

Library Research Topics

1. Research the role of hormones in treatment of non-hormone-related disorders.

2. Research the inheritance aspect of certain hormones (such as diabetes mellitus, and certain thyroid gland disorders).

3. Research the role of prostaglandins in treatment of homeostatic imbalances.

4. Identify the various circadian rhythms in the body.

5. Research the methods used to test the levels of hormones in blood.

6. Define diabetes insipidus. How is this type of diabetes related to insulin-related diabetes?

7. Research the various diseases of the pituitary, and discuss what body effects will be produced.

Audio-Visual Aids/Computer Software

Slides

1. The Endocrine System and Its Function (EIL, Slides). This diverse and complex system of hormone-producing organs is viewed anatomically and functionally with emphasis on hormone action and interaction.

2. The Chemistry of Life: Hormones and the Endocrine System, Parts 1-3 (HRM)

Videotapes

1. Animal Hormones I: Principles and Functions (EIL, VHS or BETA, 1984). A detailed review of how hormone systems function and are regulated.

2. Animal Hormones II: Regulation of the Human Menstrual Cycle and of Insect Metamorphosis and Growth (EIL, VHS or BETA, 1984). Describes two well-studied hormone systems. Presents two types of mammalian reproduction cycles, then details the human female menstrual cycle and how hormonal regulation coordinates ovulation and uterine pregnancy.

3. The Living Body: Messengers (FHS, VHS or BETA). Describes the role of hormones including the role hormones play in response to sudden emergency.

4. Your Body, Part 3: Endocrine system (PLP, CH-140203, VHS)

Computer Software

1. Dynamics of the Human Endocrine System (EIL, Apple II or IBM PC, 1988). View of endocrine organs both anatomically and functionally, with emphasis on hormone actions and interactions.

2. Hormones (EIL, Apple II, 1987). Emphasizes the basic principles of hormone action rather than exhaustively surveying hormone effects.

3. Blood Sugar (PLP, Apple II, IIE, II+, IIC 48K). Gives students a greater understanding of the principles of homeostasis by using a computer model to investigate the complexities of the control of blood sugar level in humans.

4. Diabetes and Hypoglycemia (PLP, Apple II, IIE, II+, IIC 48K). Topics include symptoms, causes, treatments, tests, complications, and role of diet in treatment of diabetes and hypoglycemia.

5. Biochemistry of Hormones (CDL, Apple, 1988). This user-friendly program emphasizes the basic principles of hormone action rather than surveying hormone effects.

See Guide to Audio-Visual Resources on page 339 for key to AV distributors.

LECTURE ENHANCEMENT MATERIAL

Clinical and Related Terms

1. Adrenogenital syndrome — a group of symptoms associated with changes in sexual characteristics as a result of increased secretion of adrenal androgens.

2. Exophthalmos — an abnormal protrusion of the eyes.

3. Hirsutism — excessive growth of hair.

4. Hyperglycemia — higher than normal levels of glucose in the blood.

5. Hyperplasia — increase in the number of cells due to an increase in the frequency of cell division.

6. Hypocalcemia — a deficiency of blood calcium.

7. Hypoglycemia — lower than normal levels of glucose in the blood.

8. Hypoplasia — defective development of tissue.

9. Iatrogenic disorder — physician-induced disorder, such as inducing Cushing's syndrome due to excessive use of corticosteroids.

10. Pheochromocytoma — a type of tumor found in the adrenal medulla and usually accompanied by high blood pressure.

11. Polyphagia — excessive eating.

12. Virilization — masculinization of a female; feminization of a male.

Disorders/Homeostatic Imbalances

Pituitary Gland Disorders

1. Acromegaly — a hyperfunction disorder of adults; produces thickening and coarsening of bones and generalized enlargement of the viscera; individuals show coarse facial features, large prominent jaws, and large spade-like hands.

2. Amenorrhea — cessation of menstrual periods due to hypersecretion of prolactin.

3. Diabetes Insipidus — a rare disease characterized by failure of the posterior lobe of the gland to secrete ADH, the antidiuretic hormone; results in the excretion of a large volume of extremely dilute urine; a hypofunction disorder.

4. Galactorrhea — spontaneous secretion of milk from the breasts due to hypersecretion of prolactin.

5. Gigantism — a disorder of children, resulting from hyperfunction of the pituitary gland, resulting in excess height; usually due to a hormone-secreting tumor of the gland.

6. Panhypopituitarism — failure of the pituitary gland anterior lobes to secrete any hormones; a hypofunction disorder.

7. Pituitary Dwarfism — a disorder due to deficiency of growth hormone, characterized by retarded growth and development.

8. Pituitary Infarction — may be due to cardiogenic shock from hemorrhage during childbirth or from intracranial hemorrhage; results in progressive loss of hormonal function.

9. Secondary Hypofunction of All Trophic Hormones — functions of the thyroid gland, adrenal glands, and gonads are impaired due to loss of trophic hormone stimulation.

Thyroid Gland Disorders

1. Exophthalmos — protrusion of the eyes due to hyperthyroidism; other symptoms include weight loss, nervousness, heat intolerance, and increased metabolic activity.

2. Goiter — an enlargement of the thyroid gland; types include a diffuse goiter (gland is uniformly enlarged), a nodular goiter (gland shows multiple nodules of proliferating thyroid tissue), a toxic goiter (gland produces an excessive amount of hormone), and a nontoxic goiter (gland does not increase hormone output). Nontoxic goiters are generally due to iodine deficiency, enzyme deficiency or malfunction, increased hormone requirements, or inadequate hormone output. Toxic goiters are due to an antithyroid antibody, which stimulates the gland.

3. Myxedema — a hypothyroid condition which causes metabolic slowing; symptoms include bloated face; thickened, dry skin; sparse, coarse hair; muscular weakness; mental lethargy, cold intolerance; and hoarseness.

4. Thyroiditis — a condition in which autoantibodies destroy thyroid tissue and cause hypothyroidism.

Parathyroid Gland Disorders

1. Hypoparathyroidism — a condition usually due to accidental removal of the parathyroid glands during thyroid surgery; results in muscle tetany.

2. Hyperparathyroidism — a relatively common disorder resulting from a hormone-secreting parathyroid hormone-secreting tumor; results in hypercalcemia, fragile bones due to calcium loss, formation of renal calculi, and calcium being deposited in the tissues.

Adrenal Gland Disorders

1. Addison's Disease (Chronic Adrenal Insufficiency) — a hypofunction disorder resulting from gradual destruction of the adrenal glands; characterized by hyperpigmented skin; muscular weakness, weight loss, and hypotension; causes include infectious diseases such as histoplasmosis, autoimmunity, and metastatic cancer of the adrenal gland.

2. Aldosteronism (Conn's Disease) — a disorder due to hypersecretion of adrenal cortex hormones; results in muscle weakness, hypertension, excessive urination, and high sodium and low potassium serum levels; causes include an adenoma or hyperplasia of the adrenal cortex.

3. Cushing's Disease — an adrenal cortex hypersecretion disorder characterized by disturbed carbohydrate, fat, and protein metabolism (due to excess glucocorticoids) and increased blood volume and blood pressure (due to excess mineralocorticoids); symptoms include a "moon" face and obesity of the trunk, sparing the distal limbs; hyperglycemia, hypertension, osteoporosis, muscular weakness, hirsutism, and psychological disturbances are commonly found.

4. Pheochromocytoma — a benign adrenal medullary tumor of adults; symptoms include secretion of excess catecholamines with resulting pronounced cardiovascular effects.

5. Virilization — a condition due to overproduction of adrenal sex hormones; causes masculinization in females with excess male sex hormones and feminization in males with excess female sex hormone production.

Pancreatic Disorders

1. Diabetes Mellitus — a disorder of the pancreas resulting from failure of the pancreatic islets to secrete insulin or from inefficient utilization of insulin. A major symptom is hyperglycemia. Two types exist: insulin-dependent diabetes (Type I or juvenile onset diabetes) and non-insulin-dependent diabetes (Type II or adult onset diabetes).

Other Disorders

1. Ectopic Hormone Disorders — disorders due to nonendocrine tumors that secrete hormones that produce the same symptoms as tumors arising from endocrine glands; most such tumors are malignant and are a result of carcinomas of the lung, kidney, pancreas, or malignant connective tissue tumors.

2. Obesity — excess weight due to excessive body fat; in rare cases, due to endocrine or metabolic disturbances; usual cause is caloric intake exceeding body requirements.

ANSWERS TO END-OF-CHAPTER QUESTIONS

Multiple Choice/Matching

1. b
2. (1)c; (2)b; (3)f; (4)d;
   (5)a; (6)g; (7)a; (8)h;
   (9)e,b; (10)a
3. a
4. c
5. d
6. d

7. c
8. b
9. d
10. b
11. d
12. b
13. c
14. d

Short Answer Essay Questions

15. Hormone — a chemical substance, secreted by cells into the extracellular fluids, that regulates the metabolic functions of specific body cells. (p. 548)

16. The anterior pituitary is connected to the hypothalamus by a stalk of tissue called the infundibulum. It produces the following hormones: growth hormone, thyroid-stimulating hormone, adrenocorticotropic hormone, follicle-stimulating hormone, luteinizing hormone and prolactin. (pp. 553-561)

The pineal gland hangs from the roof of the third ventricle within the diencephalon. Melatonin is its major secretory product. (p. 576)

The pancreas is located partially behind the stomach in the abdomen. The islets of Langerhans of the pancreas produce glucagon and insulin, as well as small amounts of somatostatin. (pp. 571-575)

The ovaries are located in the female's abdominopelvic cavity. They produce estrogens and progesterone. (p. 575)

The male testes are located in an inferior extra-abdominal skin pouch called the scrotum. They produce the hormone testosterone. (p. 575)

The adrenal glands are perched atop the kidneys. The adrenal glands produce the adrenocortical hormones mineralocorticoids, glucocorticoids and gonadocorticoids; and the adrenal medullary hormones epinephrine and norepinephrine. (pp. 566-571)

17. Endocrine regions that are important in stress response are the adrenal medulla and adrenal cortex. The adrenal medulla produces hormones that mimic the effects of neurotransmitters of the sympathetic division of the autonomic nervous system. The adrenal cortex produces the glucocorticoids (and mineralocorticoids) important in stress response. The adrenal medulla hormones function in the alarm reaction; the adrenal cortex hormones, in the resistance stage. (pp. 570-571)

18. The release of anterior pituitary hormones is controlled by hypothalamic-releasing (and -inhibiting) factors. (pp. 553-554)

19. The posterior pituitary gland is composed largely of glial cells and nerve fibers. It releases ready-made neurohormones that it receives via nerve fibers from the hypothalamus. It serves as a hormone storage area. (p. 558)

20. A lack of iodine (required to make functional T3 and T4) causes a colloidal, or endemic, goiter. (p. 561)

21. Problems that elderly people might have as a result of decreasing hormone production include the following:

    a. chemical or borderline diabetes

    b. lessening of basal metabolic rate

    c. increase in body fat

    d. osteoporosis (pp. 577-580)

Critical Thinking and Application Questions

1. It is not unusual to find them in other regions of the neck or even the thorax. The adjacent neck regions should be checked first. (pp. 565-566)

2. Insulin should be administered because symptoms are indicative of diabetic shock. (p. 573)

3. The hypersecreted hormone is growth hormone. The disorder is gigantism. (p. 556)

Key Figure Questions

Figure 17.1 See Table 17.4. **Figure 17.2** Phototransduction in the eye. **Figure 17.3** In the one illustrated here, the hormone (first messenger) enters the cell nucleus and directly stimulates the desired effect by acting on DNA. In Figure 17.2, the hormone interacts with a plasma membrane receptor and its effects are mediated by an intracellular second messenger. **Figure 17.12** In the zona glomerulosa, the cells are in clusters. In the zona fasciculata, the cells are in parallel cords. In the zona reticularis, the cells have a netlike arrangement. **Figure 17.13** The renin-angiotensin mechanism.

SUGGESTED READINGS

1. Allman, William F. "Steroids in Sports: Do They Work?" *Science* 4 (Nov. 1983): 14.

2. Axelrod, J., and T.D. Reisine. "Stress Hormones: Their Interaction and Regulation." *Science* 224 (1984): 452.

3. Berridge, M.J. "The Molecular Basis of Communication Within the Cell." *Scientific American* 253 (1985): 142-152.

4. Benson, M.A. and N.K. Macklaren. "What Causes Diabetes?" *Scientific American* 263 (July 1990): 62-71.

5. Bower, B. "Sizing Up Sadness According to Latitude." *Science News* 136 (Sept. 1989): 198.

6. Carmichael, S.W., and H. Winkler. "The Adrenal Chromaffin Cell." *Scientific American* 253 (Aug. 1985): 40-49.

7. Crapo, L. *Hormones: Messengers of Life*. New York: W.H. Freeman & Co., 1985.

8. Donham, J. "The Weakness of Steroids." *American Journal of Nursing* 8 (Aug. 1986): 917-919.

9. "Estrogen Use Raises Questions." *Science News* 128 (1985): 279.

10. Ezrin, D., J.O. Godden, and R. Volpe. *Systematic Endocrinology*. 2d ed. New York: Harper & Row, 1979.

11. Fellman, B. "A Clockwork Gland." *Science* 85 (May 1985).

12. Gibbons, B. "The Intimate Sense of Smell." *National Geographic* 170 (3) (Sept. 1986): 324-360.

13. Gilman, A.G., and M.E. Linder. "G Proteins." *Scientific American* (Sept. 1992): 70.

14. Goldfinger, S.E. (ed). "The Diseases Called Sugar Diabetes." (Part I) *The Harvard Medical School Health Letter* (Feb. 1985).

15. "Insulin and the Diabetic." *The Harvard Medical School Health Letter* (Mar. 1985).

16. "Diabetes: The Long Run." *The Harvard Medical School Health Letter* (Apr. 1985).

17. Martin C. *Endocrine Physiology*. New York: Oxford Univ. Press, 1985.

18. Murray, Thomas. "The Growing Danger from Gene-Spliced Hormones." *Discover* 8 (Feb. 1987): 88.

19. Notkins, A.L. "The Causes of Diabetes." *Scientific American* (Nov. 1979).

20. Orci, L., et al. "The Insulin Factory." *Scientific American* 259 (Sept. 1988): 85-94.

21. Pennisi, E. "Diabetes Stopped Before It Starts." *Science News* 149 (Nov. 1993): 292.

22. Raloff, J. "Fats May Influence Insulin Sensitivity." *Science News* 143 (Jan. 1993): 68.

23. Rasmussen, H. "The Cycling of Calcium as an Intracellular Messenger." *Scientific American* 261 (Oct. 1989): 66-73.

24. Sapolsky, R.M. "Stress in the Wild." *Scientific American* 262 (Jan. 1990): 116-123.

25. Shodell, M. "The Prostaglandin Connection." *Science* 83 (Mar. 1983): 78-82.

26. Stryer, L. and H.R. Bourne. "G Proteins: A Family of Signal Transducers." *Annual Review of Cell Biology* 2 (1986): 391-419.

27. Wechsler, Rob. "Unshackled from Diabetes." *Discover* 1 (Sept. 1986): 77.

28. Williams, R. *Textbook of Endocrinology*. 7th ed. Philadelphia: Saunders, 1985.

29. Wurtman, R.J. and J.J. Wurtman. "Carbohydrates and Depression." *Scientific American* 260 (Jan. 1989): 68-75.

18

Blood

Chapter Preview

The cardiovascular system includes the blood, heart, and blood vessels. This chapter deals with the fluid of the system: blood. Blood is the medium that transports all substances that must be carried from one place to another within the body, such as nutrients, wastes, respiratory gases, and hormones. Blood consists of two components: (1) plasma, a fluid consisting of a salt-water solution with inorganic and organic constituents and (2) formed elements: the erythrocytes, leukocytes, and platelets. This chapter describes the composition and functions of this life-sustaining fluid.

INTEGRATING THE PACKAGE

Suggested Lecture Outline

I. Overview: Composition and Functions of Blood (pp. 585-586)
    A. Components (Fig. 18.1, p. 585)
        1. Formed Elements
        2. Plasma
    B. Physical Characteristics and Volume (p. 585)
    C. Functions (pp. 585-586)
        1. Distribution
        2. Regulation
        3. Protection
II. Blood Plasma (Table 18.1, p. 586)
III. Formed Elements (pp. 587-598)
    A. Erythrocytes (pp. 587-593)
        1. Structural Characteristics (Fig. 18.3, p. 588)
        2. Function
            a. Hemoglobin and Oxygen
            b. Hemoglobin and CO_2
        3. Production of Erythrocytes
            a. Location
            b. Phases of Erythropoiesis
        4. Regulation and Requirements for Erythropoiesis (Fig. 18.6, p. 590)
            a. Hormonal Controls
            b. Dietary Requirements: Iron and B-Complex Vitamins
        5. Fate and Destruction of Erythrocytes (Fig. 18.7, p. 592)

6. Erythrocyte Disorders
   a. Anemias
      1. Insufficient Numbers of RBCs
         a. Hemorrhagic Anemias
         b. Hemolytic Anemias
         c. Aplastic Anemias
      2. Decreases in Hemoglobin Content
         a. Iron-Deficiency Anemia
         b. Pernicious Anemia
      3. Abnormal Hemoglobin
         a. Thalassemias
         b. Sickle-Cell Anemia
   b. Polycythemia
      1. Polycythemia Vera
      2. Secondary Polycythemias
      3. Blood Doping
B. Leukocytes (pp. 594-598)
   1. General Structural and Functional Characteristics
   2. Types (Fig. 18.8, p. 594)
      a. Granulocytes
         1. Neutrophils
         2. Eosinophils
         3. Basophils
      b. Agranulocytes
         1. Lymphocytes
         2. Monocytes
         3. Production and Life Span of Leukocytes (Fig. 18.9, p. 597)
         4. Leukocyte Disorders
            a. Leukemias
            b. Infectious Mononucleosis
C. Platelets (Fig. 18.10, p. 598)
IV. Hemostasis (pp. 598-604)
   A. Vascular Spasms (p. 599)
   B. Platelet Plug Formation (Fig. 18.11, p. 600)
   C. Coagulation (p. 601)
   D. Clot Retraction and Repair (p. 601)
   E. Fibrinolysis (p. 602)
   F. Factors Limiting Clot Growth or Formation (p. 603)
      1. Factors Limiting Normal Clot Growth
      2. Factors Preventing Undesirable Clotting
   G. Disorders of Hemostasis (pp. 603-604)
      1. Thromboembolytic Conditions

Review Items

1. Diffusion (Chapter 3, p. 66)

2. Osmosis (Chapter 3, p. 67)

3. Tissue repair (Chapter 4, p. 128)

4. Hematopoietic tissue (Chapter 6, p. 160)

Cross References

1. The role of the heart in blood delivery is detailed in Chapter 19.

2. Vasoconstriction as a mechanism of blood flow control is explained in Chapter 20, pp. 650-660.

3. A general overview of arteries, capillaries, and veins is given in Chapter 20, pp. 645-650.

4. The role of the spleen in the removal of old red blood cells is described in Chapter 21, p. 700.

5. Granulocyte function in nonspecific resistance is detailed in Chapter 22, pp. 708-712.

6. Lymphocyte function (T and B cells) in specific immune response is explained in Chapter 22, pp. 719-727.

7. The role of monocytes (macrophages) in the immune response in explained in detail in Chapter 22, pp. 709, 719-720.

8. AIDS is discussed in Chapter 22, p. 738.

9. Antigen-antibody interaction is covered in Chapter 22, pp. 715-718.

10. Diapedesis is explained in more detail in Chapter 22, p. 712.

11. Gas exchange between blood, lungs, and tissues is described in Chapter 23, pp. 744-750.

12. Chapter 23, pp. 750-754, has a complete description of respiratory gas transport.

13. Vitamin B_{12} absorbance is described in Chapter 24, p. 815.

14. Production of vitamin K in the large intestine is examined in Chapter 24, p. 815.

15. The role of blood in body temperature regulation is explained in Chapter 25, p. 863.

16. The control of water and ion balance is detailed in Chapter 27, pp. 930-940.

17. Chapter 27, pp. 940-950, contains a detailed description of acid-base balance.

18. Erythropoietin related to renal function is described in Chapter 26, p. 896.

19. Plasma filtration is detailed in Chapter 26, pp. 904-908.

20. Chemotaxis is explained in Chapter 22, p. 712.

21. Macrophages are mentioned again in Chapter 21, p. 698.

Laboratory Correlations

1. Marieb, E. N. *Human Anatomy and Physiology Laboratory Manual: Cat and Fetal Pig Versions.* 3rd. ed. Benjamin/Cummings, 1989.

   Exercise 29: Blood

2. Marieb, E. N. *Human Anatomy and Physiology Laboratory Manual: Brief Version.* 3rd. ed. Benjamin/Cummings, 1992.

   Exercise 26: Blood

Transparencies Index

18.1   Components of whole blood

18.4   Structure of hemoglobin

18.6   Erythropoietin mechanism for regulating the rate of erythropoiesis

18.11   Events of platelet plug formation and blood clotting

INSTRUCTIONAL AIDS

Lecture Hints

1. Give a brief introduction to the heart and blood vessels so that students realize the closed nature of the system. It is worthwhile to point out that there are two circulatory routes, systemic and pulmonary.

2. Emphasize that the hematocrit is an indirect measurement of the O_2 carrying capacity of the blood. More red cells mean more O_2 carried by the same volume of blood.

3. Emphasize that simple diffusion gradients cause the loading and unloading of respiratory gases and other substances. It may be of benefit to ask the students pointed questions about respiratory gas diffusion during lecture to be sure the class has mastered this concept.

4. As a point of interest, mention that well-oxygenated blood is bright red, normal deoxygenated blood (at the tissue level) is dark red, and that under hypoxic conditions, hemoglobin becomes blue.

5. Spend some time with the feedback loop involved in erythropoiesis. This is a typical negative feedback mechanism that allows the application of critical thought processes.

6. Mention that serum is essentially plasma without clotting proteins.

7. Point out the delicate balance between clotting and prevention of unwanted clotting. We want to be sure that hemorrhage is arrested, but at the same time, we need to prevent clot formation in unbroken blood vessels.

8. Emphasize that ABO incompatibility does not require sensitization by a previous blood transfusion but that Rh incompatibility does.

Demonstrations/Activities

1. Audio-visual materials of choice.

2. Use models to exhibit blood cells.

3. Use a magnet and iron filings to demonstrate the attraction of antibodies to antigens when discussing blood grouping and typing.

4. Display equipment used to perform a hematocrit, sedimentation rate, and cell counts. Describe how these tests are performed and the information they yield. Run a hematocrit so that students can see the difference in volume of plasma and formed elements.

5. Set up a stained blood smear to illustrate as many types of white blood cells as possible.

6. Provide blood-typing sera and have the students type their own blood. All lancets and disposable items are to be placed immediately in a disposable autoclave bag after use, and used slides are to be placed in a solution of freshly prepared 10% bleach and soaked for at least two hours. Both the autoclave bag and the slides are to be autoclaved for 15 min. at 121°C, 15 lbs. pressure to ensure sterility. After autoclaving, the autoclave bag may be discarded in any disposable container; the glass slides may be washed with laboratory detergent and reprepared for use.

7. Provide a sample of centrifuged animal blood so that students can examine consistency, texture, and color of plasma. Have pH paper available so that students can determine its pH. Use this activity as a lead-in to a discussion about the composition and importance of plasma.

Critical Thinking/Discussion Topics

1. Discuss the fears and facts associated with blood donation and transfusion and AIDS.

2. Discuss the problems associated with IV drug use (i.e., hepatitis, AIDS, necrosis of tissue, other blood related disorders).

3. Discuss the procedure of autologous transfusion.

4. Discuss why gamma globulin injections are painful.

5. Why do red blood cells lack a nucleus? Why is this an advantage?

6. How can you explain that an incompatible ABO blood group will generate a transfusion reaction the first time a transfusion is given, while Rh incompatibility creates a problem the second time a transfusion is given?

Library Research Topics

1. Research the blood disorders associated with IV street drug use.

2. Research the role of blood in the AIDS epidemic.

3. Research inherited blood disorders.

4. Research the blood antigens other than A, B, and Rh.

5. Research the various blood immunoglobulins, their functions, and how they are made (i.e., stimulus required).

6. Research the various uses of donated blood; i.e., packed red cells, platelets, etc.

7. Research which diseases are transmitted by blood and why these diseases are increasing in incidence. Why is careful handling of blood in the clinical agency vitally important?

Audio-Visual Aids/Computer Software

Slides

1. Visual Approach to Histology: Blood and Bone Marrow (FAD, 19 Slides)

2. Hematology (EIL, 99 Slides)

Videotapes

1. The Living Body: Life Under Pressure (FHS, VHS or BETA). This program follows the journey of a red blood cell around the circulatory system to demonstrate the efficiency and elegance of design that delivers oxygen and food to each part of the body and removes wastes before they can do harm.

2. Blood (PLP, CH-460755, VHS). A thorough and up-to-date look at the cellular functions of blood.

3. What Can Go Wrong: (Circulatory System) (PLP, CH-140501, VHS)

4. Blood: River of Life, Mirror of Health (EI, 600-2365V, VHS)

5. Circulation (EI, 475-2354V, VHS). In three parts: blood; the heart; and lymphatics.

Computer Software

1. The Heart (QUE, Apple II's, 64K). Tutorial introducing the structure and function of the human heart.

   See Guide to Audio-Visual Resources on page 339 for key to AV distributors.

LECTURE ENHANCEMENT MATERIAL

Clinical and Related Terms

1. Anemia — reduced oxygen carrying capacity; due to various causes.

2. Anisocytosis — condition in which there is an abnormal variation in the size of erythrocytes.

3. Autoantibodies — antibodies formed in response to, and reacting against, one of the individual's own normal antigenic endogenous body constituents.

4. Autologous transfusion — donating one's own blood before elective surgery to ensure an abundant supply and reduce transfusion complications.

5. Christmas disease — a hereditary bleeding disease due to a deficiency in a clotting factor; also called hemophilia B.

6. Coagulopathy — any disorder of blood coagulation.

7. Direct (immediate) transfusion — transfer of blood directly from one person to another without exposing the blood to air.

8. Exchange transfusion — removing blood from the recipient while simultaneously replacing it with donor blood.

9. Hemorrhage — loss of blood from the body either from blood vessels into tissues or from blood vessels directly to the surface of the body.

10. Hemorrhagic telangiectasia — a hereditary disorder in which there is a tendency to bleed from localized lesions of capillaries.

11. Indirect (mediate) transfusion — transfer of blood from a donor to a container and then to the recipient.

12. Pancytopenia — deficiency of all cell elements of blood; aplastic anemia.

13. Purpura — a disease characterized by spontaneous bleeding into the tissues and through the mucus membrane.

14. Reciprocal transfusion — transfer of blood from a person who has recovered from a contagious infection into the vessels of a patient with the same infection.

15. Spherocytosis — a hereditary form of hemolytic anemia characterized by the presence of spherical erythrocytes.

Disorders/Homeostatic Imbalances

Anemias

1. Hereditary Spherocytic Anemia — an autosomal dominant, inherited disease in which the RBC membrane proteins are abnormal. As a result, the RBCs become rigid spheres, which are readily destroyed by the spleen.

2. Iron-Deficiency Anemia — may result from a diet lacking iron or blood loss due to chronic bleeding (as in excessive menstruation, GI hemorrhage due to peptic ulcers, colon carcinoma, hookworm infestation, or hemorrhoids. Treatment involves identifying and correcting the condition responsible for the deficiency of iron.

3. Macrocytic Normochromic Anemia — anemia associated with a variety of diseases, such as cirrhosis and hemolytic anemia, and deficiencies of vitamin B_{12} and folic acid.

4. Microcytic Normochromic Anemia — anemia due to chronic disease and malignancy or toxic agents; also known as hemoglobin red cell mass deficit.

5. Microcytic Hypochromic Anemia — anemia due to iron deficiency, chronic lead poisoning, thalassemia, and miscellaneous factors.

6. Normocytic Normochromic Anemia — anemia associated with sudden loss of blood; increased destruction of RBCs (hemolytic anemias); decreased production of RBCs (hemoglobin red cell mass deficit) due to chronic disease, malignancy, toxic agents, and bone marrow failure; and compensated hemolytic anemia.

7. Pernicious Anemia — a special form of macrocytic anemia caused by inadequate secretion of intrinsic factor or from autoantibodies that inactivate intrinsic factor.

8. Sickle-Cell Anemia — an autosomal recessive, inherited disease resulting from the substitution of valine for glutamic acid in the sixth position in the hemoglobin molecule. The hemoglobin molecules become rigid in hypoxic conditions. The normally disc-shaped RBCs become sickle shaped.

9. Thalassemia — an autosomal recessive disease resulting from absent or depressed synthesis of one of the globin chains; characterized by the presence of very thin, fragile erythrocytes.

Polycythemia

1. Polycythemia Vera — a special form of polycythemia (excessive number of RBCs) that can lead to leukemia. Platelets, WBCs, and RBCs all increase in number. Thrombosis, as well as clogged vessels due to numbers of cells, are common complications.

Leukocyte Disorders

1. Benign Reactive Lymphoproliferation Disorders — enlargement of lymph nodes due to infectious mononucleosis, various viral and bacterial infections, and toxoplasmosis; a nonmalignant disorder.

2. Eosinophilia — leukocytosis associated with allergic disorders such as asthma and hay fever, parasitic infections, and chronic myelogenous leukemia and Hodgkin's disease.

3. Hodgkin's Disease — a type of lymphoma characterized by enlarged lymph nodes, fever, itching of the skin, and fatigue.

4. Leukopenia — refers to reduced WBCs below 4000/cm^3 of blood. It is often induced by viral infections. It may accompany leukemia and radio- or chemotherapy treatment, due to neutropenia or aplastic anemia.

5. Lymphocytosis — leukocytosis associated with certain acute infections such as infectious mononucleosis, whooping cough, mumps, and measles; certain chronic infections such as tuberculosis and hepatitis; and lymphocytic leukemia.

6. Lymphoma — refers to a malignancy of lymphoid tissue. Lymphomas arise in the immune and reticuloendothelial tissue, with the majority of malignancies arising from malignant B lymphocytes.

7. Monocytosis — leukocytosis associated with chronic infections such as tuberculosis and endocarditis, Hodgkin's disease, and monocystic leukemia.

8. Multiple Myeloma — a malignancy of plasma cells (B lymphocytes) that occurs most frequently in elderly persons. Due to the pressure of malignant plasma cells on the nerves in the periosteum of bone, bone pain in the back or hip may be the initial complaint. Complications include severe infections and myeloma kidney (a blocked kidney leading to renal failure).

9. Neutrophilia — leukocytosis associated with acute infections, toxic metabolic states and poisoning, tissue necrosis (as in acute myocardial infarction), and myelogenous leukemia.

Clotting Disorders

1. Acquired Plasma Disorders — clotting disorders resulting from liver disease or damage, since clotting factors are presumed to be made principally or entirely in the liver.

2. Hemophilia A — an X-linked recessive disorder characterized by a factor VIII activity of less than 5 percent. Affected individuals bruise easily and have spontaneous nosebleeds.

3. Hemophilia B — an X-linked recessive disorder in which factor IX is missing. Hemophilia signs are basically the same as for hemophilia A.

4. Hemorrhage — loss of blood from the cardiovascular system. Hemorrhage can be divided into two major groups and many subgroups. The major causes include surgical or traumatic and nonsurgical. Nonsurgical types of hemorrhage include coagulopathy and thrombocytopenia. Types of coagulopathy include disseminated intravascular coagulation, liver disease, vitamin K deficiency, anticoagulant (heparin) use, factor VIII deficiency, factor IX deficiency, and von Willebrand Syndrome.

5. Von Willebrand Syndrome — an autosomal dominant bleeding disorder characterized by platelet abnormalities and a low level of factor VIII.

Platelet Disorders

1. Thrombocytopenia — decreased platelets caused by increased destruction and decreased production of platelets.

2. Thrombocytopenia — autoantibody production to platelets, caused by certain drugs, infectious agents, or unknown factors, resulting in clotting disorders.

ANSWERS TO END-OF-CHAPTER QUESTIONS

Multiple Choice/Matching

1.	c	5.	d	9.	c
2.	c	6.	a	10.	d
3.	d	7.	a		
4.	b	8.	b		

Short Answer Essay Questions

11. a. The formed elements are living blood cells (p. 585). The major categories of formed elements are erythrocytes, leukocytes, and platelets. (p. 587)

  b. The least numerous of the formed elements are the leukocytes. (p. 587)

  c. The buffy coat in a hematocrit tube comprises the white blood cells and platelets. (p. 585)

12. Hemoglobin is made up of the protein globin bound to the red pigment heme. Each molecule contains four polypeptide chains (globins) and four heme groups, each bearing an atom of iron in its center. Its function is to bind oxygen to each iron atom. When oxygen is loaded (bound to hemoglobin) the hemoglobin becomes bright red. When oxygen is unloaded from the iron, the hemoglobin becomes dark red. (p. 588)

13. With a high hematocrit, you would expect the hemoglobin determination to be high, since the hematocrit is the percent of blood made up of RBCs. (p. 585)

14. The nutrients needed for erythropoiesis, in addition to carbohydrates for energy and amino acids needed for protein synthesis, are iron and certain B vitamins. (pp. 589-590)

15. a. In the process of erythropoiesis, a hemocytoblast is transformed into a proerythroblast, which gives rise to early, then late erythroblasts, normoblasts and reticulocytes.

  b. The immature cell type released to the circulation is the reticulocyte.

  c. The reticulocyte differs from a mature erythrocyte in that it still contains some rough ER. (p. 589-590)

16. The physiological attributes which contribute to the function of white blood cells in the body include the ability to move by amoeboid action, exhibition of positive chemotaxis enabling them to pinpoint areas of tissue damage, and the ability to participate in phagocytosis. (p. 594)

17. a. With a severe infection, the WBC count would be closest to 15,000 WBC/mm^3 of blood.

  b. This condition is called leukocytosis. (p. 594)

18. a. Platelets appear as small discoid fragments of large, multinucleated cells called megakaryocytes. They are essential for the clotting process and work by clumping together to form a temporary plug to prevent blood loss.

  b. Platelets should not be called "cells" because they are only fragments of cells. (p. 598)

19. a. Literally, hemostasis is "blood standing still" because it has been clotted. It refers to the prevention of blood loss from blood vessels. (p. 599)

  b. The three major steps of coagulation include the formation of prothrombin activator by a cascade of activated procoagulants, the use of prothrombin activator enzymatically to release the active enzyme thrombin from prothrombin, and the use of thrombin to cause fibrinogen to form fibrin strands. (p. 601)

c. The intrinsic pathway depends on substances present in (intrinsic to) blood. It has many more steps and intermediates, and is slower. The extrinsic mechanism bypasses the early steps of the intrinsic mechanism and is triggered by tissue thromboplastin released by injured cells in the vessel wall or in surrounding tissues. (p. 601)

d. Calcium is essential to virtually all stages of coagulation. (p. 601)

20. a. Fibrinolysis is the disposal of clots when healing has occurred.

b. The importance of this process is that without it, blood vessels would gradually become occluded by unnecessary clots. (p. 602)

21. a. Clot overgrowth is usually prevented by rapid removal of coagulation factors and inhibition of activated clotting factors.

b. Two conditions that may lead to unnecessary (and undesirable) clot formation are roughening of the vessel wall endothelium and blood stasis. (pp. 602-603)

22. The liver can cause bleeding disorders when it cannot synthesize its usual supply of procoagulants. (p. 603)

23. a. A transfusion reaction involves agglutination of foreign RBCs, leading to clogging of small blood vessels, and lysis of the donated RBCs. It occurs when mismatched blood is transfused.

b. Possible consequences include disruption of oxygen carrying capacity, fever, chills, nausea, vomiting, general toxicity, and renal failure. (pp. 604-606)

24. Among other things, poor nutrition can cause iron-deficiency anemia due to inadequate intake of iron-containing foods or to pernicious anemia due to deficiency of vitamin B_{12}. (pp. 592-593)

25. The most common blood-related problems for the aged include chronic types of leukemias, anemias, and thromboembolytic disease. (p. 607)

Critical Thinking and Application Questions

1. Hemopoiesis is a process involving fairly rapid cell production. Since chemotherapeutics simply target cells exhibiting rapid turnover (rather than other properties of cancer cells), hemopoiesis is a target of chemotherapeutic drugs and must be carefully monitored. (pp. 588-590)

2. a. The woman would probably be given a whole blood transfusion. It is essential that she maintain sufficient O_2 carrying capacity to serve fetal needs and blood volume to maintain circulation.

b. The blood tests that would be performed include tests for ABO and Rh group antigen and cross-matching. (pp. 604-605)

3. a. Polycythemia accounts for his higher erythrocyte count because of the need to produce more RBCs to increase his O_2 binding and transport ability. Enhanced production of RBCs was due to an increased production of erythropoietin.

b. His RBC count will not stay higher than normal because the excess production of RBCs will depress erythropoietin production by the kidneys when adequate levels of O_2 are transported in the blood. (p. 593)

4. Janie's leukocytes are immature or abnormal and are incapable of defending her body in the usual way. (pp. 597-598)

Key Figure Questions

Figure 18.6 Blood doping, which increases the number of red blood cells in the blood stream and thus increases the amount of oxygen being transported, would inhibit erythropoietin release. **Figure 18.7** It would increase in the blood because liver bilirubin-processing functions are impaired.

SUGGESTED READINGS

1. Bank, A., et al. "Disorders of Human Hemoglobin." *Science* 207 (1980): 486.

2. Bennett, D.D. "Human White Blood Cells as Carcinogens." *Science News* 127 (Mar. 1985): 150.

3. Cooper, M.D. "B Lymphocytes. Normal Development and Function." *The New England Journal of Medicine* 317 (Dec. 1987): 1452-1456.

4. Davis, L. "New Ideas in Fighting Clots, Heart Attacks." *Science News* 129 (May 1986).

5. Dixon, B. "Of Different Bloods." *Science* 84 (Nov. 1984).

6. Doolittle, R.F. "Fibrinogen and Fibrin." *Scientific American* 245 (Dec. 1981): 126-135.

7. Edwards, D.D. "After the Battle, TPA Declared a Winner." *Science News* (Nov. 1987): 325.

8. Fackelmann, K.A. "Blood Substances Linked to Heart Risk." *Science News* 144 (Nov. 1993).

9. Golde, D.W., and J.C. Gasson. "Hormones That Stimulate the Growth of Blood Cells." *Scientific American* 259 (July 1988): 62-70.

10. Lawn, R.M., and G.A. Vehar. "The Molecular Genetics of Hemophilia." *Scientific American* 254 (Mar. 1986): 48-54.

11. Levine, R.A., and S.C. Wardlaw. "A New Technique for Examining Blood." *American Scientist* 76 (Nov.-Dec. 1988): 592-598.

12. Loupe, D.E. "Breaking the Sickle Cycle." *Science News* 136 (Dec. 1989): 360-362.

13. Morse, E.E. and B. Jacobs. "Blood Donation and Its After-effects." *American Scientist* 73 (Jan.-Feb. 1985): 68-69.

14. Oliwenstein, L. "Liquid Assets." *Discover* 14 (Sept. 1993): 34.

15. Perutz, M.F. "Hemoglobin Structure and Respiratory Transport." *Scientific American* 239 (Dec. 1978): 92-98, 103-125.

16. Platt, W.R. *Color Atlas and Textbook of Hematology.* 2d ed. Philadelphia: J.B. Lippincott, 1979.

17. Silberner, J. "Blood Change Linked to Cancer." *Science News* 130 (Dec. 1986): 356.

18. Swithers, C.M. "Tools for Teaching About Anticoagulants." *RN* (Jan. 1988): 57-58.

19. Vichinsky, E.P., D. Hurst, and B. Lubin. "Sickle-Cell Disease: Basic Concepts." *Hospital Medicine* (Sept. 1983).

20. Weiss, R. "Sanguine Substitutes?" *Science News* 132 (Sept. 1987): 200-202.

21. Weiss, R. "Postponing Red-Cell Retirement." *Science News* 136 (Dec. 1989): 424-425.

22. Wickelgren, J. "Drug Shows Promise in Sickle-Cell Anemia." *Science News* 135 (Jun. 1989): 349.

23. Zucker, M B. "The Functioning of Blood Platelets." *Scientific American* 242 (June 1980): 86-103.

19

The Cardiovascular System: The Heart

Chapter Preview

The cardiovascular system includes the blood, a fluid that transports substances; the heart, a pump to move the blood; and the blood vessels. This chapter deals with the pump of the system. Using blood as the medium for transport, the heart continually propels nutrients, respiratory gases, wastes, and many other substances into the interconnecting blood vessels that pass in relatively close proximity to all body cells.

INTEGRATING THE PACKAGE

Suggested Lecture Outline

I. Heart Anatomy (pp. 613-621; Fig. 19.4, pp. 616-617)
    A. Size, Location, and Orientation (p. 613; Fig. 19.1, p. 613)
        1. Base and Apex
        2. PMI (Fig. 19.1, p. 613)
    B. Coverings (pp. 613-614)
        1. Parietal Pericardium (Fig. 19.2, p. 614)
        2. Visceral Pericardium
        3. Epicardium
    C. Heart Wall (p. 614)
        1. Myocardium
        2. Endocardium
    D. Chambers and Associated Great Vessels (p. 615)
        1. Atria
            a. Superior and Inferior Vena Cava
            b. Coronary Sinus
            c. Pulmonary Veins
        2. Functions
        3. Ventricles
            a. Pulmonary Trunk
            b. Aorta
    E. Pathway of Blood Through the Heart (pp. 615-619)
        1. Pulmonary Circuit
        2. Systemic Circuit (Fig. 19.5, p. 618)
    F. Heart Valves (pp. 619-621)

      1. Atrioventricular Valves

         a. Tricuspid

         b. Bicuspid (Mitral)

      2. Semilunar Valves

         a. Aortic

         b. Pulmonary

II. Coronary Circulation (pp. 621-622)

  A. Coronary Arteries (Fig. 19.10, p. 621)

  B. Cardiac Veins and Coronary Sinus

  C. Disorders

      1. Angina Pectoris

      2. Myocardial Infarction

III. Properties of Cardiac Muscle (pp. 622-627)

  A. Microscopic Anatomy (Fig. 19.11, p. 623)

  B. Energy Requirements (p. 626)

  C. Mechanism and Events of Contraction (pp. 626-627)

      1. Differences Between Cardiac and Skeletal Muscle

         a. All or None Law

         b. Means of Stimulation

         c. Length of Absolute Refractory Period

IV. Heart Physiology (pp. 627-638)

  A. Electrical Events (pp. 627-630)

      1. Intrinsic Conduction System of the Heart

      2. Action Potential Generated by Autorhythmic Cells

         a. Pacemaker Potentials

         b. Sequence of Excitation

            1. Sinoatrial Node

            2. Atrioventricular Node

            3. Atrioventricular Bundle

            4. Bundle Branches

            5. Purkinje Fibers

         c. Defects

            1. Arrhythmias

            2. Fibrillation

            3. Heart Block

      3. Extrinsic Innervation of the Heart

         a. Cardioacceleratory Center

         b. Cardioinhibitory Center

      4. Electrocardiography (Fig. 19.16, p. 630)

         a. P Wave

         b. QRS Complex

         c. T Wave

B. Mechanical Events: The Cardiac Cycle (pp. 631-633)

    1. Terms

        a. Systole

        b. Diastole

        c. Cardiac Cycle

            1. Period of Ventricular Filling: Mid-to-Late Diastole

            2. Ventricular Systole

            3. Isovolumetric Relaxation: Early Diastole

C. Heart Sounds (pp. 633-634)

    1. Normal

    2. Abnormal or Unusual

D. Cardiac Output (pp. 634-637)

    1. General

        a. Cardiac Reserve

    2. Regulation of Stroke Volume

        a. Frank-Starling Law of the Heart

        b. Changes Affecting Stroke Volume

    3. Preload: Degree of Stretch

    4. Contractility

    5. Afterload: Back Pressure

    6. Regulation of Heart Rate

        a. Autonomic Nervous System Regulation

            1. Sympathetic

            2. Parasympathetic

            3. Bainbridge Reflex

        b. Chemical Regulation

            1. Hormones

            2. Ions

        c. Other Factors

    7. Homeostatic Imbalance of Cardiac Output

V. Developmental Aspects of the Heart (pp. 638-640)

  A. Embryological Development (p. 638)

    1. Normal Development

    2. Congenital Heart Defects

        a. Patent Ductus Arteriosus

        b. Tetralogy of Fallot

  B. Aging Aspects of the Heart (pp. 639-640)

    1. Sclerosis and Thickening of the Valve Flaps

    2. Decline in Cardiac Reserve

    3. Fibrosis of Cardiac Muscle

    4. Atherosclerosis

Review Items

1. Ventral body cavity (Chapter 1, pp. 17-19)

2. Mediastinum (Chapter 1, p. 17)

3. Cell junctions (Chapter 3, p. 64)

4. Serous membranes (Chapter 4, p. 124)

5. Cardiac muscle (Chapter 4, p. 126)

6. Squamous epithelium (Chapter 4, p. 105)

7. Collagen (Chapter 4, p. 115)

8. Sliding filament mechanisms (Chapter 9, pp. 255-257)

9. Membrane potential (Chapter 11, pp. 351-352)

10. Medullary control of cardiac rate (Chapter 12, p. 399)

11. Vagus nerve (Chapter 13, p. 441)

12. Neurotransmitters and cardiac rate (Chapter 14, pp. 469-470)

13. General sympathetic function (Chapter 14, p. 473)

14. General parasympathetic function (Chapter 14, p. 473)

Cross References

1. Atherosclerosis is described in detail in Chapter 20, p. 648.

2. Hydrostatic pressure and fluid movement are described in Chapter 20, p. 664.

3. Fetal circulation and the modifications which occur during birth are detailed in Chapter 29, p. 1022.

4. The function of the pulmonary arteries and veins are further described in Chapter 23.

5. Cardiac output and the regulation of blood pressure are described in Chapter 20, p. 654.

6. Vasomotor centers and the control of blood pressure is examined in Chapter 20, p. 654.

7. Baroreceptors' and chemoreceptors' function in blood pressure control is described in Chapter 20, pp. 654-656.

8. Blood volume and pressure control is detailed in Chapter 20, p. 644.

9. Blood flow to the heart is mentioned in Chapter 20, pp. 656-657.

Laboratory Correlations

1. Marieb, E. N. *Human Anatomy and Physiology Laboratory Manual: Cat and Fetal Pig Versions.* 3rd. ed. Benjamin/Cummings, 1989.

   Exercise 30: Anatomy of the Heart

   Exercise 31: Electrocardiography

2. Marieb, E. N. *Human Anatomy and Physiology Laboratory Manual: Brief Version.* 3rd. ed. Benjamin/Cummings, 1992.

   Exercise 27: Anatomy of the Heart

   Exercise 28: Conduction System of the Heart and Electrocardiography

Transparencies Index

Bassett Atlas Figures Index

INSTRUCTIONAL AIDS

Lecture Hints

1. Use plenty of visual material to emphasize the gross anatomy of the heart.

2. Point out that the visceral layer of the pericardium (epicardium) is the same as the outermost layer of the heart wall.

3. Emphasize that the pericardial cavity is not a huge cavity (as some students imagine) but a thin potential space filled with a thin film of fluid.

4. Display a single diagram of both pericardium and heart wall so that students get an overall perspective of construction.

5. Give a brief introduction of the purpose of the foramen ovale as a point of interest.

6. Clearly distinguish between atrium and auricle.

7. Students often have difficulty with the idea that the right and left pumps operate in series, not parallel. A schematic block diagram (with the right and left heart depicted as boxes with the lungs between) will often help students understand that the amount of blood pumped by each side of the heart must be equal.

8. Students often picture heart valves as substantial structures, rigid and mechanical. It is shocking for most to see actual valves for the first time. Emphasize the "flimsy" nature of the valves.

9. Clearly distinguish between atrioventricular and semilunar valve construction. Point out why the construction of each is so well-suited to its location.

10. Point out that there is not a muscular wall separating the atria and ventricles, but that the AV valves serve as the structure dividing the chambers. Students often picture the AV valve as a small circular structure embedded in a thick muscular septum.

11. Point out the difference between ion flow in a neuron and that in the cardiac pacemaker cell.

12. Emphasize that the pacemaker cells are cardiac muscle cells, just modified so that they spontaneously depolarize.

13. Clearly distinguish between the basic rate set by the conduction system of the heart and the acceleratory or inhibitory controls (sympathetic and parasympathetic) set by the medulla.

14. Emphasize that the ECG is the measurement of the total electrical activity of the heart at the surface of the body. Students often wonder why the ECG does not look like an action potential.

15. When discussing ventricular systole and diastole, reinforce the definitions of root words so that students can think critically about meanings rather than memorize terminology, e.g., isovolumetric means "same volume."

Demonstrations/Activities

1. Audio-visual materials of choice.

2. Use a torso model and/or dissected animal model to show the position of the heart in the chest cavity.

3. Use a heart model and/or dissected animal model to show the structure of the heart.

4. Use a simple pump apparatus to demonstrate the work of the heart (pumping action).

5. Obtain and play a recording of normal and abnormal heart sounds to accompany your presentation of valve function and malfunction. ("Interpreting Heart Sounds" is available on free loan from local chapters of the American Heart Association.)

6. If recording equipment is available, demonstrate an EKG.

Critical Thinking/Discussion Topics

1. Relate the functioning of the heart to the functioning of a water pump. Include problems associated with low blood pressure going to the heart and high pressure leaving the heart.

2. Discuss the significance of the heart's pacemaker. Include the use and roles of intermittent pacemakers as well as other pacemaker types.

3. Discuss the signs of impending heart attack.

4. Discuss the need for CPR training.

5. Discuss the lifestyle necessary for a healthy heart.

6. Discuss the steps in recovery from a heart attack.

7. Discuss the significance of ventricular fibrillation as opposed to atrial fibrillation.

8. Discuss the role of cardiac muscle in ejecting blood from the ventricles as opposed to ejecting blood from the atria.

9. Why is the wall of the left ventricle much more substantial than the right ventricle?

10. Is there a difference in the amount of blood pumped per beat by each ventricle? Explain.

11. Would a semilunar type valve work between atria and ventricles? Why are chordae tendinae not used in the semilunar valves? Discuss why the atrioventricular and semilunar valves are constructed the way they are.

12. How would heart function change if cells of the AV node depolarized at a faster rate than SA node cells?

13. What would happen to the heart (and the rest of the body) over a period of time if a partial blockage of the aortic semilunar valve occurred?

Library Research Topics

1. Research the role of antihypertensive drugs on the action of the heart.

2. Research the alternatives to coronary bypass operations.

3. Research the known effects of street drugs on heart activity.

4. Research the effect of smoking on the heart and its function.

5. Research the criteria used for heart transplants and their success rate.

6. Research the status of artificial hearts and their problems.

7. Research the effect of exercise on heart function.

Audio-Visual Aids/Computer Software

Videotapes

1. The Living Body: Two Hearts That Beat as One (FHS, VHS or BETA). Describes the structure and functioning of the heart. Analyzes the three basic components of the heart: muscle, valves, and pacemaker, and shows how each one contributes to the demands of daily life.

2. Your Body (PLP, CH-140202, VHS)

3. Circulation — Blood; the heart; arteries and veins; lymphatics (EI, 475-2354V)

Computer Software

1. The Heart Stimulator (PLP, Apple). Uses high-resolution color graphics and animation to depict blood flowing through the heart as it beats. It is a student interactive program.

2. Heart Abnormalities and EKG's: A Simulation (PLP, Apple). Two modes of function: demonstration and tutorial. Both models show normal and abnormal EKGs, and students are asked to compare them.

3. Your Heart (EIL, Apple, IBM, 1984). A self-paced tutorial program featuring crisp, colorful, high-resolution graphics introducing and reviewing the basic structures of the heart and the flow of blood through them.

4. Dynamics of the Human Circulatory System (EI, C 3053A, Apple; C 3053M, IBM)

5. The Body Electric — EEG, EMG, ECG (QUE, HRM577A, Apple)

6. Cardiac Muscle Mechanics (QUE, COM4204B, IBM)

7. Cardiovascular System — Two-part program covers the basic anatomy of the heart and blood vessels (PLP, CH-182011, Apple; CH-182012, IBM)

8. The Human Systems: Series 1 — An overview of cells, tissues, organs, and systems. Also covered are the digestive system, blood, and the circulatory system. (PLP, CH-920001, Apple; CH-920004, IBM)

See Guide to Audio-Visual Resources on page 339 for key to AV distributors.

LECTURE ENHANCEMENT MATERIAL

Clinical and Related Terms

1. Arrythmia — disturbance of cardiac rhythm.

2. Commissurotomy — an operation to widen the opening in a heart valve that has become narrowed by scar tissue.

3. Compensation — a change in the circulatory system to compensate for some abnormality.

4. Endocarditis — inflammation of the internal lining of the heart.

5. Heart block — impairment of conduction in heart excitation.

6. Myocardial infarction — death of heart muscle cells due to oxygen deficiency.

7. Myocardial ischemia — reduced blood supply to the heart muscle.

8. Pericarditis — inflammation of the pericardium.

9. Sinus rhythm — normal cardiac rhythm regulated by the SA node.

Disorders/Homeostatic Imbalances

1. Bacterial Endocarditis — an infection of a heart valve. It is most often seen as a complication of rheumatic heart disease, but may occur with any valvular heart disease. Fibrin thrombi may form on the surface of a valve, and bacteria become trapped. In some cases the thrombi become dislodged and are carried as emboli to other areas of the body, with the possibility of producing infarcts. The bacteria may occasionally lodge on normal valves and cause severe damage and/or destruction. Prophylactic antibiotic therapy is indicated for individuals with valvular defects prior to tooth extraction or any surgical procedure.

2. Calcific Aortic Stenosis — a condition which occurs as a degenerative change in elderly persons. The valve becomes so rigid that it is unable to open properly.

3. Congenital Bicuspid Aortic Valve — a nonrheumatic aortic stenosis in which the aortic valve shows two rather than three cusps. Such a valve can function normally for a time but eventually thickens and may even become calcified. Once calcified, the condition is called aortic stenosis, secondary to bicuspid aortic valve.

4. Congenital Heart Disease — refers to defects in the heart that occurred during embryologic and fetal development. Involves defective communication between cardiac chambers, malformation of valves, and malformation of septa.

5. Cor Pulmonale — refers to hypertrophy of the right ventricle due to hypertension in the pulmonary circulation. Disorders such as those involving the respiratory centers in the medulla and pons, lung and lung airway diseases, thorax deformities, and neuromuscular disorders can cause lung hypoxia. Tumors, aneurysms, and diseased blood vessels can cause decreased blood flow in the lungs. Reduction in blood flow and increased resistance in the lungs can bring about pulmonary hypertension. This causes increased pressure in pulmonary arteries and right ventricle hypertrophy.

6. Coronary Heart Disease — a condition resulting from arteriosclerosis of the large coronary arteries. Major risk factors for coronary heart disease are high blood lipid levels, high blood pressure, cigarette smoking, and diabetes. This condition can result in myocardial ischemia, myocardial infarction, or myocardial necrosis.

7. Cyanotic Congenital Heart Disease — refers to the inability of the individual to get adequate blood oxygenation due to extensive cardiac abnormalities. The cardiac defects cause blood to be shunted away from the lungs. An example would be the child called a "blue baby" due to its cyanotic color.

8. Heart Attack — a cessation of normal cardiac contraction (a cardiac arrest) or an actual necrosis of heart muscle (a myocardial infarction). Triggering mechanisms include sudden blockage of a coronary vessel (coronary thrombosis), hemorrhage into an atheromatous plaque, arterial spasm, or suddenly increased myocardial oxygen requirements.

9. Mitral-Valve Prolapse — refers to the stretching of the valve leaflets due to degeneration of connective tissue. This results in prolapse into the left atrium and possibly mitral insufficiency. The chordae tendineae may rupture due to the stress. The heart, under these conditions, is predisposed to arrhythmia.

10. Myocardial Infarct — the actual death (necrosis) of heart muscle due to severe ischemia. Most frequently the muscle of the left ventricle and septum is involved. Complications include disturbances of cardiac rhythm (arrhythmias), heart failure, cardiac rupture, intracardial thrombi, pericarditis, papillary muscle dysfunction, and ventricular aneurysm. Arrhythmias are due to the extreme irritability of the heart muscle adjacent to the infarct. If the conduction system of the heart is damaged, then a heart block can occur, sometimes requiring a pacemaker to maintain normal heart function.

11. Myocardial Ischemia — a condition in which the coronary arteries have become obstructed, resulting in a reduced blood supply available to the heart muscle. A clinical symptom of ischemic heart disease is angina pectoris, a pain in the chest. With severe and prolonged cardiac ischemia, a heart attack may occur. Myocardial ischemia may lead to cardiac arrhythmias due to increased irritability of myocardial tissue.

12. Rheumatic Heart Disease — A complication of rheumatic fever, due to scarring of the heart valves subsequent to the healing of a rheumatic infection. The mitral and aortic valves are primarily affected. Prevention involves prompt treatment of beta-streptococcal infection. Treatment may include valve replacement.

13. Tetralogy of Fallot — refers to a common cardiac abnormality consisting of four defects: (1) pulmonary stenosis or narrowing, (2) a ventricular septal defect, (3) an overriding aorta, and (4) hypertrophy (thickening) of the right ventricle. Surgery may correct some or all of the defects.

14. Valvular Heart Disease — involves abnormalities of the heart valves, especially the mitral and aortic. One of the leading causes of this condition is rheumatic fever, a hypersensitivity reaction to the streptococcus antigens.

15. Ventricular Fibrillation — an arrhythmia of the heart. It is the most common cause of cardiac arrest and sudden death in patients with coronary heart disease. In this condition, the ventricles are unable to force blood into the arteries due to random firing of electrical impulses to the cardiac muscles of the ventricles. As a result, the muscles contract at rapid and random intervals, rather than contracting as a unit.

16. Ventricular Septal Defect — refers to the failure of a partition to form between atria or ventricles, causing aseptal defect. A transposition of the great arteries may occur with the aorta arising from the right ventricle rather than from the left. One example is a patent ductus arteriosus.

ANSWERS TO END-OF-CHAPTER QUESTIONS

Multiple Choice/Matching

1.	a	4.	b	7.	c
2.	c	5.	b	8.	d
3.	b	6.	b	9.	b

Short Answer Essay Questions

10. The heart is enclosed within the mediastinum. It lies anterior to the vertebral column and posterior to the sternum. It tips slightly to the left. (p. 613)

11. The pericardium has two layers, a fibrous and a serous layer. The outer fibrous layer is a fibrous connective tissue that protects the heart and anchors it to surrounding structures. The inner serous layer (squamous epithelial cells) lines the fibrous layer as the parietal serous pericardium and at the base of the heart continues over the heart surface as the visceral serous pericardium. The visceral serous pericardium is the outermost layer of the heart wall, i.e., the epicardium. (pp. 613-614)

12. Blood that enters the right atrium on its way to the left atrium is in the pulmonary circuit. The path is as follows: right atrium, right ventricle, pulmonary trunk, right and left pulmonary arteries, lungs, pulmonary veins, left atrium. (pp. 615-618)

13. a. The coronary arteries are actively delivering blood to the myocardium when the heart is relaxed. The coronary vessels are compressed and ineffective in blood delivery when the ventricles are contracting.

    b. The major branches of the coronary arteries and the areas they serve are as follows. The left coronary artery runs toward the left side of the heart and divides into the anterior interv entricular artery, which supplies blood to the interventricular septum and anterior walls of both ventricles and the circumflex artery, which serves the left atrium and the posterior walls of left ventricle. The right coronary artery courses to the right side of the heart, where it divides into the marginal artery, which serves the myocardium of the lateral part of the right side of the heart and the posterior interventricular artery, which runs to the heart apex and supplies the posterior ventricular walls. (pp. 621-622)

14. A longer refractory period of cardiac muscle is desirable because it prevents the heart from going into prolonged or tetanic contractions which would stop its pumping action. (p. 626)

15. a. The elements of the intrinsic conduction system of the heart, beginning with the pacemaker, are the SA node or pacemaker, AV node, AV bundle, right and left bundle branches, and Purkinje fibers.

    b. This system functions to initiate and distribute impulses throughout the heart so that the myocardium depolarizes and contracts in an orderly, sequential manner from atria to ventricles. (pp. 627-628)

16. The P wave results from impulse conduction from the SA node through the atria. The QRS complex results from ventricular depolarization and precedes ventricular contraction. Its shape reveals the different size of the two ventricles and the time required for each to depolarize. The T wave is caused by ventricular repolarization. (p. 630)

17. The cardiac cycle includes all events associated with the flow of blood through the heart during one complete heartbeat. One cycle includes a period of ventricular filling (mid-to-late diastole at the end of which atrial systole occurs), ventricular systole, and isovolumetric relaxation (early diastole). (pp. 631-633)

18. Cardiac output is the amount of blood pumped out by each ventricle in one minute. It can be calculated by the following equation: cardiac output = heart rate x stroke volume. (p. 634)

19. The Frank-Starling Law explains that the critical factor controlling stroke volume is the degree of stretch of the cardiac muscle cells just before they contract. The important factor in the stretching of cardiac muscle is the amount of blood returning to the heart and distending its ventricles. (pp. 634-635)

20. The common function of the foramen ovale and the ductus arteriosus in a fetus, is to allow blood to bypass the pulmonary circulation. The problem that exists, if these shunts remain patent after birth, is that the opening prevents adequate gas exchange, O_2 loading and CO_2 unloading, in the pulmonary circulation. (p. 638)

Critical Thinking and Application Questions

1. Cardiac tamponade is compression of the heart due to accumulation of blood or inflammatory fluid in the pericardial sac. Such compression reduces the ability of the heart to beat and act as an effective pump, leading to inadequate blood delivery (which results in ischemia and cyanosis), and ultimately cardiogenic shock.

2. a. To auscultate the aortic valve, place the stethoscope over the second intercostal space at the right sternal margin. To auscultate the mitral valve, place the stethoscope over the heart apex, in the fifth intercostal space in line with the middle of the clavicle.

   b. These abnormal sounds would be heard most clearly during ventricular diastole for the aortic valve and during atrial systole for the mitral valve.

   c. An incompetent valve has a swishing sound after the valve has supposedly closed. A stenosed valve has a high-pitched sound when blood is being forced through it during systole just before valve closure. (pp. 633-634)

3. Failure of the left ventricle (which pumps blood to the body) can result in chest pain due to dying or dead ischemic cardiac cells; pale, cold skin due to lack of circulation of blood from blocked ventricular contraction; and moist sounds in the lower lungs due to high pressure and pooling of blood in the pulmonary circulation because of nonfunction of the ventricle. (p. 637)

Key Figure Questions

Figure 19.1 The medial cavity of the thorax within which the heart, great vessels, and trachea are found. **Figure 19.4** Left ventricle. It is the systemic pump which has to pump blood through the entire systemic circulation against high resistance. **Figure 19.9** Both valves operate in response to blood attempting to flow backwards. In the AV valves, rising ventricular pressure forces the valve flaps upward into a closed position, a condition maintained by the chordae tendonae. In the semilunar valves, blood flowing back toward the heart from the large vessels fills the valve cusps, closing the valves. No device is used to hold the valve in the closed position. **Figure 19.14** Calcium entry.

SUGGESTED READINGS

---

1. Baas, L., and C. Kretten. "Valvular Heart Disease." *RN* (Nov. 1987): 30-37.

2. Bendit, E.P. "The Origin of Atherosclerosis." *Scientific American* (Feb. 1977).

3. Berne, R.M., and M.N. Levy. *Cardiovascular Physiology*, 4th ed. St. Louis: C.V. Mosby Co., 1981.

4. Cantin, M., and J. Genest. "The Heart as an Endocrine Gland." *Scientific American* 254 (Feb. 1986): 76.

5. Edwards, D.D. "Aspirin Cuts Risk of First Heart Attack." *Science News* 133 (Jan. 1988): 68.

6. Fackelmann, K.A. "Herpes and Heart Disease." *Science News* 143 (Apr. 1993): 216.

7. Fackelmann, K.A. "Sex and the Risk of Heart Attack." *Science News* (Nov. 1993): 332.

8. Grady, Denise. "Can Heart Disease Be Reversed?" *Discover* (March 1987): 55.

9. Hammond, Cecile. "ECGs Made Easier Than Ever." *RN* 42 (Oct. 1979): 55.

10. Harken, A.M. "Surgical Treatment of Cardiac Arrhythmias." *Scientific American* 269 (July 1993): 68.

11. Jarvik, Robert K. "The Total Artificial Heart." *Scientific American* 244 (Jan. 1981): 74.

12. Johansen, Kaj. "Aneurysms." *Scientific American* 247 (July 1982): 110.

13.   Johnson, G.T. "Heart Arrhythmias." *The Harvard Medical School Health Letter*, Jan. 1982.

14.   "Heart Attacks and Counter Attacks." *The Harvard Medical School Health Letter,* Feb. 1982.

15.   Peterson, Ivars. "Heart Flow." *Science News* 130 (Sept. 1986): 204.

16.   Raloff, J. "Vitamin E Appears to Cut Heart Disease Risk." *Science News* 143 (May 1993): 327.

17.   Robinson, T.F., S.M. Factor, and E.H. Sonnenblick. "The Heart as a Suction Pump." *Scientific American* 254 (June 1986): 84.

18.   Silberner, J. "Anatomy of Atherosclerosis." *Science News* (March 1985).

19.   "Cocaine Cardiology: Problems, Mysteries." *Science News* 131 (Jan. 1987): 69.

20.   "Giving Hearts Extra Muscle." *Science News* 129 (May 1986): 284.

21.   "The Heart and Heredity." *Science News* 129 (Feb. 1986): 126.

22.   Winfree, Arthur R. "Sudden Cardiac Death: A Problem in Topology." *Scientific American* 248 (May 1983): 144.

20

The Cardiovascular System: Blood Vessels

Chapter Preview

This chapter deals with the blood vessels. These structures form a closed delivery system that begins and ends at the heart. Blood vessels are dynamic structures that constrict, relax, and even proliferate, as demanded by the changing needs of the body. This chapter examines the structure and function of the circulatory pathways of the body.

INTEGRATING THE PACKAGE

Suggested Lecture Outline

I. Overview of Blood Vessel Structure and Function (pp. 643-650)
   A. Structure of Blood Vessel Walls (Fig. 20.1, p. 644)
   B. Arterial System (p. 645)
      1. Elastic Arteries
      2. Muscular Arteries
      3. Arterioles
   C. Capillaries (Fig. 20.1, p. 644)
      1. Types of Capillaries
         a. Continuous Capillaries
         b. Fenestrated Capillaries (Fig. 20.2, p. 646)
         c. Sinusoids
      2. Capillary Beds (Fig. 20.3, p. 647)
   D. Venous System (pp. 647-650)
      1. Venules
      2. Veins
      3. Varicose Veins
   E. Vascular Anastomoses (p. 650)
      1. Arterial Anastomoses
      2. Arteriovenous Anastomoses
      3. Venous Anastomoses

II. Physiology of Circulation (pp. 650-667)
   A. Introduction to Blood Flow, Blood Pressure, and Resistance (pp. 650-651)
      1. Blood Flow
      2. Blood Pressure

3. Peripheral Resistance
   a. Blood Viscosity
   b. Blood Vessel Length
   c. Blood Vessel Diameter
4. Relationship Between Blood Flow, Blood Pressure, and Resistance

B. Systemic Blood Pressure (pp. 650-653)
1. Arterial Blood Pressure
2. Capillary Blood Pressure
3. Venous Blood Pressure
   a. Factors Aiding Venous Return
      1. Respiratory Pump
      2. Muscular Pump (Fig. 20.5, p. 653)
4. Maintaining Blood Pressure
   a. Cardiac Output
   b. Peripheral Resistance
   c. Blood Volume
5. Short-Term Mechanisms: Neural Controls
   a. Role of the Vasomotor Center
      1. Vasomotor Fibers
   b. Baroreceptor-Initiated Reflexes (Fig. 20.7, p. 655)
   c. Chemoreceptor-Initiated Reflexes
   d. Higher Brain Centers
6. Short-Term Mechanisms: Chemical Controls
   a. Adrenal Medulla Hormones
   b. Atrial Natriuretic Factor
   c. Antidiuretic Hormone
   d. Endothelium-Derived Factors
   e. Nitric Oxide (NO)
   f. Alcohol
   g. Inflammatory Chemicals
7. Long-Term Mechanisms: Renal Regulation
8. Monitoring Circulatory Efficiency
   a. Taking a Pulse
   b. Measuring Blood Pressure
9. Variations in Blood Pressure
   a. Hypotension
   b. Hypertension

C. Blood Flow (pp. 660-667)
1. Velocity of Blood Flow (Fig. 20.12, p. 661)
2. Autoregulation: Local Regulation of Blood Flow
   a. Metabolic Controls
   b. Myogenic Controls
   c. Long-Term Autoregulation

    3. Blood Flow in Special Areas

        a. Skeletal Muscles

        b. The Brain

        c. The Skin

        d. The Lungs

        e. The Heart

    4. Capillary Dynamics

        a. Exchanges of Respiratory Gases and Nutrients

        b. Fluid Movements

            1. Hydrostatic Pressures

            2. Osmotic Pressures

            3. Hydrostatic-Osmotic Pressure Interactions

    5. Circulatory Shock

        a. Hypovolemic Shock

        b. Vascular Shock

        c. Cardiogenic Shock

        d. Orthostatic Intolerance

III. Circulatory Pathways (pp. 667-690; Table 20.2, pp. 670-671)

  A. Pulmonary Circulation (Fig. 20.15, p. 670)

  B. Systemic Circulation (Fig. 20.16, p. 671)

    1. Systemic Arteries

        a. Aorta (Table 20.2, p. 673)

        b. Arteries of the Head and Neck (Fig. 20.18, p. 675)

        c. Arteries of the Upper Limbs and Thorax (Fig. 20.19, p. 677)

        d. Arteries of the Abdomen (Fig. 20.20, p. 679)

        e. Arteries of the Pelvis and Lower Limbs (Fig. 20.21, p. 681)

    2. Systemic Veins

        a. Venae Cavae and Their Major Tributaries

        b. Major Veins of the Systemic Circulation (Fig. 20.22, p. 683)

        c. Veins of the Head and Neck (Fig. 20.23, p. 685)

        d. Veins of the Upper Limbs and Thorax (Fig. 20.24, p. 687)

        e. Veins of the Abdomen (Fig. 20.25, p. 689)

        f. Veins of the Pelvis and Lower Limbs (Fig. 20.26, p. 690)

IV. Developmental Aspects of the Blood Vessels (p. 667)

  A. Embryological Development

    1. Angioblasts

    2. Foramen Ovale

    3. Ductus Arteriosus

  B. Aging Problems

Review Items

1. Tight junctions (Chapter 3, p. 64)

2. Diffusion (Chapter 3, p. 66)

3. Osmosis (Chapter 3, p. 67)

4. Simple squamous epithelium (Chapter 4, p. 105)

5. Dense connective tissue (Chapter 4, p. 118)

6. Elastic connective tissue (Chapter 4, p. 120)

7. Smooth muscle (Chapter 9, pp. 275-278)

8. Sensory receptors (Chapter 13, pp. 426-429)

9. Blood-brain barrier (Chapter 12, pp. 405-406)

10. Medulla (Chapter 12, p. 399)

11. Hypothalamus (Chapter 12, p. 395)

12. Sympathetic control (Chapter 14, p. 472)

13. Epinephrine, norepinephrine (Chapter 14, p. 470)

14. Atrial natriuretic factor (Chapter 17, p. 568)

15. Blood characteristics (Chapter 18, p. 585)

16. Cardiac output (Chapter 19, pp. 634-637)

17. Cardioinhibitory and cardioacceleratory centers (Chapter 19, p. 629)

Cross References

1. Antidiuretic hormone and aldosterone function in blood pressure control is described in detail in Chapter 26.

2. The effects of alcohol on antidiuretic hormone (and, therefore, blood pressure) is explained in Chapter 26, p. 916.

3. The renal mechanism of electrolyte balance is covered in Chapter 27, pp. 934-940.

4. Fetal shunts are described in Chapter 29, p. 1022.

5. Varicose veins and the effects on the pregnant mother are detailed in Chapter 29, p. 1019.

6. The function of fenestrated capillaries in glomerular filtration is detailed in Chapter 26, pp. 900, 904.

7. The function of arterioles in the control of glomerular filtration is presented in Chapter 26, pp. 907-908.

8. An example of vascular resistance and autoregulation of blood flow is presented in Chapter 26, pp. 903, 906-908.

9. A specific body example of capillary dynamics (in this case, filtration) is detailed in Chapter 26, pp. 906-908.

10. The renin-angiotensin mechanism is further explained in Chapters 26, p. 908, and 27, p. 934.

11. Blood flow regulation in the control of body temperature is described in Chapter 25, pp. 887-890.

12. Splanchnic circulation is further described in Chapter 24.

13. The relationship between blood capillaries and lymphatic capillaries is examined in Chapter 21, pp. 694-695.

14. Factors affecting fluid movement through capillary membranes is further explained in Chapter 21, pp. 694-695.

15. The factors that aid venous return are the same as the factors that aid lymph return described in Chapter 21, p. 697.

16. Edema is covered in Chapter 27, pp. 932-934.

Laboratory Correlations

1. Marieb, E. N. *Human Anatomy and Physiology Laboratory Manual: Cat and Fetal Pig Versions.* 3rd. ed. Benjamin/Cummings, 1989.

   Exercise 32: Anatomy of Blood Vessels

   Exercise 33: Human Cardiovascular Physiology — Blood Pressure and Pulse Determinations

   Exercise 34: Frog Cardiovascular Physiology

2. Marieb, E. N. *Human Anatomy and Physiology Laboratory Manual: Brief Version.* 3rd. ed. Benjamin/Cummings, 1992.

   Exercise 29: Anatomy of Blood Vessels

   Exercise 30: Human Cardiovascular Physiology — Blood Pressure and Pulse Determinations

   Exercise 31: Frog Cardiovascular Physiology

Transparencies Index

Bassett Atlas Figures Index

Slide Number	Figure Number	Description
62	6.10	Triceps of the arm
72	7.4A,B	Inguinal area
73	7.5	Inguinal area
74	7.6A,B	Thigh, posterior view
76	7.8	Popliteal fossa
84	7.16A,B	Foot, medial view

INSTRUCTIONAL AIDS

Lecture Hints

1. Emphasize the smooth transition in wall structure from artery → arteriole → capillary → venule → vein.

2. Rather than strictly lecturing about muscular and elastic arteries, capillaries, etc., try giving basic wall structure and ask the class what the logical functions of these vessels could be, given their construction.

3. Point out that the specific construction of the capillary wall determines permeability, and that, depending on location, capillaries can be highly selective to the substance that passes (e.g., in the brain), or relatively nonselective (in the kidney).

4. When discussing blood flow, introduce each factor (pressure, resistance, etc.), then relate all together logically so that students can see the dynamic nature of blood transport. Students often treat the vascular system as being a rigid set of tubes.

5. Spend some time with a series of diagrams depicting the muscular pump as a factor aiding venous return. Students are often confused as to how this mechanism works.

6. Work through the respiratory factor aiding venous return in conjunction with the split of the second heart sound. Students will grasp the concept more easily if the two ideas are related together.

7. Emphasize the importance of vasomotor tone. Students should be made aware that if the arteries were not in a constant state of partial contraction (during normal activity), there would be no mechanism to allow for vasodilation.

8. It is effective to diagram the reflex from baroreceptor to effector to illustrate the negative feedback loop involved in blood pressure control.

9. Mention the word "alcohol" and you instantly have class attention. Use this to your advantage to point out the effects of alcohol on ADH and blood pressure, then gradually work in a discussion of the renal mechanism of blood pressure control.

10. Since hypertension and its risks are of interest to most individuals, use a discussion of this topic to bring together important concepts regarding cardiac and vascular function, and mechanisms of blood pressure control.

11. Students often have difficulty with the idea that capillaries have more total cross-sectional area than arteries. It often helps to diagram a single artery about 5 mm in diameter and 10-20 capillaries about 1 mm in diameter to show the relationship between total area and the total number of vessels.

12. A discussion of the pulmonary circulation is a good place to recall the relative size of the left and right heart. Point out the short-distance, low-pressure route from right ventricle to left atrium.

13. It is difficult for some students to recognize that coronary circulation is highest when cardiac muscle is in diastole. Emphasize the elastic recoil of the aorta as the driving force, i.e., the aorta stores the force of ventricular contraction in its elastic connective tissue.

14. A thorough understanding of fluid movement at the capillary level is crucial for a complete understanding of how other systems function. Refer students to Chapter 3 for a review on osmosis and emphasize that this topic will be seen again in future systems.

15. Emphasize that arteries always carry blood away from the heart and veins always return blood to the heart. In other words, the name artery or vein is in relation to coming or going from the heart, not whether oxygenated or deoxygenated blood is carried. You may wish to mention the exceptions to the rule that arteries carry oxygenated blood (pulmonary trunk, arteries and pulmonary veins; umbilical arteries and umbilical vein).

16. By this time, most important aspects of cardiovascular function have been covered. The function of the cardiovascular system is simple to master if schematic diagrams are used. Test student knowledge by asking what possible consequences are likely if partial blockages occur in different parts of the system. For example, ask "If a partial blockage of the aorta developed (or aortic stenosis), what would be the possible systemic consequences?" Students should be able to describe reduced blood flow to the systemic circulation, LV hypertrophy and eventual failure, bicuspid failure, pulmonary edema, eventual venous congestion, etc. This is an excellent exercise in demonstrating the circular nature of this closed system, and that effects at one point in the system are likely to go full circle if not checked.

Demonstrations/Activities

1. Audio-visual materials of choice.

2. Use a torso model, circulatory system model, and/or dissected animal model to exhibit major circulatory pathways.

3. Use a short piece of cloth-wrapped garden hose to show the layers of the veins and arteries.

4. Use a short piece of soaker hose to illustrate a capillary as the functional unit of the circulatory system.

5. Use a small-mouth jar with a two-hole rubber stopper in it. Put a solid glass rod in one hole and a flexible rubber hose in the other, to illustrate the differences between arteries and veins.

6. Use a hand-held pinball machine to illustrate the one-way nature of valves in veins.

7. Use a long, not completely blown up, balloon to illustrate the effect of altering pressure in a closed system.

8. Demonstrate how apical and radial pulse are taken and have students practice on one another.

9. Have students practice finding important pressure points.

10. Demonstrate the auscultatory method of determining arterial blood pressure, and provide the necessary equipment (sphygmomanometers and stethoscopes) so that the students can practice on each other.

11. Use a model of human vasculature or an unlabeled acetate and ask students to call out the names of the vessels indicated.

Critical Thinking/Discussion Topics

1. Discuss the need for blood pressure monitoring in regard to hyper- or hypotension.

2. Discuss the significance of proper diet in maintaining normal blood flow.

3. Discuss why some coronary bypass surgeries have to be repeated.

4. Discuss how an artery loses its elasticity.

5. In terms of the mechanics of blood flow, discuss why a pulse is evident on the arterial side but not the venous side of circulation.

6. What factors retard venous return?

7. Why do water and dissolved solutes leave the bloodstream at the arteriole end of the capillary bed and enter the bloodstream at the venous end?

8. Why is the elasticity of the large arteries so important (or why is arteriosclerosis such a threat)?

9. Why does a body region develop edema after the lymphatics are moved from that area?

Library Research Topics

1. Research the role of diet in the clearance or obstruction of blood vessels.

2. Research the possible congenital defects of circulation resulting from the differences in the fetal and adult circulation.

3. Research the effect of untreated hypertension on kidney function.

4. Research the procedures and types of valves currently used in valve replacement surgery.

5. Research the various types of heart blockages and their significance.

6. Research the risk factors implicated in heart disease and what can be done to minimize the risk.

Audio-Visual Aids/Computer Software

Videotapes

1. Your Body, Part 2: Your Circulatory System (PLP, CH-140202, VHS)

2. The Human Body: What Can Go Wrong? — Circulatory System (PLP, CH-140501, VHS)

Computer Software

1. Cardiovascular Fitness Lab (PLP, Apple, IBM, C64). Using sophisticated software and heart rate sensor (included) to record pulse, students can perform numerous fitness experiments.

2. Osmo: Osmosis in Red Blood Cells (EI, C 4042A, Apple; C4024M, IBM)

3. Osmosis and Diffusion (EI, C4022, Apple)

4. Disk X: The Human Body, Part 1 (QUE, COM4239A, Apple; COM4239M, Macintosh; COM4239B, IBM)

5. Body Language: Study of Anatomy and Physiology — Cardiovascular System (PLP, CH-182011, Apple; CH-182012, IBM)

    See Guide to Audio-Visual Resources on page 339 for key to AV distributors.

LECTURE ENHANCEMENT MATERIAL

Clinical and Related Terms

1. Angiospasm — a muscular spasm in the wall of a blood vessel.

2. Arteriosclerosis — hardening and thickening of the walls of the smaller arteries.

3. Atheroma — a mass of plaque of degenerated, thickened arterial intima occurring in atherosclerosis.

4. Atherosclerosis — a common form of arteriosclerosis in which deposits of yellowish plaques (atheromas) containing cholesterol, lipoid material, and lipophages are formed within the intima and inner media of large- and medium-sized arteries.

5. Blood stasis — a stoppage or diminution of the flow of blood.

6. Claudication — pain and lameness or limping caused by defective circulation of the blood in the vessels of the limbs.

7. Cyanosis — a bluish discoloration, especially discoloration of skin and mucus membranes due to excessive concentration of reduced hemoglobin in the blood.

8. Embolectomy — removal of an embolus through an incision in a blood vessel.

9. Endarterectomy — removal of the inner wall of an artery to reduce an arterial occlusion.

10. Hypotensive — low blood pressure; most commonly used to describe an acute drop in blood pressure, as occurs in circulatory shock.

11. Normotensive — characterized by normal blood pressure.

12. Thrombophlebitis — usually a disorder of deep veins of lower extremities in which thrombi form (a blood clot formed in response to venous inflammation). Postoperative and bed patients are predisposed to the disorder.

Disorders/Homeostatic Imbalances

1. Aneurysm — a dilatation of the wall of an artery or an outpouching of a portion of the wall usually occurring as a result of arteriosclerosis.

2. Arterial Thrombosis — clots formed when platelets adhere to roughened surfaces on the arterial wall. The roughened surfaces are usually due to arteriosclerosis. Depending on the vessel involved, heart attacks, strokes, or gangrene of an extremity can occur.

3. Edema — swelling of tissues due to increased capillary permeability with low plasma protein levels. Causes include excess protein loss due to kidney failure, malnutrition, heart failure, localized venous obstruction, and lymphatic obstruction.

4. Intracardiac Thrombosis — a clot in the heart itself. Usual sites include atrial appendage (in heart failure), heart valves (secondary to valve injury), and left ventricles (secondary to infarct of dead muscle).

5. Pulmonary Embolism — obstruction of a pulmonary vessel due to a traveling blood clot, fat plaque, or some other obstructive material. Usually the lung is not infarcted if a large vessel is blocked since collateral blood flow is provided by bronchial arteries. Collateral circulation is not adequate in small vessels and lung infarction may occur if the vessels are blocked. Symptoms include chest pain, cough, and bloody sputum secondary to infarct.

6. Septic Pulmonary Emboli — emboli which occur in pelvic veins following pelvic surgery. Bacteria invade the thrombi which are then carried to the lungs and cause pulmonary infarct. The bacteria in the clot invade the infarct, forming a lung abscess.

7. Thrombi Due to Increased Blood Coagulability — clots formed due to an increase in blood coagulation factors resulting in blood that clots more readily. Estrogen in contraceptive pills have been shown to stimulate clotting factor synthesis increasing the possibility of both venous and arterial thrombi.

8. Varicose Veins — veins that have a weakened wall resulting in pooling of blood in the region. They are usually due to congenital weakness of a vein wall or valves predisposing to varicosities. They may occur if deeper veins are blocked or valves are damaged by thrombophlebitis, which diverts more blood flow to the superficial veins. Complications include stasis ulcers, rupture with bleeding, and thrombophlebitis. Varicose veins occur in the rectum (hemorrhoids) and accompanying cirrhosis of the liver (esophageal varices).

9. Venous Thrombosis — a clot in a vein. Predisposing factors include stasis of blood in the veins, varicose veins, and increased blood coagulability. Stasis of blood in veins is likely to occur during periods of long bed rest, or after a cramped position has been maintained for a long time. Weakness of vein walls (varicose veins) will also contribute to the problem. The major complication of venous thrombosis is the detachment of the clot from the vein wall and its subsequent movement to block major vessels and blood flow.

ANSWERS TO END-OF-CHAPTER QUESTIONS

Multiple Choice/Matching

1.	d	5.	e	9.	b
2.	b	6.	d	10.	a
3.	d	7.	c	11.	b
4.	c	8.	b	12.	c

Short Answer Essay Questions

13. Capillary anatomy is suited to the exchange of material between blood and interstitial fluid because its walls are very thin and devoid of muscle and connective tissue. (pp. 645-647)

14. Elastic arteries are the large, thick-walled arteries close to the heart. They have generous amounts of elastic tissue in all tunics, but especially in the tunica media. This elastic tissue enables them to withstand large pressure fluctuations by expanding when the heart contracts, forcing blood into them. They recoil as blood flows forward into the circulation during heart relaxation. They also contain substantial amounts of smooth muscle but are relatively inactive in vasoconstriction.

    Muscular arteries are medium- and smaller-sized arteries farther along in the circulatory pathway, that carry blood to specific body organs. Their tunica media contains proportionately more smooth muscle and less elastic tissue than that of elastic arteries, but they typically have an elastic lamina on each face of the media. They are more active in vasoconstriction and are less distensible.

    Arterioles are the smallest of the arterial vessels. The smallest, terminal arterioles, feed directly into the capillary beds. The larger arterioles exhibit all three tunics and their tunica media is chiefly smooth muscle with a few scattered elastic fibers. The walls of the smaller arterioles are little more than smooth muscle cells that coil around the tunica intima lining. When arterioles constrict, the tissues served are largely bypassed; when the arterioles dilate, blood flow into the local capillaries increases dramatically. (p. 645)

15. The equation showing the relationship between peripheral resistance, blood flow, and blood pressure is as follows: blood flow equals a change in blood pressure between two points in the circulation divided by the resistance. (p. 650)

16. a. Blood pressure is the force per unit area exerted on the wall of a blood vessel by its contained blood. Systolic pressure is the pressure that occurs during systole when the aortic pressure reaches its peak. Diastolic pressure is the pressure that occurs during diastole when aortic pressure drops to its lowest level. (pp. 650-652)

    b. The normal blood pressure for a young adult is between 110 and 140 mm Hg systolic and between 75 and 80 mm Hg diastolic. (p. 659)

17. The neural controls responsible for controlling blood pressure operate via reflex arcs chiefly involving the following components: baroreceptors and the associated afferent fibers, the vasomotor center of the medulla, vasomotor (efferent) fibers, and vascular smooth muscle. The neural controls are directed primarily at maintaining adequate systemic blood pressure and altering blood distribution to achieve specific functions. (pp. 654-656)

18. Changes in the velocity in different regions of the circulation reflect the cross-sectional area of the vascular tubes to be filled. Since the cross-sectional area is least in the aorta and greatest in the capillaries, the blood flow is fastest in the aorta and slowest in the capillaries. (pp. 660-661)

19. Nutrient blood flow to the skin is controlled by autoregulation in response to the need for oxygen; whereas blood flow for regulating body temperature is controlled by neural intervention, i.e., the sympathetic nervous system. (pp. 660-663)

20. When one is fleeing from a mugger, blood flow is diverted to skeletal muscles from other body systems not in direct need of large volumes of blood. Blood flow increases in response to acetylcholine release by sympathetic vasodilator fibers and/or epinephrine binding to beta receptors of vascular smooth muscles in the skeletal muscles, and virtually all capillaries open to accommodate the increased flow. Systemic adjustments, mediated by the sympathetic vasomotor center, occur to ensure that increased blood volume reaches the muscles. Strong vasoconstriction of the digestive viscera diverts blood away from those regions temporarily, ensuring that an increased blood supply reaches the muscles. Blood-borne epinephrine enhances blood glucose levels, alertness, and metabolic rate. The major factor determining how long muscles can continue vigorous activity is the ability of the cardiovascular system to deliver adequate oxygen and nutrients. (pp. 660-662)

21. Nutrients, wastes, and respiratory gases are transported to and from the blood and tissue spaces by diffusion. (p. 664)

22. a. The veins draining the digestive viscera contribute to the formation of the hepatic portal circulation. The most important of these are the superior and inferior mesenteric veins and the splenic.

    b. The function of the hepatic portal circulation is to deliver blood laden with nutrients from the digestive organs to the liver.

    c. The portal circulation is a "strange" circulation because it consists of veins draining into capillaries which drain into veins again. (p. 688)

23. a. The text states that postcapillary venules function "more like capillaries" (p. 647), meaning that exchanges of small molecules between the blood and the surrounding tissue fluid occur across these venules. Furthermore, inflammatory fluid and leukocytes leave the postcapillary venules just as they exit the capillaries.

    b. Whereas capillaries consist only of an endothelium, postcapillary venules have scattered fibroblasts on their endothelium layer.

Critical Thinking and Application Questions

1. His fears are unfounded because there is no artery in the exact midline of the anterior forearm. Most likely the median vein of the forearm has been cut but it is a fairly small superficial vein and venous bleeding is much slower than arterial bleeding. (pp. 686-687)

2. The compensatory mechanisms of Mrs. Johnson induce an increase in heart rate and an intense vasoconstriction, which allows blood in various blood reservoirs to be rapidly added to the major circulatory channels. (pp. 665-666)

3. If the sympathetic nerves are severed, vasoconstriction in the area will be reduced and vasodilation will occur. Therefore, blood flow to the area will be enhanced. (p. 654)

4. An aneurysm is a balloon-like outpocketing of a blood vessel that places the vessel at risk for rupture. In this case, the aneurysm was so large that it was pressing on the brain stem and cranial nerves, threatening to interfere with the functions of these structures. The surgeons removed the ballooned section of artery and sewed a section of strong tubing in its place.

Key Figure Questions

Figure 20.1 It forms an exceedingly thin diffusion membrane for the exchange of substances between the blood and the interstitial fluid. **Figure 20.7** Arterial blood pressure initially drops, resulting in stimulation of the vasomotor and cardioacceleratory centers. The result is increased heart rate and vasoconstriction, which increases arterial blood pressure, restoring homeostasis.

SUGGESTED READINGS

1. Arehart-Treichel, Joan. "Eating Your Way out of High Blood Pressure." *Science News* 123: 233.

2. Benditt, Earl P. "The Origin of Atherosclerosis." *Scientific American* 236 (Feb. 1977): 74.

3. Bower, B. "Signs and Sounds of High Blood Pressure." *Science News* 129 (Feb. 1986): 116.

4. Bredt, D.S., and S.H. Snyder. "Biological Roles of Nitric Oxide." *Scientific American* (1993): 22.

5. Brown, M.S., and J.L. Goldstein. "How LDL Receptors Influence Cholesterol and Atherosclerosis." *Scientific American* (Nov. 1984).

6. Donald, D.E., and J.T. Shepherd. "Autonomic Regulation of the Peripheral Circulation." *Annual Review of Physiology* 42 (1980): 419.

7. Edwards, D.D. "Lipoprotein Findings May Solve One Riddle . . . and Lose Another." *Science News* 132 (Nov. 1987): 311.

8. Fackelmann, K.A. "Exercise May Help To Keep Arteries Stringy." *Science News* 144 (Oct. 1993): 246.

9. Fackelmann, K.A. "Nitirc Oxide Flow Found in Hypertension." *Science News* 143 (May 1993): 327.

10. Gore, R.W., and P.F. McDonagh. "Fluid Exchange Across Single Capillaries." *Annual Review of Physiology* 42 (1980): 337.

11. Hazzard, William R. "Atherosclerosis: Why Women Live Longer Than Men." *Geriatrics* 40 (Jan. 1985): 1042.

12. Hilton, S.M., and K.M. Spyer. "Central Nervous Regulation of Vascular Resistance." *Annual Review of Physiology* 42 (1980): 399.

13. Katzir, A. "Optical Fibers in Medicine." *Scientific American* 260 (May 1989): 120-125.

14. Lancaster, J.R. "Nitric Oxide in Cells." *American Scientist* 80 (May-June 1992): 248.

15. Loscalzo, J., D.S. Singel, and J.S. Stamler. "Biochemistry of Nitric Oxide and Its Redox-Activated Forms." *Science* 258 (Dec. 1992): 189.

16. Olsson, R.A. 1981. "Local Factors Regulating Cardiac and Skeletal Muscle Blood Flow." *Annual Review of Physiology* 43 (1981): 385.

17. Silberner, J. "Artery Clogging and APO-B." *Science News* 131 (Feb. 1987): 90.

18. Snyder, S.H. "Nitric Oxide: First in a New Class of Neurotransmitters." *Science* 257 (July 1992): 494.

19. Weisburd, S. "Magnesium Plays a Role in Hypertension." *Science News* 125 (Mar. 1984): 182.

20. Wood, J. Edwin. "The Venous System." *Scientific American* 218 (Jan. 1968).

21. Zweifach, Benjamin W. "The Microcirculation of the Blood." *Scientific American* (Jan. 1959).

21

The Lymphatic System

Chapter Preview

This chapter introduces the lymphatic system. Lymphatic vessels are essential for removing excess interstitial fluid from tissues and returning the fluid to venous circulation. Lymphatic structures and vessels are important in immune system function. The organs of this system house cells that function in specific and nonspecific resistance to disease.

INTEGRATING THE PACKAGE

Suggested Lecture Outline

I. Lymphatic Vessels (pp. 694-697)
    A. Distribution and Structure of Lymphatic Vessels (pp. 694-697)
        1. Lymph Capillaries (Fig. 21.1, p. 695)
        2. Lymphatic Collecting Vessels
    B. Lymph Transport (p. 697)

II. Lymphoid Cells, Tissues, and Organs: An Overview (pp. 697-698)
    A. Lymphoid Cells (p. 697)
    B. Lymphoid Tissue (p. 698)
    C. Lymphoid Organs (p. 698, Fig. 21.4, p. 698)

III. Lymph Nodes (pp. 699-700)
    A. Structure (pp. 699-700)
    B. Circulation (pp. 699-700)

IV. Other Lymphoid Organs (pp. 700-704)
    A. Spleen (Fig. 21.6, p. 700)
        1. Functions
        2. Structure
    B. Thymus (Fig. 21.7, p. 701)
        1. Function
        2. Structure
    C. Tonsils (Fig. 21.8, p. 704)
    D. Aggregates of Lymphoid Nodules (p. 704)

V. Developmental Aspects of the Lymphatic System (p. 704)
    A. Embryological Development

Review Items

1. Interstitial fluid (Chapter 3, p. 65)
2. Reticular connective tissue (Chapter 4, p. 118)
3. Blood capillaries (Chapter 20, pp. 645-647)

4. Hydrostatic and osmotic pressure related to fluid movement (Chapter 20, pp. 664-665)

5. Factors that aid venous return (Chapter 20, pp. 647-648)

6. Agranulocytes (Chapter 18, p. 596)

7. Leukocyte production and lifespan (Chapter 18, pp. 596-597)

8. Granulocytes (Chapter 18, p. 594)

Cross References

1. Lacteal function is examined in Chapter 24, p. 815.

2. The function of lymphatic organs in immunity is detailed in Chapter 22.

3. The palatine tonsils are further described in Chapter 24, p. 795.

4. Lymphatic tissue related to the digestive system (Peyer's patches) is explained in Chapter 24, p. 818.

5. Tonsils are further described in Chapter 23, p. 747.

6. The thymus gland and hormone production is examined in Chapter 17, p. 576.

7. The relationship of the thymus to cell-mediated immunity is described in Chapter 22, pp. 727-731.

Laboratory Correlations

1. Marieb, E. N. *Human Anatomy and Physiology Laboratory Manual: Cat and Fetal Pig Versions.* 3rd. ed. Benjamin/Cummings, 1989.

   None in this chapter.

2. Marieb, E. N. *Human Anatomy and Physiology Laboratory Manual: Brief Version.* 3rd. ed. Benjamin/Cummings, 1992.

   None in this chapter.

Transparencies Index

21.1   Distribution and special features of lymphatic capillaries
21.2   The lymphatic system
21.5   Structure of a lymph node

Bassett Atlas Figures Index

Slide Number	Figure Number	Description
18	3.2	Anterior mediastinum
22	3.6A,B	Lateral view of mediastinum
32	4.5	Stomach, liver, and spleen in situ
33	4.6A,B	Stomach, liver, and spleen, partial dissection
34	4.7A,B	Stomach, open
37	4.10A,B	Small bowel mesentery, dissected
38	4.11A,B	Small bowel mesentery, dissected, detail
58	6.6A,B	Brachial plexus and axilla

INSTRUCTIONAL AIDS

Lecture Hints

1. Emphasize the difference between lymphatic capillaries and blood capillaries. Students often raise the question: "Since both are capillaries, why does interstitial fluid flow from blood capillary to lymphatic capillary?"

2. Point out that the same factors that help venous return also aid lymph movement. Since both are structurally similar, and follow the same general pathways, logic dictates that factors affecting fluid movements should also be similar.

3. Mention that bubos (inflamed lymph nodes) are the "swollen glands" seen in infectious processes.

4. Point out that the structure and locations of lymph nodes is ideal for filtering the interstitial fluid (containing tissue proteins, metabolic wastes, and pathogenic microorganisms) flushed from tissue spaces.

Demonstrations/Activities

1. Audio-visual materials of choice.

2. Use a torso model and/or dissected animal model to exhibit lymph organs.

3. Use a short piece of cloth-wrapped garden hose to show the layers of the lymph vessels and the open-ended nature of the vessels.

4. Use a hand-held pinball machine to show the one-way nature of valves in lymph vessels.

Critical Thinking/Discussion Topics

1. Explain why and how lymphedema occurs after a modified radial mastectomy or other such surgery.

2. Discuss the reason behind a physician checking for swollen lymph nodes in the neck when examining a patient who shows respiratory symptoms.

3. Discuss the ramification of spleen removal.

4. Discuss the ramification of tonsillar tissue removal.

5. Discuss, briefly, the role of the thymus gland in the body's immune response.

Library Research Topics

1. Research the causes, effects, and treatment of lymphedema.

2. Research the differences between lymph nodes swollen due to disease such as a viral infection and those swollen due to cancer.

3. Research the changes that appear in the thymus gland with age and relate those changes to the body's immune response.

Audio-Visual Aids/Computer Software

Slides

1. Visual Approach to Histology: Lymphatic System (FAD, 18 Slides)

2. The Immune Response (EIL, Slides). High-quality artwork provides a detailed look at antibody formation and antigen-antibody interactions. Explores the lymphatic system and passive and active immunity.

Videotapes

1. The Living Body: Internal Defenses (FHS, VHS or BETA). This program deals with events when the body is under attack. It shows the roles of the spleen, the lymphatic system and the WBCs, and explains the body's production of antibodies.

   See Guide to Audio-Visual Resources on page 339 for key to AV distributors.

LECTURE ENHANCEMENT MATERIAL

Clinical and Related Terms

1. Lymphadenectomy — surgical removal of lymph nodes.

2. Lymphadenitis — an inflammation of the lymph nodes characterized by a transitory enlargement and swelling of affected lymph nodes. As the infection travels through lymph channels, acute inflammation can occur in the node.

3. Lymphadenotomy — an incision into a lymph node.

4. Lymphedema — chronic unilateral or bilateral edema of the extremities due to accumulation of interstitial fluid as a result of stasis of lymph, which is secondary to obstruction of lymph vessels or disorders of the lymph nodes.

5. Lymphocytopenia — an abnormally low concentration of lymphocytes in the blood.

6. Lymphocytosis — an abnormally high concentration of lymphocytes in the blood.

7. Lymphoma — a tumor composed of lymphatic tissue.

8. Lymphosarcoma — a cancer within the lymphatic tissue.

9. Splenomegaly — an abnormal enlargement of the spleen.

Disorders/Homeostatic Imbalances

1. Congenital Lymphedema — an isolated limb abnormality which occurs due to failure of normal lymphatics to develop.

2. Infectious Mononucleosis — a common viral disease caused by the Epstein-Barr virus. Enlargement and tenderness of nodes is common. The virus affects the B lymphocytes and is destroyed by the T lymphocytes and antibodies. Symptoms include extreme lethargy, swollen nodes, possible fever, and possible swollen spleen. The swollen spleen can rupture and cause a fatality.

3. Filariasis — lymphedema caused by parasites entering through the skin, finding their way to regional lymph nodes, and causing obstructive fibrosis of lymphatics and nodes, resulting in edema of the extremity — elephantiasis.

4. Lymphadenitis — an inflammation of the lymph nodes characterized by a transitory enlargement and swelling of affected lymph nodes. As the infection travels through lymph channels, acute inflammation can occur in the node.

5. Milroy's Disease (Heredofamilial Lymphedema) — lymphedema that occurs following a familial inheritance pattern. It is similar in appearance to congenital lymphedema.

6. Neoplasms — cancers of lymph tissue. Three types of neoplasms affect lymph tissue. In metastatic carcinoma, the tumor spreads from primary site to regional nodes. Malignant lymphoma and lymphatic leukemia are discussed in a previous chapter.

7. Obstruction Lymphedema — lymphedema due to obstruction of lymphatic vessels or lymph nodes by spread of metastatic tumors.

8. Post Inflammatory Fibrosis Lymphedema — lymphedema following soft-tissue infection.

9. Surgical Disruption Lymphedema — lymphedema due to surgical disruption of lymph channels, particularly with excision of lymph node groups (i.e., radical mastectomy with axillary lymph node dissection).

ANSWERS TO END-OF-CHAPTER QUESTIONS

Multiple Choice/Matching

1.	c	4.	c	7.	a
2.	c	5.	a	8.	b
3.	a	6.	b	9.	d

Short Answer Essay Questions

10. Blood, the carrier of nutrients, wastes, and gases, circulates within blood vessels through the body, exchanging materials with the interstitial fluid. Interstitial fluid, formed by filtration from blood, is the fluid surrounding body cells in the tissue spaces. Lymph is the protein-containing fluid that enters the lymphatic capillaries (from the tissue spaces) hence its composition is the same as that of the interstitial fluid. (pp. 694-695)

11. Lymph nodes are very small bean-shaped structures consisting of both a medulla and cortex, which act as filters to cleanse lymph before it is allowed to reenter the blood. Each node is surrounded by a dense fibrous capsule from which connective tissue strands called trabeculae extend inward to divide the node into a number of compartments. The basic internal framework or stroma is provided by a soft, open network of reticular fibers that physically support lymphocytes and macrophages. The outer cortex contains densely packed spherical collections of lymphocytes called follicles, which frequently have lighter staining centers called germinal centers. Cord-like extensions of the cortex, called medullary cords, invade the medulla. Macrophages are located throughout the node but are particularly abundant lining the sinuses of the medulla. The spleen is the largest lymphoid organ. It functions to remove aged or defective blood cells, platelets, and pathogens from the blood and to store some of the breakdown products of RBCs or release them to the blood for processing by the liver. The spleen is surrounded by a fibrous capsule, and has trabeculae. It contains lymphocytes, macrophages, and huge numbers of erythrocytes. Venous sinuses and other regions that contain red blood cells and macrophages are referred to as red pulp, whereas areas composed mostly of lymphocytes suspended on reticular fibers are called white pulp. The white pulp clusters around small branches of the splenic artery within the organ and serves the immune functions of the organ. (pp. 698-699, 700-701)

12. a. The anatomical characteristic which ensures that the flow of lymph through a lymph node is slow is that there are fewer efferent vessels draining the node than afferent vessels feeding it.

    b. This feature is desirable to allow time for the lymphocytes and macrophages to perform their protective functions. (p. 699)

Critical Thinking and Application Questions

1. a. With removal of the lymphatic vessels, fluid has built up in the tissues and drains very slowly back to the bloodstream.

   b. Yes, she can expect to have relief since lymphatic drainage is eventually reestablished by regrowth of the lymphatic vessels. (pp. 695-697)

2. Her swollen "glands" are inflamed cervical lymph nodes. Bacteria have spread from lymph vessels that drain the region of the cut in her face, and have lodged in the lymph nodes of the neck, infecting these nodes. (p. 699)

Key Figure Questions

Figure 21.1 The arteries. It's not a problem, because lymphatic fluid is formed by the blood capillary beds and is only moved in one direction, toward the heart. **Figure 21.5** It causes lymphatic fluid to back up or accumulate somewhat in lymph nodes, allowing more time for its cleansing. **Figure 21.6** The red pulp is the blood processing region; the white pulp has immune functions.

SUGGESTED READINGS

1. Marrach, P., and J. Kappler. "The T Cell and Its Receptor." *Scientific American* 254 (Feb. 1986): 36.

2. Mayerson, H.S. "The Lymphatic System." *Scientific American* (June 1963): 158.

3. Tonegawa, S. "The Molecules of the Immune System." *Scientific American* (Oct. 1985).

22

Nonspecific Body Defenses and Immunity

Chapter Preview

This chapter describes the functional, rather than structural, system of the body: the immune system. Its organs are: (1) trillions of individual immune cells, which inhabit lymphatic tissues and circulate in body fluids, and (2) a diverse array of molecules. The immune system is designed to eliminate almost any type of pathogen that intrudes into the body, protect the body from transplanted organs or grafts, and to protect the body against the body's own cells that have turned against it.

Specific defense mechanisms usually require prior exposure to each foreign substance before a response is mounted against that substance, whereas the nonspecific defenses respond immediately to protect the body from all foreign substances. This chapter discusses both specific and nonspecific defenses of the body.

INTEGRATING THE PACKAGE

Suggested Lecture Outline

I. Nonspecific Body Defenses (pp. 708-715)

    A. Surface Membrane Barriers (p. 708)

        1. Skin

        2. Mucous Membranes

    B. Nonspecific Cellular and Chemical Defenses (pp. 708-715)

        1. Phagocytes (Fig. 22.1, p. 709)

        2. Natural Killer Cells

        3. Inflammation: Tissue Response to Injury: (Fig. 22.2, p. 711)

            a. Vasodilation and Increased Vascular Permeability

            b. Phagocyte Mobilization

        4. Antimicrobial Proteins

            a. Complement

            b. Interferon

        5. Fever

II. Specific Body Defenses: Immunity (pp. 715-720)

    A. Aspects of the Immune Response (p. 715)

        1. Antigen-Specific

        2. Systemic

        3. Memory

        4. Humoral Immunity

        5. Cell-Mediated Immunity

B. Antigens (pp. 716-718)
    1. Complete Antigens and Haptens
    2. Antigenic Determinants
        a. Self Antigens
C. Cells of the Immune System (pp. 718-720)
    1. Lymphocytes
    2. Macrophages

III. Humoral Immune Response (pp. 720-727)
A. Clonal Selection and Differentiation of B Cells (p. 720)
    1. Clonal Selection
    2. Plasma Cells
B. Immunological Memory (p. 721)
    1. Primary Immune Response
    2. Secondary Immune Response
    3. Immunological Memory
C. Active and Passive Humoral Immunity (p. 722)
    1. Active Humoral Immunity
        a. Vaccines
    2. Passive Humoral Immunity
D. Antibodies (pp. 722-727)
    1. Basic Antibody Structure (Fig. 22.11, p. 723)
        a. Heavy and Light Chains
        b. Variable and Constant Regions
    2. Antibody Classes (Table 22.3, p. 724)
    3. Mechanisms of Antibody Diversity (Fig. 22.12, p. 725)
    4. Antibody Targets and Functions (Fig. 22.13, p. 726)
        a. Complement Fixation and Activation
        b. Neutralization
    5. Monoclonal Antibodies

IV. Cell-Mediated Immune Response (pp. 727-734)
A. Introduction (p. 727)
    1. Effector Cells
        a. Cytotoxic T Cells
    2. Regulatory Cells
        a. Helper T Cells—CD4 Cells
        b. Supressor T Cells—CD8 Cells
B. Clonal Selection and Differentiation of T Cells
    1. Antigen Recognition and MHC Restriction
        a. Class I MHC
        b. Class II MHC
        c. MHC Restriction
    2. T Cell Activation
        a. Step 1: Antigen Binding
        b. Step 2: Costimulation

3. Cytokines

C. Specific T Cell Roles (pp. 730-731)

1. Helper T Cells (Fig. 22.15, p. 729)

2. Cytotoxic T Cells (Fig. 22.16, p. 731)

3. Suppressor T Cells

4. Summary of T Cell Roles

D. Organ Transplants and Prevention of Rejection (pp. 731-734)

1. Grafts

a. Autografts

b. Isografts

c. Allografts

d. Xenografts

2. Immunosuppressive Therapy

V. Homeostatic Balance of Immunity (pp. 734-737)

A. Types (pp. 734-736)

1. Immunodeficiencies

2. Hypersensitivities

a. Immediate Hypersensitivities (Fig. 22.18, p. 735)

b. Anaphylaxis

c. Atopy

3. Subacute Hypersensitivities

a. Antibody-Dependent Cytotoxic (Type II) Reactions

b. Immune-Complex (Type III) Hypersensitivity

4. Delayed Hypersensitivities

a. Allergic Contact Dermatitis

B. Autoimmune Diseases (pp. 736-737)

1. Inefficient/Ineffective Lymphocyte Programming

2. Appearance of Self Proteins in the Circulation

3. Cross-Reaction of Antibodies

VI. Developmental Aspects of the Immune System (p. 737)

A. Embryologic Development

B. Aging

Review Items

1. Protein structure (Chapter 2, pp. 47-53)

2. Cilia (Chapter 3, pp. 83-84)

3. Lysosomes (Chapter 3, pp. 80-81)

4. Lysozyme Chapters 16, p. 508)

5. Mechanical and chemical protection of the skin (Chapter 5, p. 145)

6. Langerhan's cells (Chapter 5, p. 137)

7. Granulocytes (Chapter 18, pp. 594-596)

8. Agranulocytes (Chapter 18, p. 596)

9.   Chemotaxis (Chapter 18, p. 594)

10. Diapedesis (Chapter 18, p. 594)

11. Thymus (Chapters 17, p. 576; 21, p. 701)

Cross References

1.   Protection of the mucous barrier in the stomach is described in Chapter 24, p. 806.

2.   The role of saliva in protection of mucous barriers is detailed in Chapter 24, p. 798.

3.   Kupfer cells are detailed in Chapter 24, p. 821.

4.   Body temperature regulation is further explained in Chapter 25, pp. 885-890.

5.   Antibody protection of the fetus due to maternal antibodies is mentioned in Chapter 29, p. 1022.

6.   Inflammatory processes involving respiratory tissues are presented in Chapter 23, pp. 778-780.

Laboratory Correlations

1.   Marieb, E. N. *Human Anatomy and Physiology Laboratory Manual: Cat and Fetal Pig Versions.* 3rd. ed. Benjamin/Cummings, 1989.

     None in this chapter.

2.   Marieb, E. N. *Human Anatomy and Physiology Laboratory Manual: Brief Version.* 3rd. ed. Benjamin/Cummings, 1992.

     Exercise 32: The Lymphatic System and Immune Responses

Transparencies Index

22.3   Events of phagocyte mobilization

22.4   Events and results of complement activation

22.5   Antigenic determinants

22.7   Macrophage's role in immunity

22.8   Simplified version of clonal selection of a B cell stimulated by antigen binding

22.11 Basic Antibody structure

22.13 Mechanisms of antibody action

22.14 Clonal selection of helper and cytotoxic T cells involves simultaneous recognition of self and antiself

22.15 The central role of helper T cells

22.18 Mechanism of an acute allergic response

Bassett Atlas Figures Index

Slide Number	Figure Number	Description
18	3.2	Anterior mediastinum
22	3.6A,B	Lateral view of mediastinum
32	4.5	Stomach, liver, and spleen in situ
33	4.6A,B	Stomach, liver, and spleen, partial dissection
34	4.7A,B	Stomach, open
37	4.10A,B	Small bowel mesentery, dissected
38	4.11A,B	Small bowel mesentery, dissected, detail
58	6.6A,B	Brachial plexus and axilla

INSTRUCTIONAL AIDS

Lecture Hints

1. Although specific and nonspecific defense mechanisms are treated as separate entities, emphasize that there is much overlap of function, e.g., in antibody-mediated complement lysis, formation of the antibody that labels a cell is due to specific processes, but the actual lysis of that cell is accomplished nonspecifically by complement fixation.

2. Point out that the body has several lines of defense — mechanical, chemical, and cellular. It is helpful to orient students toward the idea that even with all of its complexity, the immune system has one underlying theme: rid the body of unwanted substances/life forms.

3. Students often have problems deciding what mechanisms are specific and which are nonspecific. Mention that specific processes are something like enzyme-substrate interactions, highly specific to a single substance.

4. Emphasize the logic behind the four cardinal signs of inflammation. For example, to bring the large quantities of oxygen and nutrients for repair processes, blood supply to an area must be increased. Redness results as vasodilation increases; heat, as warm blood is delivered; swelling as capillary walls become more permeable; pain as pressure due to swelling is transmitted to nerve endings.

5. Point out that neutrophils are seen early in an infection, but that macrophages are characteristic of chronic infection.

6. To illustrate the action of pyrogens on the hypothalamus, use the example of resetting the thermostat to a higher temperature in a home. As always, relating a physiological concept to something students can relate to will help reinforce the idea.

7. Mention that cytotoxic T cells must come in contact with the invader, but that B cells send out antibodies from sometimes remote locations to target specific antigens.

8. Emphasize the difference between antigens and haptens, and that size is the cause of the distinction between the two. Essentially, one can think of a hapten as an antigenic determinant if it is bound to a carrier molecule or, alternatively, if an antigenic determinant were not part of a large molecule, it would be a hapten.

9. To reinforce the idea of clonal selection, point out that a single B cell could not possibly produce enough antibody to neutralize a large quantity of antigen.

10. Stress the difference between active and passive immunity, and that the body does not care where antibodies come from. If the students understand the concept, the question should arise: "Why don't the foreign antibodies generate a response in the recipient?" If no one asks, ask this of the class to generate discussion.

11. In the discussion of complement, ask the class "What would happen if an antibody bound to a completely normal body cell?" Use this as a lead-in to the topic of complement recognition of the constant region of the antibody.

12. It is often difficult for students to grasp the concept of somatic recombination in the generation of antibody diversity. Ask plenty of questions during lecture to reinforce concepts.

13. Clearly distinguish between different types of allergy. As the material is being covered, ask students which arm of the immune system is responsible for what type of hypersensitivity—immediate, subacute, or delayed. Students should be able to make the connection that a cell-mediated response will take time (delayed hypersensitivity).

Demonstrations/Activities

1. Audio-visual materials of choice.

2. Use a lock and several keys (with only one that fits the lock) to demonstrate the specificity of antigens and antibodies.

3. Use a magnet and iron filings to demonstrate the attraction of antibodies to antigens.

4. Use a film clip of covered wagons in a circle, surrounded by charging Indians, with the cavalry coming to the rescue to illustrate the body's rush to defend itself against invaders (microorganisms, etc.).

5. To open a discussion of the inflammatory process, ask if anyone has a cut or injury that is in the process of healing. If so, have the class members observe it, describe all obvious signs, and provide the underlying reason for the signs seen.

6. Ask students to come prepared to discuss the following questions during a subsequent lecture:

    a. Explain why vaccination provides long-term protection against a particular disease while passive immunization provides only temporary protection.

    b. What is the important difference between natural killer cells and cytotoxic T cells?

    c. Why can T helper cells be called the "managers" of the immune system?

Critical Thinking/Discussion Topics

1. Discuss the pros and cons surrounding the use of immunizations for mumps, measles, etc.

2. Discuss the rationale for the demise of the Martian invaders in H.G. Wells's *War of the Worlds*.

3. Discuss the autoimmune diseases, how they occur, symptoms, prognosis, and treatment.

4. Discuss why some individuals are sensitive (allergic) to drugs, etc., from one source, but are not so sensitive to drugs from another source.

5. Discuss the social implications of immunity disorders such as AIDS, ARC, SCID, etc.

6. Discuss the role of the Epstein-Barr virus in immunity and immunity disorders.

7. Discuss the effects of AIDS both immunologically and socially.

8. Explain why we need specific resistance mechanisms if nonspecific resistance mechanisms attack all foreign substances (i.e., why is specific resistance necessary at all?).

9. Explain what the body's immune response is to an antitoxin or other passive immunization.

10. Discuss why chemotherapeutics attached to monoclonal antibodies are an advantage over injection of the chemical agent alone. Could there be any drawbacks to this therapy?

Library Research Topics

1. Research some of the opportunistic diseases that often accompany AIDS.

2. Research the difficulties involved in transplant surgeries.

3. Research the causes of several known autoimmune diseases.

4. Research the signs, symptoms, and treatment of anaphylactic shock.

5. Research the possible side effects of vaccines.

Audio-Visual Aids/Computer Software

Slides

1. The Immune Response (EIL, Slides). High-quality artwork provides a detailed look at antibody formation and antigen-antibody interactions. Explores the lymphatic system and passive and active immunity.

2. Immunoglobulin Structure and Function (EIL, 98 Slides). Excellent introduction to the role of antibodies in health and disease. Begins with simple structure and demonstrates various properties, actions of immunoglobulin classes, molecules, and gamma globulin. Compares systemic versus local immune responses, which in turn result in either protection or pathology.

3. Antigens and Immunogens (EIL, 55 Slides). Defines immune response, immunologic recognition and memory. Diagrams and enumerates antigenic and immunogenic properties. Gives rationale, methodologies, and results of immunization.

4. Mounting Immune Responses (EIL, 98 Slides). Defines general nature of immune response and diagrams the lymphatic system's role in it, using graphics as well as color photomicrographs. Various influences on the immune response and dose tolerance are studied in detail.

5. Antigen-Antibody Reactions (EIL, 98 Slides). Terms are defined and illustrated. Nature, detection, and measurement of antigen-antibody reactions are then revealed through diagrams, tables, and photomicrographs of affected tissues. Painstakingly details clinical methodology for determining various reactions.

6. The Complement Systems (EIL, 65 Slides). Diagrams and differentiates between classical and alternate pathways. Describes various biological properties of complement components and their still uncertain roles in health. Step-by-step analysis of complement component measurement and complement fixation testing.

7. Anaphylactic Hypersensitivity (EIL, 80 Slides). Introduces the five types of immunopathology. Concentrates on Type I Hypersensitivity—the terms, the target cell and the antibody/antigen, the mediators of anaphylaxis—its modification and methods of testing.

8. Cytotoxic Hypersensitivity (EIL, 47 Slides). Type II sensitivity in detail. Combines tables for terms and comparisons, artwork diagramming processes involved, and enhanced photography of affected tissues. A complete, yet concise presentation.

9. Immune Complex Disease (EIL, 65 Slides). Involves Type III and V Hypersensitivity. Arthus reaction, serum sickness and chronic systemic immune complex diseases, what forms them and how, as well as their investigation and differentiation.

10. Lymphocyte Mediated Hypersensitivity (EIL, 63 Slides). Type IV Hypersensitivity. Cell-mediated immunity, resulting in damage; antigens involved, resultant lesions, effect on bacteria, methodology of immunologic investigation. Includes enhanced photography of affected tissues.

11. Hyper-Immunoglobulinemias (EIL, 51 Slides). Outlines immunoglobulin production and measurement and details the differences between polyclonal and oligoclonal hyper-immunoglobulinemias and their resultant conditions. Emphasizes diagnosis and management.

12. Immunologic Deficiency States (EIL, 110 Slides). Infantile hypogammaglobulinemia, defects in the inflammatory response that lead to immune deficiency, defects in stem cells, defects in T cells, defects in B cells, defective immunoglobulin synthesis, alterations in the fate of immunoglobulins, evaluation of an immunologic defect. An effective combination of clear, precise artwork, medical photomicrographs, and unadorned, concise narrative make an indispensible teaching tool.

Videotapes

1. The Human Immune System: The Fighting Edge (FHS, VHS or BETA, 52 min., C. 1988). This program tells the stories of four individuals whose immune system failed to function properly. It provides the very latest information on how the immune system works, and explains why some treatments work and others do not.

2. AIDS: Our Worst Fears (FHS, VHS or BETA, 57 min., C, 1988). This program, updated with the latest statistics, explains what we do and don't know about AIDS; who is most susceptible and most at risk; and what preventive actions and precautions can be taken.

3. Transplants: The Immune System at Risk (PLP, CH-460769, VHS)

See Guide to Audio-Visual Resources on page 339 for key to AV distributors.

LECTURE ENHANCEMENT MATERIAL

Clinical and Related Terms

1. Acquired immune deficiency syndrome (AIDS) — a viral disease that compromises the immune system leaving the individual susceptible to opportunistic diseases.

2. Acquired immunity — specific immunity attributable to the presence of antibody and to heightened reactivity of antibody-forming cells, specifically immune lymphoid cells responsible for cell-mediated immunity, and of phagocytosis.

3. AIDS-related complex (ARC) — a condition that may be developed by individuals carrying the AIDS virus; symptoms include fever, diarrhea, malaise, and a generalized swelling of lymph nodes; may progress to full-blown AIDS.

4. Allergy — a response to antigens such as those of ragweed, other plant pollens, and various other antigens. Individuals form specific IgE antibodies. Treatment usually involves the use of antihistamine drugs to relieve some of the allergic symptoms. A more specific method of treatment involves desensitization.

5. Autoimmune disease — any of a group of disorders in which tissue injury is associated with humoral or cell-mediated responses to body constituents; they may be systemic (systemic lupus erythematosus) or organ specific (autoimmune thyroiditis).

6. Cytoxic drug — a drug that has the action of a cytotoxin, a toxin or antibody that has a specific toxic action upon cells of special organs.

7. Hypersensitivity — a state of altered reactivity in which the body reacts with an exaggerated response to a foreign agent.

8. HLA antigens — histocompatibility antigens on the surface of nucleated cells determined by a single major chromosomal locus, the HLA locus; they are important in cross-matching for transplantation procedures.

9. Immunologic surveillance — the constant monitoring of the body tissues by the immune system for abnormal cells.

10. Malaise — a vague feeling of bodily discomfort.

11. Serum sickness — a hypersensitivity reaction occurring eight to 12 days following a single, relatively large injection of foreign serum and marked by urticarial rashes, edema, adenitis, joint pains, high fever, and prostration; serum disease; serum intoxication.

12. Severe combined immunodeficiency disease (SCID) — a congenital defect in which the patient is unable to mount a defense against disease due to a lack of the stem cell progenitor of T cells and B cells.

Disorders/Homeostatic Imbalances

1. Acquired Immune Deficiency Syndrome (AIDS) — a primary disease caused by HIV (human immunodeficiency virus). HIV attacks CD4 (helper T) cells, resulting in inhibition of the roles of helper T cells in the immune response. Individuals who are exposed to the virus may carry the virus throughout their lifetimes without further symptoms. They are known as carriers. Some people who carry the AIDS virus will develop the AIDS-related complex (ARC). This condition may include fever, diarrhea, malaise, and a generalized swelling of lymph nodes. ARC may resolve, persist for a long period, or progress to full-blown AIDS. Full-blown AIDS symptoms include malaise; a low-grade fever or night sweats; coughing; shortness of breath; sore throat; extreme fatigue; muscle aches; unexplained weight loss; enlarged lymph nodes in the neck, axilla, and groin; and blue-violet or brownish spots on the skin, usually on the lower extremities. The deterioration of the body's defenses allows for the development of opportunistic infections such as

Kaposi's sarcoma, Pneumocystis carinii pneumonia, a form of herpes that attacks the CNS, and a bacterial infection that usually causes TB in chickens and pigs. AIDS transmission is via semen, blood, and possibly other body fluids. Treatment of AIDS is currently not available, except for the use of experimental drugs. Novacine is currently available. Blood screening (for transfusions), safe sex (monogamy and/or use of condoms) or abstinence, and nonsharing of contaminated needles by drug addicts, are currently the only means of preventing the disorder.

2. Autoimmune Blood Disease — blood diseases due to autoantibodies formed against blood cell antigens and resulting in anemia, leukopenia, or thrombocytopenia.

3. Diffuse Toxic Goiter — an autoimmune disease due to autoantibodies mimicking thyroid-stimulating hormone (TSH), causing increased output of thyroid hormone (hyperthyroidism).

4. Glomerulonephritis — an autoimmune disorder due to antibodies formed against glomerular basement membrane or to antigen-antibody complexes trapped in glomeruli, resulting in inflammation of renal glomeruli.

5. Graft Rejection — rejection of grafts from another person due to a difference in HLA antigens between the donor and the recipient. The body recognizes the antigens as being foreign and rejects the transplant organ. In such cases immunosuppressive drugs are given to affect the rejection.

6. Hodgkin's Disease — a malignant disorder of lymph nodes, with later involvement of the liver, spleen, bone marrow, and lungs. It is accompanied by early and profound systemic immunological dysfunction. Cell-mediated immunological responses are most severely depressed. Patients are, therefore, particularly susceptible to viral, mycobacterial, and fungal infections.

7. Hypersensitivity — an increased responsiveness to a foreign antigen. Immediate hypersensitivity is associated with the presence of antibodies in the serum. Subsequent contact with the foreign antigen leads to an antigen-antibody reaction called anaphylactic shock or to a chronic reaction called serum sickness. Delayed hypersensitivity develops after exposure to foreign antigens. The individual has been sensitized to the antigen so that subsequent contact with the antigen leads to an acute inflammatory reaction.

8. Lupus Erythematosus — an autoimmune collagen disease due to various antinuclear antibodies, which cause injury to several organs and result in systemic disease.

9. Rheumatoid Arthritis — an autoimmune disease due to antibodies formed against serum gamma globulin that results in a systemic disease with inflammation and degeneration of joints.

10. Rheumatic Fever — an autoimmune disorder due to antistreptococcal antibodies cross-reacting with antigenic heart muscle, heart valves, and other tissues, resulting in inflammation of heart and joints.

11. Thyroiditis — an autoimmune disease due to antithyroid antibody causing injury and inflammatory cell infiltration of the thyroid gland. It results in hypothyroidism.

12. Severe Combined Immunodeficiency (SCID) — a rare immunodeficiency disease in which both B cells and T cells are missing or inactive in providing immunity. As a result, the patient is unable to mount a defense against disease.

ANSWERS TO END-OF-CHAPTER QUESTIONS

Multiple Choice/Matching

1.	c	4.	d	7.	b
2.	a	5.	a	8.	c
3.	d	6.	d	9.	d

Short Answer Essay Questions

10. Mucus provides a sticky mechanical barrier that traps pathogens. Mucosae are found on the outer surface of the eye, and in the linings of all body cavities open to the exterior, i.e., the digestive, respiratory, urinary, and productive tracts.

    Lysosome, an enzyme that destroys bacteria, is found in saliva and lacrimal fluid.

    Keratin, a tough waterproofing protein in epithelial membranes, presents a physical barrier to microorganisms on the skin, as well as being resistant to most weak acids and bases and to bacterial enzymes and toxins.

    The acid pH of skin secretions inhibits bacterial growth. Vaginal secretions and urine (as a rule) are also very acidic. Hydrochloric acid is secreted by the stomach mucosa and acts to kill pathogens.

    Cilia of the upper respiratory tract mucosae sweep dust and bacteria-laden mucus superiorly toward the mouth, restraining it from entering the lower respiratory passages. (p. 708)

11. Attempts at phagocytosis are not always successful because to accomplish ingestion, the phagocyte must first adhere to the particle. Complement proteins and antibodies coat foreign particles, providing binding sites to which phagocytes can attach, making phagocytosis more efficient. (p. 709)

12. Complement refers to a heterogenous group of at least 20 plasma proteins that normally circulate in an inactive state. Complement is activated by one of two pathways (classical or alternative) involving the plasma proteins. Each pathway involves a cascade in which complement proteins are activated in an orderly sequence leading to the cleavage of C3. Once C3b is bound to the target cell's surface, it enzymatically initiates the remaining steps of complement activation, which incorporates C5 through C9 (MAC) into the target cell membrane, ensuring lysis of the target cell.

    Other roles of complement include opsonization, inflammatory actions such as stimulating mast cells and basophils to release histamine (which increases vascular permeability), and attracting neutrophils and other inflammatory cells to the area. (pp. 712-713)

13. Interferons are secreted by virus-infected cells. They diffuse to nearby cells and bind to their membrane receptors, interfering with the ability of viruses to multiply within these cells. Cells that form interferon include macrophages, lymphocytes, and other leukocytes. (pp. 714-715)

14. Humoral immunity is provided by the antibodies in the body's fluids. Cell-mediated immunity is provided by non-antibody-producing lymphocytes. (pp. 720-722, 727)

15. T cells are best suited for cell-to-cell interactions, and most of their direct attacks on antigens are mounted against body cells infected by viruses, bacteria, and intracellular parasites, against abnormal or cancerous body cells, and infused or transplanted foreign cells. Lymphokines released by T cells help to amplify and regulate both the humoral and cellular immune response as well as the nonspecific defense responses. (p. 727)

16. Immunocompetence is the ability of the immune system's cells to recognize foreign substances (antigens) in the body by binding to them. Acquisition is signaled by the appearance of a single, unique type of cell surface receptor protein on each T or B cell that enables the lymphocyte to recognize and bind to a specific antigen. (p. 718)

17. Helper T cell activation involves a double recognition: a simultaneous recognition of the antigen and an MHC II membrane glycoprotein of an antigen-presenting cell (macrophage). (p. 727)

18. A primary immune response results in cellular proliferation, differentiation of mature effector and memory lymphocytes, and the synthesis and release of antibodies. Secondary immune response results in huge numbers of antibodies flooding into the blood stream within hours after recognition of the antigen, as well as an amplified cellular attack. Secondary responses are faster because the immune system has been primed to the antigen and sizable numbers of sensitized memory cells are already in place. (p. 720)

19. An antibody is a soluble protein secreted by sensitized B cells and plasma cell offspring of B cells in response to an antigen. (Fig. 22.11, p. 723)

20. The variable regions of an antibody combine to form an antigen binding site that is shaped to "fit" a specific antigenic determinant or an antigen. The constant regions of an antibody determine what class of antibody will be formed, how the antibody class will carry out its immune roles in the body, and what cell types or chemicals with which the antibody will bind. (pp. 722-723)

21. The antibody classes and their probable locations in the body include the following:

    Class IgD — virtually always attached to B cells; B cell receptor

    Class IgM — monomer attached to B cells; pentamer free in plasma

    Class IgG — in plasma

    Class IgA — some in plasma, most in secretions such as saliva, tears, intestinal juice, and milk

    Class IgE — secreted by plasma cells in skin, mucosae of gastrointestinal and respiratory tracts and tonsils

    (p. 724; Table 22.3, p. 724)

22. Antibodies help defend the body by complement fixation, neutralization, agglutination, and precipitation. Complement fixation and neutralization are most important in body protection. (pp. 724-725)

23. Vaccines produce active humoral immunity because most contain dead or extremely weakened pathogens which have the antigenic determinants necessary to stimulate the immune response but are generally unable to cause disease. Passive immunity is less than satisfactory because neither active antibody production nor immunological memory is established. (p. 722)

24. Helper T cells function to chemically or directly stimulate the proliferation of other T cells and of B cells that have already become bound to antigen. Suppressor T cells function to temper the normal immune response by dampening the activity of both T cells and B cells by releasing lymphokines that suppress their activity. Cytotoxic T cells function to kill virus-invaded body cells and cancer cells and are involved in rejection of foreign tissue grafts. (pp. 730-731)

25. Lymphokines are soluble glycoproteins released by activated T cells. They enhance the defensive activity of T cells, B cells, and macrophages. Specific lymphokines and their role in the immune response are summarized in Table 22.1, p. 732.

26. Hypersensitivity is an antigen-induced state that results in abnormally intense immune responses to an innocuous antigen. Immediate hypersensitivities include anaphylaxic and atopy. Subacute hypersensitivities include cytotoxic and immune complex hypersensitivities. All of these involve antibodies. Delayed hypersensitivities include allergic contact dermatitis and graft rejection. These hypersensitivities involve T cells. (pp. 734-736)

27. Autoimmune disease results from changes in the structure of self antigens, from appearance of self proteins in the circulation that have not been previously exposed to the immune system, and by cross-reaction of antibodies produced against foreign antigens with self antigens. (pp. 736-737)

28. Declining efficiency of the immune system with age probably reflects genetic aging. (p. 737)

Critical Thinking and Application Questions

1. a. Jenny had severe combined immunodeficiency disease (SCID) in which T cells and B cells fail to develop. At best there are only a few detectable lymphocytes. Bone marrow transplant is the treatment of choice; however, these are unsuccessful in some cases. The transplanted cells may not survive, or, in other cases, mount an immune response against the recipient's tissues (graft versus host response).

   b. Bone marrow transplants is the only chance for survival. It was hoped that by replacing marrow stem cells, the populations of T cells and B cells would approach normal.

   c. Jenny's brother had the closest antigenic match since both children were from the same parents.

   d. Epstein-Barr virus is the etiologic agent of infectious mononucleosis, usually a self-limiting problem with recovery in a few weeks. Rarely, the virus causes the formation of cancerous B cells — Burkitt lymphoma.

   e. SCID is a congenital defect in which there is a lack of the common stem cell that develops into T cells and B cells. AIDS is the result of an infectious process by a viral microorganism that selectively incapacitates the CD4 (helper) T cells. Both result in a severe immuno-deficiency that leaves the individual open to opportunistic pathogens and body cells that have lost normal control functions (cancerous). (p. 734)

2. IgA is found primarily in mucous and other secretions that bathe body surfaces. It plays an important role in preventing pathogens from entering the body. Lack of IgA would result in frequent major/minor infections of the sinuses or respiratory tract infections. (p. 724)

3. The mechanisms for the cardinal signs of acute inflammation involve the entire inflammatory process. The inflammatory process begins as a host of inflammatory chemicals are released into the extracellular fluid. They promote local vasodilation, allowing more blood to flow into the area, causing a local hyperemia that accounts for the redness and heat of an inflamed area. The liberated chemicals also increase the permeability to local capillaries and large amounts of exudate seep from the bloodstream into the tissue space, causing local edema or swelling. The excessive fluid in the extracellular space presses on adjacent nerve endings, contributing to a sensation of pain. (pp. 710-712)

Key Figure Questions

Figure 22.2 This allows more filtrate containing oxygen, nutrients, blood-clotting proteins, and antibodies to enter the inflamed area where these substances are needed for defensive and repair processes. **Figure 22.6** It allows them to come into contact with a broader spectrum of antigens and other body cells. **Figure 22.13** Neutralization ties up the binding sites on the antigenic substance or pathogen so it can't bind body cells. Agglutination forms large clumps of the antigen which then precipitate and can't go anywhere, making it fair game for the phagocytes.

SUGGESTED READINGS

1. Ayala, F.J., J. Klein, and N. Takahata. "MHC Polymorphism and Human Origins." *Scientific American* 269 (Dec. 1993): 78.

2. Bazer, F.W., M.A. Jarpe, H.M. Johnson, and B.E. Szente. "How Interferons Fight Disease." *Scientific American* 270 (May 1994): 68.

3. Beyers, A.D., and A.F. Williams. "At Grips with Interactions." *Nature* 356 (Apr. 1992): 746.

4. Boon, T. "Teaching the Immune System to Fight Cancer." *Scientific American* 266 (Mar. 1993): 82.

5. Ada, G.L., and Sir Gustav Nossal. "The Clonal Selection Theory." *Scientific American* 257 (Aug. 1987): 62-69.

6. Buisseret, P.D. "Allergy." *Scientific American* (Aug. 1982).

7. Cohen, I.R. "The Self, the World and Autoimmunity." *Scientific American* 258 (Apr. 1988): 52-60.

8. Cooper, M.D. "B Lymphocytes. Normal Development and Function." *New England Journal of Medicine* 317 (Dec. 1987): 1452-1456.

9. Cooper, M.D., and I.L. Weissman. "How the Immune System Develops." *Scientific American* 269 (Sept. 1993): 64.

10. Crowley, G., et al. "AIDS: The Next Ten Years." *Newsweek* (June 25, 1990): 20-27.

11. Curren, J. W., et al. "The Epidemiology of AIDS." *Science* 229 (1985): 1352.

12. Edelson, R.L., and J.M. Fink. "The Immunologic Function." *Scientific American* (June 1985).

13. Edwards, D.D. and L. Beil. "Pessimistic Outlook in AIDS Reports." *Science News* 133 (June 1989): 372.

14. Fackelmann, K.A. "AIDS Predictors." *Science News* 136 (Nov. 1989): 298-299.

15. Fackelmann, K.A. "Herpes Virus Decimates Immune-Cell Soldiers." *Science News* 144 (Apr. 1993): 215.

16. Fackelmann, K.A. "HIV and Drug Abuse." *Science News* 135 (Mar. 1989): 168-169, 171.

17. Gallo, R. "The AIDS Virus." *Scientific American* 256 (Jan. 1987): 47.

18. Gray, H.M., A. Sette, and S. Buus. "How T Cells See Antigen." *Scientific American* 261 (Nov. 1989): 56-64.

19. Greene, W.C. "Aids and the Immune System." *Scientific American* 269 (Sept. 1993): 98.

20. Imagawa, D.T., et al. "Human Immunodeficiency Virus Type I Infection in Homosexual Men Who Remain Seronegative for Prolonged Periods." *New England Journal of Medicine* (June 1989).

21. Jaret, P. "The Wars Within." *National Geographic* 169 (6) (June 1986): 702-735.

22. Koffler, D. "Systemic Lupus Erythematosus." *Scientific American* 243 (July 1980): 52.

23. Lancaster, J.R. "Nitric Oxide in Cells." *American Scientist* 80 (May-June 1992): 2480.

24. Lichtenstein, L.M. "Allergy and the Immune System." *Scientific American* 269 (Sept. 1993): 116.

25. McKenzie, A. "An AIDS-Associated Microbe Unmasked." *Science News* 136 (Dec. 1989): 356.

26. Miller, J.A. "Nerve Chemicals Direct Immunity." Science News 126 (Dec. 1984): 357.

27. Old, L.J. "Tumor Necrosis Factor." *Scientific American* 258 (May 1988): 59-75.

28. Oldstone, M.B.A. "Viral Alteration of Cell Function." *Scientific American* 261 (Aug. 1989): 42-48.

29. Paul, W.E. "Infectious Diseases and the Immune System." *Scientific American* 269 (Sept. 1993): 90.

30. Pennisi, E. "HIV Also Kills Developing White Blood Cells." *Science News* 143 (June 1993): 143.

31. Rennie, J. "Who's the Dealer? What Controls Gene Shuffling in the Immune System?" *Scientific American* 262 (3) (Mar. 1990): 30-31.

32. Rose, N.R. "Autoimmune Diseases." *Scientific American* (Feb. 1981).

33. Rosen, F.S., et al. "The Primary Immunodeficiencies." *New England Journal of Medicine* 311 (1984): 235.

34. Schwartz, R.H. "T Cell Anergy." *Scientific American* 269 (Aug. 1993): 62.

35. Smith, K.A. "Interleukin-2." *Scientific American* 262 (3) (Mar. 1990): 50-57.

36. Steinman, L. "Autoimmune Disease." *Scientific American* 269 (Sept. 1993): 106.

37. Vaughan, C. "The Depression Stress Link." *Science News* 134 (Sept. 1988): 155.

38. Weiss, R. "Immune Molecule's 3-D Structure Revealed." *Science News* 132 (Oct. 1987): 228.

39. Weiss, R. "HIV: More Tricks Up Its Sleeve." *Science News* 134 (Oct. 1988): 244.

40. "What Science Knows About AIDS." *Scientific American* 259 (Oct. 1988): a single topic issue.

41. Young, J.D.E., and Z.A. Cohn. "How Killer Cells Kill." *Scientific American* 258 (1) (Jan. 1988): 38-44.

23

The Respiratory System

Chapter Preview

This chapter deals with the system designed to exchange respiratory gases between air and the body cells. In order to accomplish that task, four distinct events must occur: (1) pulmonary ventilation, (2) external respiration, (3) respiratory gas transport, and (4) internal respiration.

Pulmonary ventilation is the process of taking air into and out of the lungs. Taking air in is called inspiration (inhalation). Letting air out is called expiration (exhalation). The total process is commonly called ventilation. External respiration is the process of gas exchange between the air in lung alveoli and the blood within lung capillaries. Respiratory gas transport is the process of the movement of oxygen and carbon dioxide in the blood. Internal respiration is the exchange of gases between body cells and the blood in the surrounding capillaries.

INTEGRATING THE PACKAGE

Suggested Lecture Outline

I. Introduction (pp. 743-744)
    A. Function of the Respiratory System (p. 744)
    B. Events of Respiration (p. 744)
        1. Pulmonary Ventilation
        2. External Respiration
        3. Transport Respiratory Gases
        4. Internal Respiration
II. Functional Anatomy of the Respiratory System (pp. 744-756; Table 23.1, p. 756)
    A. The Nose (pp. 744-746)
        1. Functions
        2. Structure
            a. External Nose
            b. Nasal Cavities
            c. Paranasal Sinuses
            d. Disorders
                1. Rhinitis
                2. Sinusitis
    B. Pharynx (p. 747)
        1. Function
        2. Structure

     a. Nasopharynx

     b. Oropharynx

     c. Laryngopharynx

C. The Larynx (pp. 747-749)

   1. Location

   2. Function

   3. Associated Structures (Fig. 23.4, p. 748)

     a. Thyroid Cartilage

     b. Cricoid Cartilage

     c. Epiglottis

     d. Vocal Ligaments

     e. Vocal Folds

   4. Voice Production

     a. Speech

     b. Loudness

     c. Quality of Sound

   5. Sphincter Function of the Larynx

D. The Trachea (pp. 749-751)

   1. Location

   2. Structure

   3. Function

E. The Bronchial Tree (pp. 751-754)

   1. The Conducting Zone (Fig. 23.7, p. 751)

     a. Right and Left Primary Bronchi

     b. Secondary (Lobar) Bronchi

     c. Tertiary (Segmented) Bronchi

     d. Bronchioles

     e. Terminal Bronchioles

   2. The Respiratory Zone (Fig. 23.8, p. 752)

     a. Respiratory Bronchioles

     b. Alveolar Ducts

     c. Alveolar Sacs

     d. Alveoli

     e. Respiratory Membrane (Fig. 23.9, p. 753)

       1. Type I Cells

       2. Type II Cells

   3. Alveolar Macrophages

F. Lungs and Pleural Coverings (pp. 754-756)

   1. Gross Anatomy of the Lungs (p. 754)

     a. Surface Attachments and Associations

     b. Lobes

     c. Bronchopulmonary Segments

2. Blood Supply and Innervation of the Lungs (p. 754)

   a. Pulmonary Arteries and Veins

   b. Pulmonary Capillary Network

   c. Bronchial Arteries and Veins

3. The Pleura (p. 754)

   a. Parietal Pleura

   b. Visceral Pleura

   c. Pleural Fluid

III. Mechanics of Breathing (pp. 756-763)

  A. Pressure Relationships in the Thoracic Cavity (pp. 756-757)

    1. Intrapulmonary Pressure (Fig. 23.11, p. 757)

    2. Intrapleural Pressure

    3. Factors Holding the Lungs to the Thorax Wall

    4. Factors Pulling the Lungs away from the Body Wall

  B. Pulmonary Ventilation (pp. 757-759)

    1. Boyle's Law

    2. Inspiration

    3. Expiration

  C. Physical Factors Influencing Pulmonary Ventilation (pp. 759-761)

    1. Respiratory Passageway Resistance

    2. Lung Compliance

    3. Lung Elasticity

    4. Alveolar Surface Tension Forces

  D. Respiratory Volumes and Pulmonary Function Tests (pp. 761-763)

    1. Respiratory Volumes and Capacities (Fig. 23.15, p. 762)

      a. Respiratory Volumes

        1. Tidal

        2. Inspiratory Reserve

        3. Expiratory Reserve

        4. Residual Volume

      b. Respiratory Capacities

        1. Respiratory Capacities

          a. Inspiratory Capacity

          b. Functional Residual Capacity

          c. Vital Capacity

          d. Total Lung Capacity

    2. Dead Space

    3. Pulmonary Function Tests

      a. Spirometer

      b. Minute Respiratory Volume

      c. Forced Vital Capacity

      d. Forced Expiratory Volume

     3. Influence of Higher Brain Centers
        a. Hypothalamic Controls
        b. Cortical Controls
     4. Chemical Factors
        a. Influence of PCO_2
           1. Hypercapnia
           2. Hyperventilation
           3. Hypocapnia
           4. Apnea
        b. Influence of PO_2
        c. Influence of Arterial pH
        d. Summary of Interactions of PCO_2, PO_2, and Arterial pH

VII. Respiratory Adjustments During Exercise and at High Altitudes (pp. 777-778)
  A. Effects of Exercise (pp. 777-778)
     1. Increase in Ventilation
     2. Decrease in Ventilation
     3. Rise in Lactic Acid Levels
  B. Effects of High Altitudes (p. 778)
     1. Acclimatization
     2. Decreased Hemoglobin Saturation Levels

VIII. Homeostatic Imbalances of the Respiration System (pp. 778-780)
  A. Chronic Obstructive Pulmonary Disorders (p. 779)
     1. Obstructive Emphysema
     2. Chronic Bronchitis
  B. Tuberculosis (p. 779)
  C. Lung Cancer (p. 779)
     1. Squamous Cell Carcinoma
     2. Adenocarcinoma
     3. Small Cell Carcinoma

IX. Developmental Aspects of the Respiratory System (p. 780)
  A. Upper Respiratory Structures
  B. Lower Respiratory Organs
  C. Birth Defects
     1. Cleft Plates
     2. Cystic Fibrosis
  D. Patterns of Breathing Movements

Review Items

1. Mediastinum (Chapter 1, p. 17)

2. Acids and bases (Chapter 2, pp. 40-42)

3. Diffusion (Chapter 3, pp. 66-70)

4. Hyaline cartilage (Chapter 4, p. 119)

5. Elastic cartilage (Chapter 4, p. 120)

6. Squamous epithelium (Chapter 4, p. 105)

7. Cuboidal epithelium (Chapter 4, p. 106)

8. Pseudostratified epithelium (Chapter 4, p. 106)

9. Serous and mucous glands (Chapter 4, pp. 110-112)

10. Bones of the skull (Chapter 7, pp. 179-193)

11. Muscles of respiration (Chapter 10, pp. 306-307)

12. Medulla and pons (Chapter 12, pp. 397-399)

13. Cortex (Chapter 12, pp. 384-390)

14. Chemoreceptors, proprioceptors (Chapter 13, p. 426)

15. Sympathetic effects (Chapter 14, p. 471)

16. Auditory tube (Chapter 16, p. 528)

17. Great vessels (Chapter 19, pp. 615-619)

18. Autoregulation of blood flow (Chapter 20, pp. 661-662, 664)

19. Pulmonary circulation (Chapter 20, p. 667)

20. Tonsils (Chapter 21, pp. 701-702)

21. Inflammation (Chapter 22, pp. 710-712)

22. Macrophages (Chapter 22, p. 709)

23. Lysozyme (Chapters 16, p. 508)

Cross References

1. Acid-base balance of the blood is discussed in detail in Chapter 27, pp. 940-950.

2. The role of acidosis in initiating fetal respirations is covered in Chapter 29, p. 1022.

Laboratory Correlations

1. Marieb, E. N. *Human Anatomy and Physiology Laboratory Manual: Cat and Fetal Pig Versions.* 3rd. ed. Benjamin/Cummings, 1989.

    Exercise 36: Anatomy of the Respiratory System

    Exercise 37: Respiratory System Physiology

2. Marieb, E. N. *Human Anatomy and Physiology Laboratory Manual: Brief Version.* 3rd. ed. Benjamin/Cummings, 1992.

    Exercise 33: Anatomy of the Respiratory System

    Exercise 34: Respiratory System Physiology

Transparencies Index

Bassett Atlas Figures Index

INSTRUCTIONAL AIDS

Lecture Hints

1. Stress the difference between ventilation and respiration.

2. Show slides or acetates of the bones of the skull followed by the Bassett Atlas slides of the sagittal section of the head to illustrate the relationship between bony and soft tissue structures.

3. The conducting airways of the head are usually the most confusing of the respiratory structures. Spend some time with diagrams and photographs reinforcing the three-dimensional anatomy of the upper airway structures.

4. Point out the characteristics of the epithelia lining in the conducting airways, and why those epithelia are the correct choice for that particular area. This will reinforce epithelial types and gradually establish an intuitive sense in the students so that they can predict epithelia for any location in the body.

5. During a discussion of the trachea, ask students why the cartilage rings are "C" shaped rather than continuous.

6. Be sure the class does not confuse the respiratory membrane with subcellular level membrane structures (plasma membrane, etc.).

7. Remind students that pulmonary vessels are exceptions to the rule that arteries = oxygenated, and veins = deoxygenated, blood. Students should not confuse bronchial artery (oxygenated blood) with pulmonary artery (deoxygenated blood).

8. Draw a diagram during lecture of the position of the ribs before and after contraction of the external intercostal muscles to illustrate the volume increase of the thorax.

9. Emphasize elastic recoil as the main mechanism of normal expiration.

10. A complete understanding of diffusion is necessary for comprehension of respiratory gas movement at lung and body tissue levels. Refer the class to the section on diffusion in Chapter 3 the session before the discussion of respiratory physiology is planned.

11. Mention that cellular respiration is not the same as internal or external respiration, but that cellular respiration involves the pathways of glucose catabolism.

12. Emphasize the increasing difficulty with which successive oxygens are removed from a hemoglobin molecule. This explains why hemoglobin is not unsaturated when returned to the lungs.

13. Point out that the carbon dioxide transport (bicarbonate buffering) system is the most important mechanism of maintaining pH of the blood.

14. Mention that, unlike the heart which has its own intrinsic basic rate control, the respiratory system relies on continuous stimulation by brain stem control centers. Therefore, if innervation is lost to inspiratory muscles, breathing stops.

15. Students often have the misconception that oxygen level is the principal stimulant of respiration. Emphasize that carbon dioxide level is the most important factor via changing CSF pH.

Demonstrations/Activities

1. Audio-visual materials of choice.

2. Use a torso model, respiratory system model, and/or dissected animal model to exhibit the respiratory system and related organs.

3. Use two glass slides with water between them to demonstrate the cohesive effect of the serous fluid between the chest cavity wall and the lungs via the pleura and its parts. (Note: Due to this force, chest cavity movement results in lung movement since the lungs cannot pull away from the chest wall under normal conditions.)

4. Use an open-ended bell jar with balloons inside to demonstrate the changing pressures as the diaphragm contracts and relaxes. (Note: Top of bell jar should have a one-hole stopper with a glass Y tube extending into the jar; to the Y tube will be attached two small balloons; the bottom of the jar will be covered with a flexible elastic sheeting.)

5. Use a stringed instrument to demonstrate the effect of vibration and thickness on sound production.

6. Use laundry detergent in a glass of water with some cloth in it to demonstrate the role of surfactant in the lungs for reducing water surface tension and as attraction for other water molecules.

7. Use a freshly opened soft drink to demonstrate and explain Henry's law.

8. Demonstrate the location of the sinuses using a complete or Beauchene's skull.

9. Obtain a fresh lamb or calf pluck (lungs plus attached trachea and heart) from a slaughterhouse. Insert a rubber hose snugly into the trachea and attach the hose to a source of compressed air.

Alternately inflate the lungs with air and allow them to deflate passively to illustrate the huge air capacity and elasticity of the lungs.

10. Obtain some animal blood and bubble air through the blood via a small section of tubing to demonstrate the color change that occurs when blood is well oxygenated.

11. Provide tape measures so that students can measure the circumference of the rib cage before and after inspiration.

12. Provide stethoscopes so that students can listen to respiratory (breathing) sounds over various regions of a partner's thorax. For example, bronchial sounds are produced by air rushing through the large passages (trachea and bronchi) whereas the more muffled vesicular breathing sounds are heard over the smallest airways and alveoli.

13. Using hand-held spirometers, have students measure their respiratory volumes, particularly tidal volume and vital capacity.

14. Provide straws, beakers of water, and pH paper. Have students use the straws to blow into the water in the beakers. Since exhaled air contains a significant amount of CO_2, the water should become acidic. Have them measure the pH of the water at intervals to follow the pH change.

Critical Thinking/Discussion Topics

1. Discuss why athletes would want to train at high altitudes if their competition was to be at a high altitude (relate to USA's Denver, Colorado, Olympic training site), or even if their competition was to be at a lower altitude.

2. Discuss why an individual who has never smoked can get lung cancer.

3. Discuss the changes in respiratory volumes with obstructive or congestive disorders.

4. Discuss the relationship between oxygen debt and muscle fatigue and an elevated respiratory rate after exercise.

5. Discuss the logic behind the structure of the conducting airways, i.e., Why are cartilage rings necessary? Why are smooth muscle in the walls of the conducting tubes necessary?

6. Discuss the relationship between intrapulmonary pressure and intrapleural pressure. What happens to intrapulmonary pressure relative to intrapleural pressure when the Valsalve maneuver is performed?

7. Why are only slightly higher atmospheric levels of carbon monoxide gas dangerous?

Library Research Topics

1. Research and list the respiratory diseases caused by inhalation of toxic particles associated with specific occupations such as coal mining, etc.

2. Research the incidence of cancer in smokers versus nonsmokers, in individuals working in respiratory hazard areas versus individuals working in relatively safe respiratory areas.

3. Research the current status of heart-lung transplants, and why such a transplant would be considered.

4. Research the causes, known and supposed, of sudden infant death syndrome.

5. Research the respiratory problems a newborn infant might face.

Audio-Visual Aids/Computer Software

Slides

1. Visual Approach to Histology: Respiratory System (FAD, 11 Slides)

2. Respiration (EIL, 71 Slides)

3. Respiratory System and Its Function (EIL, Slides). The respiratory tree is viewed macroscopically and microscopically, augmented with discussion of blood oxygenation, CO_2 release, and thoracic pressure changes.

Videotapes

1. The Living Body: Breath of Life (FHS, VHS or BETA, 26 min., C). Explains why the body needs regular supplies of air and how it gets them. The camera follows the process of breathing through the ultrathin membrane of the lung into the blood, showing how the varying demand for O_2 is met by the exchange of information between the brain and the chest muscles and how the body rids itself of CO_2.

Computer Software

1. Respiratory Diseases and Disorders (PLP, Apple). Covers prevention, symptoms, and treatment of bronchitis, cystic fibrosis, emphysema, tuberculosis, pneumonia, asthma, shortness of breath.

2. Dynamics of the Human Respiratory System (EIL, C 3057A, Apple II, C 3057M, IBM, 1988). Contains a straightforward, concise text with engaging animated graphics, sound effects, and an interactive quiz to instruct and challenge.

3. Body Language: Study of Human Anatomy, Respiratory System (PLP, CH-182009, Apple; CH-182008, IBM)

4. The Human Systems: Series 3. The respiratory, excretory, and reproductive systems (PLP, CH-140220, Apple)

See Guide to Audio-Visual Resources on page 339 for key to AV distributors.

LECTURE ENHANCEMENT MATERIAL

Clinical and Related Terms

1. Anoxia — an absence or deficiency of oxygen within tissues.

2. Apnea — temporary cessation of breathing.

3. Asphyxia — condition in which there is a deficiency of oxygen and an excess of carbon dioxide in tissues and the blood.

4. Dyspnea — difficulty in breathing.

5. Eupnea — normal breathing.

6. Hemothorax — presence of blood in the pleural cavity.

7. Hyperpnea — increased rate and depth of breathing.

8. Hypoxemia — a deficiency in the oxygen concentration of blood.

9. Pneumoconiosis — a condition characterized by the accumulation of particles from the environment in the lungs and the reaction of the tissue to their presence; pulmonary fibrosis produced by inhalation of injurious dust or other particulate matter.

Disorders/Homeostatic Imbalances

Disorder Symptoms

1. Chest Pain — may be a symptom of respiratory disorder. It may be due to a variety of conditions, ranging from a transient and insignificant event to a most serious and life-threatening medical problem.

2. Cough — a respiratory action primarily intended to maintain airway patency by eliminating materials accumulated or deposited on the mucosa of the respiratory tract. Such materials include tracheobronchial secretions, aspirated substances, blood and other foreign bodies.

3. Dyspnea — an uncomfortable awareness of breathing due to increased work of ventilation out of proportion to the level of exertion. It may range in intensity from a mild discomfort to extreme distress. Causes of dyspnea include abnormalities in chemical stimulation and pulmonary vein dilation. Abnormal chemical stimulation disorders include a depressed respiratory center with periodic dyspnea (Cheyne-Stokes respiration); decreased oxygen availability (as in high altitudes); decreased cardiac output (as in valvular stenosis and constrictive pericarditis); congenital heart defects; pulmonary hypoxia (embolism); severe congestive heart failure; severe anemia; diabetic acidosis or uremia; pulmonary destruction, obstruction, or compression (emphysema, pneumonia, atelectasis, pleural effusion). Pulmonary vein dilation disorders include congenital heart defects causing increased blood flow to the lungs and the following disorders that cause increased venous pressure in the lungs: mitral valve defects (stenosis, insufficiency); constriction of the left heart, left atrial tumor or thrombus; pericardial effusion; and left ventricular failure.

4. Expectoration — the act of coughing up and spitting out sputum. Sputum, the material raised from the respiratory tract following a cough, usually consists of material secreted by the mucous glands and goblet cells of the tracheobronchial tree.

5. Hemoptysis — means spitting blood and generally refers to expectoration of blood that originates from the respiratory tract below the pharynx. Blood-tinged or blood-streaked sputum is not usually called hemoptysis. Thus hemoptysis signifies coughing up of a certain quantity of blood, pure or mixed with sputum. The amount of bleeding indicates the severity of hemoptysis. It can be caused by infectious diseases (bacterial pneumonia and tuberculosis); neoplasms (laryngeal and bronchogenic carcinomas); trauma; or cardiovascular diseases (pulmonary embolism with infarction or mitral valve narrowing with congestive heart failure).

6. Hoarseness — an indication of laryngeal diseases such as varying stages of laryngitis. It will vary from roughening of the voice to its total loss.

7. Wheezing — a whistling sound made by breathing. It is a usual symptom of asthma but may also be present in such conditions as acute bronchitis and other causes of bronchial narrowing.

Congenital and Inherited Disorders

1. Antitrypsin Deficiency — an autosomal recessive disease which can cause cirrhosis in homozygous children or emphysema in homozygous adults.

2. Cystic Fibrosis — an autosomal recessive disorder resulting in hypersecretion of NaCl, chronic respiratory tract infections, and destruction of secretory glands (liver and pancreas). Untreated, it leads to chronic debilitation and early death.

3. Infant Respiratory Distress Syndrome (Hyaline Membrane Disease) — a condition caused by inadequate production of surfactant, which impedes normal lung expansion and promotes collapse of the alveoli. Premature infants, infants born by cesarean section, and infants of diabetic mothers are predisposed to this syndrome.

Mechanical and Traumatic Disorders

1. Atelectasis — refers to a collapse of parts of the lung. Types include obstructive atelectasis, which results from bronchial obstruction, and compression atelectasis, which is the result of external compression of the lung. Treatment involves removing the obstruction or material compressing the lung.

2. Chronic Obstructive Lung Disease — a combination of emphysema and chronic bronchitis. Emphysema is characterized by enlargement of the air spaces distal to the terminal bronchioles and possible destruction of their walls. Usually there is an associated chronic inflammation of the terminal bronchiole. As a result, emphysema and chronic bronchitis are usually referred to as a single entity, chronic obstructive pulmonary disease. Clinical symptoms include dyspnea and cyanosis. Treatment cannot restore damaged lung tissue but can prevent further damage and may even improve pulmonary function.

3. Lung Carcinoma (Cancer of the Lung) — usually a smoking-related neoplasm; however, individuals who have never smoked may also develop the disorder. There are several histological types, each of which differs in its prognosis. Treatment depends on histological type. Other cancers of the respiratory tract also occur. Cancers of the lip, mouth, nasopharynx, and larynx are common, especially in smokers beyond the age of 50.

4. Pneumothorax — entrance of air into the space between the pleural membranes, followed by collapsing of the lung. It may be due to a chest or lung injury that permits air to escape into the pleural cavity. Symptoms include pain and dyspnea. Treatment includes inserting a tube into the pleural cavity through an incision in the chest wall. The tube prevents accumulation of air in the pleural cavity and aids reexpansion of the lung.

5. Pulmonary Fibrosis — a fibrous thickening of alveolar septa which makes the lungs increasingly rigid and restricts normal respiratory functions. Causes include exposure to irritating substances in the air such as silicon and asbestos. Some types of collagen diseases may also lead to pulmonary fibrosis. Pulmonary fibrosis produced by inhalation of injurious dust or other particulate matter is called pneumoconiosis. Inhalation of rock results in silicosis. Inhalation of asbestos fibers is called asbestosis. Inhalation of coal dust is called anthracosis or black lung disease. Chemical pneumonitis is due to inhalation of irritating gases such as ammonia, chlorine, nitrogen dioxide, and ozone. Exposure to metallic fumes from cadmium, beryllium, lead, and zinc can also damage the lung. Allergies can develop from inhaling organic dust such as molds.

Inflammatory and Infectious Disorders

1. Asthma — a response to inhaled antigens resulting in spasmodic contraction of bronchial smooth muscle and narrowing of air passageways.

2. Bronchiectasis — the weakening of bronchial walls in parts of the lung as a result of severe inflammation or other such factors, with the involved bronchi becoming markedly dilated. Since the distended bronchi tend to retain secretions, the patient usually has a chronic cough associated with production of large amounts of purulent sputum. The only effective treatment is surgical resection of the involved segments of lung tissue.

3. Chronic Bronchitis — a condition which results from a chronic inflammatory process in the bronchi. The inflammation causes excess mucus production, provoking coughing.

4. Colds — disorder caused by many different viruses, especially rhinoviruses. Symptoms include mucus flow, fever, and headache. No cure is available to date. Treatment usually involves treating symptoms, if any treatment other than time is used.

5. Epiglottitis — an infection of the epiglottis. It is a life-threatening complication of the common cold. It is caused by the bacterium *Hemophilus influenza* and is characterized by swelling and blockage of the airway opening.

6. Legionnaires' Disease — a type of pneumonia. Symptoms include malaise, muscle aches, headache, abdominal pain, rapidly rising fever and chest discomfort. Treatment is the same as for any form of pneumonia.

7. Lung Abscesses — can be a complication of pneumonia or bronchiectasis, with anaerobic gram negative or staphylococcal bacteria most often responsible for the abscesses.

8. Otitis Media — a bacterial infection of the middle ear causing earache and fever. It is most common in children. Lack of treatment can result in damage to the hearing apparatus and also possibly brain damage.

9. Pneumonia — can be caused by an etiologic agent (bacterial, viral, mycoplasma, or fungal), by anatomic distribution of inflammation in the lung, and by predisposing factors. Symptoms include cough, chest pain, fever, and other manifestations of systemic infection. Complications of pneumonia include fluid accumulation in the chest cavity (pleural effusion), pus in the thoracic cavity (empyema), blood poisoning (bacteremia), and on occasion, inflammation of the covering of the brain (meningitis) or inflammation of the heart valves (endocarditis).

10. Tracheobronchitis — an inflammation of the trachea. The inflamed trachea can collapse, blocking passage of air.

11. Tuberculosis — an acute and/or chronic communicable disease caused by *Mycobacterium tuberculosis* and is characterized by necrosis and giant cell production resulting in granulomatous inflammation. Treatment is a combination of drug therapy, rest, and proper nutrition.

ANSWERS TO END-OF-CHAPTER QUESTIONS

Multiple Choice/Matching

1. b	5. b	9. c, d	13. b
2. a	6. d	10. c	14. c
3. c	7. d	11. b	15. b
4. c	8. b	12. b	16. b

Short Answer Essay Questions

17. The route of air from the external nares to an alveolus and the organs involved are as follows: conducting zone structures — external nares, nasal cavity, pharynx (nasopharynx, oropharynx, laryngopharynx), larynx, trachea, and right and left primary bronchi, secondary bronchi, tertiary bronchi and successive bronchi orders, bronchioles, and terminal bronchioles; respiratory zone structures — respiratory bronchioles, alveolar ducts, alveolar sacs, and alveoli. (pp. 744-753)

18. a. The trachea is reinforced with cartilage rings to prevent the trachea from collapsing and to keep the airway patent despite the pressure changes that occur during breathing.

   b. The advantage of the rings not being complete posteriorly is that the esophagus is allowed to expand anteriorly during swallowing. (pp. 749-750)

19. The adult male larynx as a whole is larger and the vocal cords are longer than those of women or boys. These changes occur at puberty under the influence of rising levels of testosterone. (pp. 747-749)

20. a. The elastic tissue is essential both for normal inspiration and expiration; expiration is almost totally dependent on elastic recoil of the lungs when the inspiratory muscles relax.

   b. The passageways are air conduits used to transport air. (pp. 757-759)

21. The volume of gas flow to and from the alveoli is directly proportional to the difference in pressure between the external atmosphere and the alveoli. Very small differences in pressure are sufficient to produce large volumes of gas flow. (pp. 757-759)

22. The walls of the alveoli are composed of a single layer of squamous epithelium underlain by a flimsy basement membrane which is fused to the endothelium of the pulmonary capillaries. The thinness of the respiratory membrane allows gas diffusion to occur very rapidly across the membrane. (pp. 753-754)

23. Pulmonary ventilation is influenced by respiratory passageway resistance in that gas flow is equal to the pressure gradient divided by the resistance. Gas flow changes inversely with resistance. Lung compliance is assessed by measuring the increase in lung volume resulting from an increase in intrapulmonary pressure. The greater the volume increase for a given rise in pressure, the greater the compliance. The ability of lung tissue to distend and recoil, called lung elasticity, is essential for normal inspiration and expiration. Due to surfactant, the surface tension of alveolar fluid is reduced, and less energy is needed to overcome surface tension forces to expand the lungs. (pp. 759-761)

24. a. Minute respiratory volume is the total amount of gas that flows into and out of the respiratory tract in one minute. Alveolar ventilation rate takes into account the amount of air wasted in dead space areas and provides a measurement of the concentration of fresh gases in the alveoli at a particular time. (pp. 761-763)

    b. Alveolar ventilation rate provides a more accurate measure of ventilatory efficiency because it considers only the volume of air actually participating in gas exchange. (p. 763)

25. Dalton's law of partial pressure states that the total pressures exerted by a mixture of gases is the sum of the pressure exerted independently by each gas in the mixture. Henry's law states that when a mixture of gases is in contact with a liquid, each gas will dissolve in the liquid in proportion to its partial pressure and its solubility in the liquid. (pp. 763-764)

26. Partial pressure of a gas is the pressure exerted by each gas in a gas mixture and is a reflection of the relative concentrations of the gas in the gaseous mixture. (p. 764)

27. a. Hyperventilation is deep breathing which flushes carbon dioxide rapidly out of the blood.

    b. When you hyperventilate, you expel more carbon dioxide.

    c. Hyperventilation increases blood pH. (pp. 774-775)

28. Age-related changes include a loss of elasticity in the lungs and a more rigid chest wall. These factors result in a slowly decreasing ability to ventilate the lungs. Accompanying these changes is a decrease in blood oxygen levels and a reduced sensitivity to the stimulating effects of carbon dioxide. (p. 781)

Critical Thinking and Application Questions

1. Hemoglobin is almost completely (98%) saturated with oxygen in arterial blood. Hence, hyperventilation will increase the oxygen saturation very little, if at all. However, hyperventilation will flush CO_2 out of the blood, ending the stimulus to breathe and possibly causing (1) cerebral ischemia due to hypocapnia, and (2) O_2 decrease to dangerously low levels, resulting in fainting. (pp. 777-778)

2. a. The lung penetrated by the knife collapsed because the intrapleural pressure became equal to the atmospheric pressure, allowing the pleural membranes to separate.

    b. Only the penetrated lung collapsed, because it is isolated from the remaining mediastinal structures (and the other lung) by the pleural membranes. (p. 738)

3. a. Normally, during swallowing, the soft palate reflects superiorly to seal the nasopharynx and prevent food or drink from entering the nasal cavity. During giggling, however, this sealing mechanism sometimes fails to operate (because giggling demands that air be forced out of the nostrils), and swallowed fluids may enter the nasal cavity, then exit through the nostrils.

    b. Even though standing on his head, the boy made certain that he swallowed carefully, so that his soft palate correctly sealed the entrance to his nasal cavity. Then, his swallowing muscles directed the milk through his esophagus to the stomach, against gravity. (p. 749)

4. Adjacent bronchopulmonary segments are separated from one another by partitions of dense connective tissue, which no major vessels cross. Therefore, it is possible for a surgeon to dissect adjacent segments away from one another. The only vessels that had to be cauterized were the few main vessels to each bronchopulmonary segment. (p. 754)

Key Figure Questions

Figure 23.1 The lungs. **Figure 23.3** Because with increased turbulence, more debris (dust, bacteria) will be hurled into contact with the sticky mucus and prevented from going farther. **Figure 23.4** The epiglottis. **Figure 23.13** They both decrease. **Figure 23.18** They would be dilated. This response allows matching of blood flow to availability of oxygen. **Figure 23.20** The same direction.

SUGGESTED READINGS

1. Blakeslee, S. "Nicotine, Harder to Kick. . ." *New York Times Magazine* (Mar. 29, 1987): 22-23, 49-53.

2. Bloom, B.R., and C.J.L. Murray. "Tuberculosis: Commentary on a Reemergent Killer." *Science* 257 (Aug. 1992): 1055.

3. Carr, D.T. "Malignant Lung Disease." *Hospital Practice* (Jan. 1981).

4. Caruthers, D.D. "Infectious Pneumonia in the Elderly." *American Journal of Nursing* 90 (Feb. 1990): 56-64.

5. Collins, F.S. "Cystic Fibrosis: Molecular Biology and Therapeutic Implications." *Science* 256 (May 1992): 774.

6. DeVito, A.J., and M. Kleven. "Dyspnea: Find the Cause, Treating the Symptoms." *RN* (June 1987): 40.

7. Fackelmann. K.A. "The Double Whammy of TB and AIDS." *Science News* 137 (June 1990): 348.

8. Hart, S., and A. McKenzie. "Testing Newborns for Cystic Fibrosis." *Science News* 136 (Oct. 1989): 233.

9. Ioli, J.G., and M.J. Richardson. "Giving Surfactant to Premature Infants." *American Journal of Nursing* 90 (Mar. 1990): 59-60.

10. McNaull, F.W., et al. "Lung Cancer." *American Journal of Nursing* 87 (Nov. 1987): 1427.

11. Mayo, J.M., and J.B. Hammner. "A Nurse's Guide to Mechanical Ventilation." *RN* (Aug. 1987): 18-24.

12. "Lungs Hurt Most by Ozone-Acid Synergy." *Science News* 130 (Aug. 1986): 85.

13. Nalye, R.R. "Sudden Infant Death." *Scientific American* 242 (Apr. 1980).

14. Pennisi, E. "Gene, Biochemical Fixes Sought for CF." *Science News* 144 (Oct. 1993): 260.

15. Raloff, J. "Making Blood Share More of Its Oxygen." *Science News* 129 (Apr. 1986): 260.

16. Sarsany, S.L. "Respiratory Distress." *RN* (Apr. 1988): 47.

17. Sataloff, R.T. "The Human Voice." *Scientific American* 267 (Dec. 1992): 108.

18. Schulthesis, A.H. "When and How to Extubate in the Recovery Room." *American Journal of Nursing* 89 (Aug. 1989): 1040-1045.

19. Silberner, J. "Hyperbaric Oxygen Bounces Back." *Science News* 125 (Oct. 1985): 236.

20. Silverstein, S.C., and J.J. Wine. "ATP and Chlorine Conductance." *Nature* 360 (Nov. 1992): 18.

21. Snider, G. "Chronic Obstructive Pulmonary Disease: Advice on Dx and Rx." *Modern Medicine* (Dec. 1982).

22. Weiss, R. "TB Trouble: Tuberculosis Is on the Rise Again." *Science News* 133 (Feb. 1988): 92.

23. Weiss, R. "Surfactant Therapy." *Science News* 134 (Oct. 1988): 245.

24. Worthington, L. "What Those Blood Gases Can Tell You." *RN* 42 (Oct. 1979): 23.

24

The Digestive System

Chapter Preview

This chapter deals with the food procurement, processing, and elimination plant for the body. The organs of this system take in, process, and eliminate the wastes of the raw products needed for fuel and for the intricate workings of the human body. The steps involved in this procedure include ingestion of food, digestion (both mechanical and chemical), absorption of the end products of digestion, and elimination of the wastes, the unusable parts of the food (a process called defecation). Without this system, the body would be unable to obtain its needed raw products for fuel and for maintaining the various cells of the body.

INTEGRATING THE PACKAGE

Suggested Lecture Outline

I. Overview (pp. 789-795)
   A. Digestive System Organs (pp. 789-790)
      1. Alimentary Canal
      2. Accessory Digestive Organs
   B. Digestive Processes (pp. 790-791)
      1. Ingestion
      2. Propulsion
      3. Mechanical Digestion
      4. Chemical Digestion
      5. Absorption
      6. Defecation
   C. Basic Functional Concepts (pp. 791-792)
   D. Digestive System Organs: Relationships and Structural Plan (pp. 792-795)
      1. Relationship of Digestive Organs and Peritoneum
         a. Visceral and Parietal Peritoneum
         b. Peritoneal Cavity
         c. Mesentery
         d. Retroperitoneal Organs
      2. Blood Supply
         a. Splanchnic Circulation

D. Small Intestine and Associated Structures (pp. 813-825)
 1. The Small Intestine
  a. Gross Anatomy
  b. Microscopic Anatomy
   1. Plicae Circulares
   2. Villi
   3. Microvilli
  c. Histology of the Wall
  d. Intestinal Juice: Composition and Control
 2. The Liver and Gallbladder
  a. Gross Anatomy of the Liver (Fig. 24.24, p. 819)
  b. Microscopic Anatomy of the Liver (Fig. 24.25, p. 820)
   1. Composition of Bile
  c. The Gallbladder
  d. Regulation of Bile Release into the Small Intestine
 3. Pancreas (Fig. 24.27, p. 823)
  a. Composition of Pancreatic Juice
  b. Regulation of Pancreatic Secretion
 4. Digestive Processes Occurring in the Small Intestine
  a. Regulation for Optimal Intestinal Digestive Activity
  b. Motility of the Small Intestine
E. Large Intestine (pp. 826-830)
 1. Gross Anatomy (Fig. 24.30, p. 827)
 2. Microscopic Anatomy (Fig. 24.31, p. 828)
 3. Bacterial Flora
 4. Digestive Processes Occurring in the Large Intestine
  a. Motility of the Large Intestine
  b. Defecation
III. Physiology of Chemical Digestion and Absorption (pp. 830-834)
 A. Chemical Digestion (pp. 830-834)
  1. Mechanism of Chemical Digestion: Enzymatic Hydrolysis
  2. Chemical Digestion of Specific Food Groups (Fig. 24.33, pp. 831-832)
   a. Carbohydrates
    1. Salivary and Pancreatic Amylase
    2. Brush Border Enzymes
     a. Dextrinase
     b. Glucoamylase

      3. Disaccharide Enzymes
         a. Maltase
         b. Sucrase
         c. Lactase
      b. Proteins
         1. Pepsin
         2. Renin
         3. Pancreatic Enzymes
            a. Trypsin
            b. Chymotrypsin
         4. Brush Border Enzymes
            a. Carboxypeptidase
            b. Aminopeptidase
            c. Dipeptidase
    c. Lipids
         1. Lipases
    d. Nucleic Acids
         1. Pancreatic Nucleases
  B. Absorption (pp. 834-837)
    1. Absorption of Specific Nutrients
      a. Carbohydrates
      b. Proteins
      c. Lipids
         1. Micelles
         2. Chylomicrons
      d. Nucleic Acids
      e. Vitamins
      f. Electrolytes
      g. Water
    2. Malabsorption of Nutrients
      a. Gluten Enteropathy
IV. Developmental Aspects of the Digestive System (pp. 837-840)
  A. Embryonic Development (Fig. 24.37, p. 840)
  B. Aging

Review Items

1. Serous membranes (Chapter 1, p. 20)
2. Enzyme function (Chapter 2, pp. 50-53)
3. Acids and bases (Chapter 2, pp. 40-42)
4. Carbohydrates, lipids, proteins, and nucleic acids (Chapter 2, pp. 43-56)

5. Microvilli (Chapter 3, p. 64)

6. Membrane transport (Chapter 3, pp. 65-73)

7. Simple columnar epithelium (Chapter 4, p. 105)

8. Areolar connective tissue (Chapter 4, p. 116)

9. Serous and mucous glands (Chapter 4, pp. 110-111)

10. Smooth muscle (Chapter 9, pp. 275-280)

11. Mastication and tongue movement (Chapter 10, p. 298)

12. Brain stem centers (Chapter 12, pp. 395-399)

13. Receptors (Chapter 13, pp. 426-429)

14. Reflex activity (Chapter 13, pp. 450-456)

15. Nerve plexuses (Chapter 13, pp. 444-450)

16. Sympathetic and parasympathetic controls (Chapter 14, p. 472)

17. Papillae and taste buds (Chapter 16, pp. 502-503)

18. Hormones (Chapter 17, pp. 548-553)

19. Pernicious anemia (Chapter 18, p. 593)

20. Lymphatic tissue (Chapter 21, p. 698)

21. Lacteals (Chapter 21, p. 695)

22. Palatine tonsils (Chapter 21, pp. 701-702)

23. Macrophages (Chapter 22, pp. 709-710)

Cross References

1. Hepatic metabolism and detoxification are further described in Chapter 25, pp. 879-881.

2. Electrolyte balance is described in Chapter 27, pp. 934-940.

3. The role of chylomicrons in lipid metabolism is presented in Chapter 25, pp. 869-871.

4. Bile formation is further explained in Chapter 25, pp. 879-881.

5. Cholesterol and lipid transport in the blood is examined in Chapter 25, pp. 879-881.

Laboratory Correlations

1. Marieb, E. N. *Human Anatomy and Physiology Laboratory Manual: Cat and Fetal Pig Versions.* 3rd. ed. Benjamin/Cummings, 1989.

   Exercise 38: Anatomy of the Digestive System

   Exercise 39: Mechanism of Food Propulsion and the Physiology of Smooth Muscle

   Exercise 40: Chemical Breakdown of Foodstuffs: Enzymatic Action

2. Marieb, E. N. *Human Anatomy and Physiology Laboratory Manual: Brief Version.* 3rd. ed. Benjamin/Cummings, 1992.

   Exercise 35: Anatomy of the Digestive System

   Exercise 36: Food Propulsion Mechanisms

   Exercise 37: Chemical Breakdown of Foodstuffs: Enzymatic Action

Transparencies Index

Bassett Atlas Figures Index

INSTRUCTIONAL AIDS

Lecture Hints

1. Emphasize that the digestive system is not only the alimentary (gastrointestinal) canal but all organs and tissues that aid in the process of digestion.

2. Point out that the gastrointestinal tract is formed of the same basic four layers through its length, but that each area is modified to the specific task involved.

3. Digestion is the process of breaking large particles into small particles. Emphasize that the overall function of the digestive system is mechanical and chemical breakdown of ingested substances followed by the absorption of those substances and elimination of undigestible materials.

4. Most students have difficulty with the serous coverings of the abdominal viscera. Use diagrams and photographs of actual tissue to reinforce descriptions of the relatively complex folded nature of these membranes.

5. Spend some time with the hepatic portal system. This is another example of blood entering a capillary bed, feeding into a vein, then into another capillary bed before being returned to general circulation.

6. When discussing the histology of the tract, ask the class: "What is the logical epithelial choice for the mouth? For the esophagus? Point out that the choice of columnar epithelium for the mucosal layer of the gastrointestinal tract is ideally suited to its function.

7. Emphasize that the esophagus is not covered by serosa, but instead has an adventitia as its outer-most coat.

8. Although teeth and bones are both constructed of calcium salts, stress that a major distinguishing characteristic is that bones are vascular and teeth are not.

9. Emphasize that the lower esophageal sphincter (gastroesophageal) is not a true sphincter.

10. As a point of interest, mention that heartburn is actually acid reflux into the lower portion of the esophagus.

11. Point out the modification of the muscularis externa in the stomach as it relates to the function of the stomach.

12. Have students note the difference between the way the mucosa in the stomach is a relatively low surface area structure as compared to the small intestine. Ask the students why this is so.

13. Points of interest: aspirin and alcohol are two commonly ingested substances that pass through the mucosa of the stomach; intrinsic factor (necessary for vitamin B_{12} absorption) is produced by gastric mucosa.

14. As each cell of the stomach mucosa is described, relate the logical function of each type to the overall function of the stomach.

15. Mention that the three areas of the small intestine are distinguishable histologically by examination of the mucosal structure.

16. Use diagrams or black line masters to demonstrate the three structural modifications of the small intestine that greatly increase the surface area for absorption.

17. When introducing the digestive function of the small intestine, lead into the topic by asking the class: "What functions must occur as chyme enters the initial part of the small intestine?" Using carefully led questioning, the class should respond: acid neutralization, further digestion of carbohydrates and proteins, and initiation of lipid digestion.

18. Students have difficulty with the pathways of flow in the liver lobule. Use two-dimentional cross sections of a lobule and indicate the directions of blood flow and bile flow. Stress the difference between hepatic portal vein and hepatic vein.

19. Ask the class why the hepatic artery is necessary, since the liver is already supplied by the portal vein. They should be able to respond that portal blood is "used" blood from the digestive tract.

20. Emphasize that the pancreas is a dual-function/structure gland, endocrine and exocrine.

21. *Taenia coli* are best explained by using a cross-sectional diagram followed by a longitudinal section.

22. Emphasize that the amount of time the contents of the large intestine are in contact with the mucosa determines fecal water content. Too little time in the large intestine means a watery stool, and too much time results in constipation.

23. Point out the logical names of digestive enzymes: prefix usually indicating the substrate, suffix "-ase" meaning enzyme. An exception, trypsin, was named before universal acceptance of the "-ase" convention.

24. Spend time on fat digestion and absorption, from emulsification to movement through the bloodstream. Point out that carbohydrates and proteins take a different (vascular) path to the liver than do the lipids.

Demonstrations/Activities

1. Audio-visual materials of choice.

2. Use a torso model and/or dissected animal model to exhibit digestive organs.

3. Use gallstones obtained from a surgeon to exhibit as you discuss the liver and gallbladder.

4. Use a long, not quite fully blown up balloon to demonstrate peristalsis.

5. Use a human skull or dentition models to demonstrate the different tooth shapes, types, and numbers.

6. Demonstrate the emulsification action of bile: first mix oil and water together and allow the layers to separate out. Then add bile salts and shake vigorously. Point out that the layer of oil has been dispersed into hundreds of tiny fat spheres by the action of the bile salts.

7. Demonstrate molecular models of carbohydrate, fat, and protein.

8. Have students calculate their total caloric intake over a 24-hour period by using a simple caloric guide obtainable in any drugstore. Have students analyze their diet with attention to what improvements could (and should) be made in their eating habits.

Critical Thinking/Discussion Topics

1. Discuss symptoms, treatment, and prognosis of a hiatal hernia.

2. Discuss why it is important to chew food properly.

3. Discuss the importance of the liver.

4. Discuss the cause, treatment, and prevention of ulcers.

5. Discuss why it is necessary for someone with ulcer-like symptoms to consult a physician rather than to just use antacids.

6. Discuss the reasons why elderly individuals should be checked for colorectal cancer.

7. Discuss the reasons for treatment and prognosis of a colostomy.

8. If a high-salt meal is ingested, why is a large amount of water not lost in the feces?

9. Discuss how people on low-carbohydrate diets have relatively constant glucose levels.

Library Research Topics

1. Research the causes and treatment of ulcers.

2. Research the benefits of fiber in the diet.

3. Research liver transplants in terms of rationale for the transplant, procedure, prognosis, etc.

4. Research the inherited metabolic disorders.

5. Research the congenital disorders that affect a newborn's ability to survive in the first days after birth.

6. Research the latest causes and treatments of hepatitis. What are the consequences of liver inflammation/infection?

7. What are malabsorption syndromes? Their causes? Their treatments?

8. Research the different types of motility disorders associated with the digestive tract. Include possible secondary complications and suggested treatments.

9. What are the common cancers of the digestive system? Are cancers limited to the gastrointestinal tract? Are they limited to the accessory structures?

Audio-Visual Aids/Computer Software

Slides

1. Visual Approach to Histology: Digestive System (FAD, 48 Slides)

2. Digestive System and Its Function (EIL, Slides). A tour of the alimentary tract brings the student to consider the basic chemical and morphological phenomena of digestion.

3. Histology of the Digestive System: Mouth to Esophagus (EIL). Presents a general plan of the digestive tract and covers the structure of the oral cavity and associated structures down to the point of junction of the esophagus with the stomach.

4. Histology of the Digestive System: Stomach, Intestine and Major Glands (EIL). The major regions of the stomach and intestine. The thickness of various layers, presence or absence of lymphatic tissue, and types and position of glands in each region are emphasized. Pancreas, liver, and gallbladder are also included.

Videotapes

1. The Living Body: Eating to Live (FHS, VHS or BETA). This program looks at appetite and hunger, and, by means of some of the most dramatic interior film in the series, shows the actions of a salivary gland, the swallowing reflex, and the powerful churning of the stomach as food is broken down and processed.

2. The Living Body: Breakdown (FHS, VHS or BETA). A family sits down to lunch. As the first morsel is put into the mouth, the camera watches from inside as the molars clamp down and the process of breakdown and transformation occurs, and then follows the food through the entire alimentary tract, showing how it is dissolved in acid, how the liver and gallbladder work, and how digestion and absorption work.

Computer Software

1. Dynamics of the Human Digestive System (EIL, Apple II, IBM, 1988). Contains straightforward, concise text with engaging animated graphics, sound effects, and an interactive quiz to instruct and challenge.

2. Digestion and Excretion (QUE, INT4153A, Apple II; INT4153B, IBM; INT4153M, Macintosh). Tutorial review of digestion and excretion.

3. Blood Sugar (QUE, ALB4605A, Apple II). Glucose homeostasis is demonstrated. Students investigate different factors involved in glucose control.

4. Body Language: Study of Anatomy, Digestive System (PLP, CH-182015, Apple; CH-182016, IBM). Anatomy tutorial.

5. The Human Systems: Series 1, (PLP, CH-920001, Apple). Covers mechanical and chemical digestion.

6. The Digestion Simulator (PLP, CH-140153, Apple). Animated graphics demonstrate the dynamics of digestion.

See Guide to Audio-Visual Resources on page 339 for key to AV distributors.

LECTURE ENHANCEMENT MATERIAL

Clinical and Related Terms

1. Achalasia — failure of the smooth muscle to relax at some junction in the digestive tube, such as that between the esophagus and stomach.

2. Achlorhydria — lack of hydrochloric acid in gastric secretions.

3. Aphagia — inability to swallow.

4. Cholecystitis — inflammation of the gallbladder; may be caused by presence of bile stones.

5. Cholelithiasis — presence of gallstones.

6. Cirrhosis — a liver condition in which the hepatic walls degenerate and the surrounding tissue becomes thickened.

7. Colitis — inflammation of the colon and rectum.

8. Diverticulitis — inflammation of small pouches (diverticula) that sometimes form in the lining and wall of the colon.

9. Dysentery — an intestinal infection characterized by diarrhea and cramps; caused by bacteria, viruses, or protozoans.

10. Dyspepsia — indigestion.

11. Dysphagia — difficulty in swallowing; may be due to a sore throat or some other inflammatory problem.

12. Enteritis — inflammation of the intestine, particularly the small intestine.

13. Gastritis — inflammation of the stomach lining.

14. Gingivitis — inflammation of the gums.

15. Hemorrhoids — enlarged veins in the lining of the anal canal.

16. Hepatitis — inflammation of the liver.

17. Pyorrhea — an inflammation of the dental periosteum accompanied by the formation of pus.

Disorders/Homeostatic Imbalances

1. Abnormalities of Tooth Development — a relatively common familial trait that follows a multifactorial pattern of inheritance. It results from failure of one or more tooth buds to develop. Tetracycline stains enamel of developing teeth and may cause abnormal tooth development.

2. Acute Enteritis — a relatively common inflammation of the intestine caused by a number of different microorganisms. Clinical symptoms include nausea, vomiting, diarrhea, and abdominal discomfort. Duration is limited.

3. Alcohol Related Disorders — can cause either a fatty liver or alcohol hepatitis. With a fatty liver, fat accumulates in liver cells due to the injury of the cells, resulting in impaired liver function. Other chemicals can also cause a fatty liver. Alcohol hepatitis results in not only fatty change in liver cells but liver necrosis as well.

4. Appendicitis — an inflammation of the appendix. The organ swells and may rupture leading to peritonitis, a potentially fatal condition. Treatment involves removal of the appendix (appendectomy) in all suspected cases.

5. Cancer of the Pancreas — adenomas are usually related to endocrine function. Carcinomas are usually fatal disorders due to obstruction of the ducts.

6. Cholelithiasis and Cholecystitis — inflammation of the gallbladder after accompanying gallstone production. Cholesterol, bilirubin, or minerals can form the gallstones. Major complications of cholecystitis and cholethiasis (gallstones) include biliary cirrhosis, shock, and acute pancreatitis. The disorder is asymptomatic in the gallbladder.

7. Chronic Enteritis — chronic inflammation of the intestine. Two types are recognized. Regional enteritis (Chron's disease) is localized to the distal ileum. It is chronic, relapsing, and thought to be due to an infectious agent, although the actual cause is unknown. Chronic ulcerative colitis is a chronic recurring inflammation of the colon with ulceration of the mucosa. Complications include anemia, arthritis, and carcinoma. Treatment includes use of antibiotics and cortisone, and, in severe cases, surgery.

8. Cirrhosis (Chronic Liver Disease) — refers to scarring in the liver regardless of cause. Causes include alcoholism, drugs, toxins, infectious agents, bile duct obstruction, metabolic and inherited abnormalities. Impaired liver function and portal hypertension result. Complications of cirrhosis include gastrointestinal hemorrhage, ascites, neuropsychiatric disorders (coma or psychosis), hepatorenal failure, and hepatocarcinoma.

9. Cleft Lip and Palate — defects in the development of the upper lip and jaw (cleft lip) or in the palate (cleft palate). Follows a multifactorial pattern of inheritance. Cleft lip may be unilateral or bilateral and may range in severity from a relatively minor defect in the mucosa of the lip to a large cleft extending deeply into the upper jaw. Surgical treatment is generally successful.

10. Cold Sore — a common infectious disease of the mouth caused by the herpes simplex virus. It is characterized by inflammation of the oral mucosa, lips, and gums. It often occurs when the individual is stressed (when immunity is depressed).

11. Cystic Fibrosis of the Pancreas — an autosomal recessive disease which, in the pancreas, causes obstruction of pancreatic ducts with thick mucus, blocking secretion of pancreatic enzymes. Eventually, the pancreas becomes a mass of cystically dilated ducts surrounded by dense fibrous tissue.

12. Dental Caries — caused by the erosion of enamel by organic acids produced by bacterial fermentation of retained food particles. Combined acid and bacterial acid destroy tooth structure, forming the cavity. Prevention, through good oral hygiene, is the best treatment.

13. Diverticulosis and Diverticulitis of the Colon — inflammation and bleeding due to food debris accumulation in outpouchings of colonic mucosa through weak areas in the wall. Predisposing factors include chronic constipation and low-residue (low-fiber) diets.

14. Esophageal Obstruction — leads to inability to swallow, often associated with regurgitation of food and saliva into the trachea, causing episodes of choking and coughing. Three common causes include carcinoma, impaction of food in the esophagus, and stricture. Carcinoma may arise anywhere in the esophagus. When it occurs, it gradually narrows the lumen and may invade the trachea, causing a tracheoesophageal fistula. Poorly chewed food may block the esophagus, causing a food impaction. Stricture occurs when the esophagus is narrowed by a scar such as caused by ingestion of corrosive chemicals.

15. "Gay Bowel" Syndrome — refers to disturbed bowel function experienced by gay men due to bacterial-parasite infection spread by sexual practices. It may be misinterpreted as irritable bowel syndrome. Specific treatment depends upon the source (bacterial or parasitic) of the inflammation.

16. Gluten Intolerance — a food intolerance disorder characterized by a hypersensitivity to wheat protein, leading to impaired intestinal absorption and atrophy of intestinal villi. Treatment involves removing wheat protein from the diet.

17. Hemorrhoids — varicose veins within the rectum. A predisposing factor is constipation and straining during a bowel movement. Hemorrhoids may become thrombosed and bleed. Treatment is conservative but may include surgery.

18. Hepatitis — inflammation and necrosis of liver tissue. Causes include infectious agents (viruses) and chemical toxins (such as alcohol). Infectious hepatitis is caused by hepatitis A virus obtained from contaminated food or liquids. Serum hepatitis is caused by hepatitis B virus obtained from blood. Other viral hepatitis infections include non-A, non-B hepatitis and hepatitis due to the Epstein-Barr virus. In all cases, anorexia, malaise, nausea, vomiting, diarrhea, and chills and fever may or may not occur. Many infected individuals are asymptomatic but are carriers of the disease.

19. Hiatal Hernia — a protrusion of the stomach through the diaphragm into the thoracic cavity. The individual feels extreme discomfort upon eating and in fact may need to maintain an upright position after eating to allow gravity to help push the stomach back into the abdominal cavity.

20. Intestinal Obstruction — blockage of the colon or small intestine. High blockage refers to the small intestine; low blockage, to the colon. Intestinal obstruction is always serious. The severity of the problem depends on the location of the blockage, its completeness, and possible blood supply blockage. Common causes of blockage include intestinal adhesions, hernia, tumor, volvulus (rotating twisting of the bowel on the fold of peritoneum that suspends the bowel from the posterior wall of the abdomen), and intussusception (the telescoping of one segment of bowel into an adjacent segment).

21. Intestinal Tumors — tumors of the small intestine are uncommon. Benign polyps may occur in the large intestine. Of much more concern are carcinomas of the colon. Carcinoma of the left half of the colon results in colon obstruction. Carcinoma of the right half of the colon usually causes chronic blood loss but no obstruction. Colon cancer is easily diagnosed and if caught early, easily treated.

22. Irritable Bowel Syndrome (Spastic Colitis, Mucous Colitis) — refers to disturbed bowel function without structural or biochemical abnormalities. Diagnosis is by exclusion, i.e., ruling out other causes of the problem. It is seen as a manifestation of emotional stress. Symptomatic treatment is usually successful.

23. Jaundice — refers to a yellowing of the skin due to excessive red cell breakdown (hemolytic jaundice) by the liver, liver cell injury (hepatocellular jaundice), and common duct obstruction by tumor or stone (obstructive jaundice).

24. Lactose Intolerance — a food intolerance due to lactase deficiency. It occurs in adults. The unabsorbed lactose raises osmotic pressure of bowel contents, leading to diarrhea. Treatment involves removing lactose (milk products) from the diet.

25. Malabsorption — faulty absorption of nutrient materials from the digestive tract. Causes include intestinal inflammation or infection, pancreatic disorders, biliary tract obstruction, or primary disease of the intestines. Malabsorption is especially serious in infants and young children.

26. Merkel's Diverticulum — a tubular outpouching of remnant vitelline duct from the distal ileum. May cause symptoms that mimic those of appendicitis.

27. Mesenteric Thrombosis — clot formation in the superior mesenteric artery that supplies the small intestine and the proximal half of the large intestine. If the artery becomes blocked by thrombus, embolus, or atheroma, infarction of the intestine can occur. Infarction of the intestine is invariably fatal.

28. Pancreatitis — inflammation of the pancreas due to escape of pancreatic juice from the ducts into the body of the pancreas. The pancreatic enzymes begin to digest pancreas tissue. Acute pancreatitis may be fatal. Chronic pancreatitis will lead to progressive destruction of the pancreas. Predisposing factors to acute pancreatitis include gallbladder disease and alcohol consumption.

29. Peptic Ulcer — a disorder caused by increased acid secretions and digestive enzymes eroding the gastric mucosa. Predisposing factors include congenital weakness of the stomach wall, excess production of acid (acid stomach), and nervous, high-strung personality. Complications include bleeding, perforations, and pyloric obstruction by scarring. Treatment involves use of antacids, cimetidine (to block secretion of gastric acid), and surgery.

30. Periodontal Disease — an infection between the roots of the teeth and the gums. Spread of infection into tooth sockets may cause teeth to loosen and eventually fall out. This is a treatable condition.

31. Salivary Gland Disorders — the most common lesion is characterized by inflammation and is called sialoadenitis. Sialoadenitis is caused by the mumps virus, obstruction of the salivary ducts by stones (sialithiasis), as well as by other means. Neoplasms are rare, with adenoma the most common types.

32. Thrush or Moniliasis — an infection of the mouth caused bythe fungus *Candida albicans*. It is most common in newborn infants and individuals treated with high doses of antibiotics or toxic drugs.

33. Tumor of the Oral Cavity — carcinoma may arise from the squamous epithelium of the lips, cheek, tongue, palate, or back of the throat. Treatment is by surgical resection or radiation therapy.

34. Tumors of Liver and Gallbladder — benign adenomas and primary carcinoma are uncommon. Metastatic carcinoma, spread from GI tract, breast, lung or other sites, is common.

ANSWERS TO END-OF-CHAPTER QUESTIONS

Multiple Choice/Matching

1. d	5. a	9. c	13. d
2. d	6. d	10. c	14. b
3. b	7. d	11. a	15. c
4. b	8. b	12. d	16. a

Short Answer Essay Questions

17. A drawing of the organs of the alimentary tube and labels can be found on page 789, Fig. 24.1.

18. The basic alimentary canal wall structure consists of four tunics: the mucosa, submucosa, muscularis, and serosa. The mucosa consists of a surface epithelium underlain by a small amount of connective tissue called the lamina propria and a scanty amount of smooth muscle fibers, the muscularis mucosae. Typically, the epithelium of the mucosa is a simple columnar epithelium rich in mucus-secreting goblet cells and other types of glands. The mucus protects certain digestive organs from being digested themselves by the enzymes working within their cavities and eases the passage of foods along the tract. In some digestive organs the mucosa contains both enzyme-secreting and hormone-secreting cells. The lamina propria, consisting of areolar connective tissue and containing lymph nodules, is important in the defense against bacteria and other pathogens. In the small intestine, the muscularis mucosae throw the mucosal tunic into a series of small folds that vastly increases its surface area for secretion and absorption.

The submucosa is areolar connective tissue containing blood vessels, lymphatic vessels, nerve endings, and epithelial glands. Its vascular network supplies surrounding tissues and carries away absorbed nutrients. Its nerve plexus is part of the enteric nerve supply of the gastrointestinal tube.

The muscularis externa mixes and propels food along the digestive tract. This muscular tunic usually has an inner circular layer and an outer longitudinal layer of smooth muscle cells, although there are variations in this pattern.

The serosa is formed of areolar connective tissue covered with mesothelium, a single layer of squamous epithelial cells. It is the protective outermost layer and the visceral peritoneum. (pp. 792-795)

19. The mesentery is a double pritoneal fold that suspends the small intestine from the posterior abdominal wall. The mesocolon is a special dorsal mesentery that secures the transverse colon to the parietal peritoneum of the posterior abdominal wall. The greater omentum is also a double peritoneal sheet that covers the coils of the small intestine and wraps the transverse portion of the large intestine. (pp. 792-793)

20. The six functional activities of the digestive system are ingestion, propulsion, mechanical digestion, chemical digestion, absorption, and defecation. (pp. 790-791)

21. The boundaries of the oral cavity include the lips, cheeks, tongue, palate, and oropharynx. The epithelium is stratified squamous epithelium, because the walls have to withstand considerable abrasion. (pp. 795-796)

22. a. The normal number of permanent teeth is 32; deciduous teeth, 20.

    b. Enamel covers the crown; cementum, the root.

    c. Dentin makes up the bulk of the tooth.

    d. Pulp is found in the central cavity in the tooth. Soft tissue structures (connective tissue, blood vessels, and nerve fibers) compose pulp. (pp. 799-800)

23. The two phases of swallowing are as follows:

    a. Buccal (voluntary) phase of swallowing: organs involved — tongue, soft palate; activities — tongue compacts food into a bolus, forces the bolus into the oropharynx via tongue contractions; the soft palate rises to close off the superior nasopharynx.

    b. Pharyngeal-esophageal (involuntary) phase organs involved — pharynx and esophagus, activities — motor impulses sent from the swallowing center to their muscles, which contract to send the food to the esophagus by peristalsis. Arrival of food/peristaltic wave at the gastroesophageal sphincter causes it to open. (pp. 802-804)

24. The parietal cells secrete hydrochloric acid and intrinsic factor. Chief cells produce pepsinogen. Mucous neck cells produce mucus that helps shield the stomach wall from damage by gastric juices. Enteroendocrine cells secrete hormones into the lamina propria. (p. 806)

25. Gastric secretion is controlled by both neural and hormonal mechanisms. The stimulation of gastric secretion involves three distinct phases: the cephalic, gastric and intestinal phases.

The cephalic phase occurs before food enters the stomach and is triggered by the sight, aroma, taste, or thought of food. Input is relayed to the hypothalamus, which stimulates the vagal nuclei of the medulla oblongata, causing motor impulses to be sent via vagal nerve fibers to the stomach. This reflex may be dampened during depression or loss of appetite.

The gastric phase is initiated by neural and hormonal mechanisms once food reaches the stomach. Stomach distension activates stretch receptors and initiates reflexes that transmit impulses to the medulla and then back to the stomach, leading to acetylcholine release. Acetylcholine stimulates the output of gastric juice. During this phase, the hormone gastrin is more important in gastric juice secretion than neural influences. Chemical stimuli provided by foods directly activate gastrin-secreting cells. Gastrin stimulates the gastric glands to spew out even more gastric juice. Gastrin secretions are inhibited by high acidity.

The intestinal phase is set into motion when partially digested food begins to fill the duodenum. This filling stimulates intestinal mucosal cells to release a hormone (intestinal gastrin) that encourages the gastric glands to continue their secretory activity briefly; but as more food enters the small intestine, the enterogastric reflex is initiated, which inhibits gastric secretion and food entry into the duodenum to prevent the small intestine from being overwhelmed. Additionally, intestinal hormones (enterogastrones) inhibit gastric activity. (pp. 808-810)

26. a. The cystic and hepatic ducts fuse to form the common bile duct, which fuses with the pancreatic ducts just before entering the duodenum. (pp. 813, 818-819)

b. The point of fusion of the common bile duct and pancreatic duct is called the hepatopancreatic ampulla. (p. 813)

27. The absence of bile (which causes fat emulsification) and/or pancreatic juice (which contains essentially the only source of lipase) causes fat absorption to be so slow as to allow most of the fat to be passed into the large intestine. (p. 821)

28. The Kupffer cells function to remove debris such as bacteria from the blood. The hepatocytes function to produce bile, in addition to their many metabolic activities. (p. 821)

29. a. Brush border enzymes are intestinal digestive enzymes; these are part of the plasma membrane of the microvilli of the intestinal absorptive cells. (p. 815)

    b. Chylomicrons are fatty droplets consisting of triglycerides combined with small amounts of phospholipids, cholesterol, free fatty acids, and coated with proteins. They are formed within the absorptive cells and enter the lacteals. (p. 835)

30. Common inflammatory conditions include appendicitis in adolescents, ulcers and gallbladder problems in middle-age adults, and constipation in old age. (p. 840)

31. The effects of aging on digestive system activity includes declining mobility, reduced production of digestive juice, less efficient absorption, and slowing of peristalsis. (p. 840)

Critical Thinking and Application Questions

1. If the agent promotes increased bowel motility without providing for increased bulk, diverticulosis is a possibility, because the rigor of the colonic contractions increases when the volume of residues is small. This increases the pressure on the colon wall, promoting the formation of diverticula. If the product irritates the intestinal mucosa, intestinal contents will be moved rapidly through both the small and large intestines, leaving inadequate time for absorption of water, which can result in dehydration and electrolyte imbalance. (p. 829)

2. This patient has the classical symptoms of a gallbladder attack in which a gallstone has lodged in the cystic duct. The pain is discontinuous and colicky because it is caused by peristaltic contractions (contract-relax-contract-relax, etc.). The stone can be removed surgically or by sound or laser treatment. If it is not removed, bile will back up into the liver, and jaundice will ensue. (p. 822)

3. The baby's food would indicate acidosis due to the intestinal juice passing through the large intestine with little or no time for reabsorption of water and substances dissolved in water by the large intestine. (pp. 829)

4. a. The vagotomy is suggested to inhibit the production of hydrochloric acid and gastric juices.

   b. Possible consequences of nontreatment could be surgical removal of the existing ulcer due to internal bleeding, or the occurrence of multiple ulcers. (pp. 807-809)

Key Figure Questions

Figure 24.1 The tongue, salivary glands, liver, gallbladder, pancreas. **Figure 24.12** The presence of the stratified squamous epithelium indicates that the esophagus is mainly a chute and, thus, must accomodate high friction, whereas the stomach has a secretory simple columnar epithelium, reflecting its role in chemical digestion. **Figure 24.14** It has a third layer of smooth muscle in its muscularis, in which the fibers run obliquely, allowing it to exhibit churning movements that physically digest the food. **Figure 24.16** It would decrease it. **Figure 24.28** They are activated in the small intestine where they work and where the cells exposed to them are protected by mucus. No such protection exists in the pancreas.

SUGGESTED READINGS

1. Bansil, R., K.R. Bhaskar, J.D. Bradley, P. Garik, J.T. LaMont, H.E. Stanley, and B.S. Turner. "Viscous Fingering of HCl Through Gastric Mucin." *Nature* 360 (Dec. 1992): 458.

2. Bennett, W.I. "Overactive Machinery." *New York Times Magazine* (May 7, 1989): 259-260.

3. Bnendia, M., and P. Tiollais. "Hepatitus B Virus." *Scientific American* 264 (Apr. 1991): 116.

4. "Bugged by an Ulcer? You Could Have a Bug." *Discover* 8 (May 1987): 10.

5. Davenport, H.W. *Physiology of the Digestive Tract.* 5th ed. Chicago: Yearbook Medical Publishers, 1982.

6. Desai, K.M., W.C. Sessa, and J.R. Vane. "Involvement of Nitric Oxide in the Reflex Relaxation of the Stomach to Accomodate Food or Fluid." *Nature* 351 (June 1991): 477.

7. Doyle, R. "On the Disease Trail." *Science* 86 (7) (May 1986): 16-17.

8. Eisenberg, S. "Biting Down on the Culprit Causing Gum Disease." *Science News* 133 (Jan. 1988): 21.

9. Erickson, R.A. "Peptic Ulcer Disease in 1985: A Diagnostic Update." *Modern Medicine* (Feb. 1985).

10. Eron, C. "One-Two Punch for Hepatitis B." *Science News* 134 (Aug. 1988): 84.

11. Fackelmann, K.A. "Nabbing a Gene for Colorectal Cancer." *Science News* 144 (Dec. 1993): 388.

12. Goodfield, J. "Vaccine on Trial." *Science* 84 (5) (Mar. 1984): 79-84.

13. Greenaugh, W.B., and N. Hirschhorn. "Progress in Oral Rehydration Therapy." *Scientific American* 264 (May 1991): 50.

14. Grossman, M.I. "Neural and Hormonal Regulation of Gastrointestinal Function." *Annual Review of Physiology* 41 (1979): 27.

15. Halmi, K.A. "Anorexia Nervosa and Bulimia." *Psychomatics* 24 (Feb. 1983): 111.

16. Jenkins, W. "Large Bowel Exam: When to Use What Test." *Modern Medicine* (Mar. 1984).

17. Johnson, G.T. (ed). "The Tragedy of Anorexia Nervosa." *The Harvard Medical School Health Letter* (Dec. 1981).

18. Kopeski, L.M. "Diabetes and Bulemia: A Deadly Duo." *American Journal of Nursing* 89 (Apr. 1989): 482-485.

19. Loupe, D.E. "Hepatitus C May Spread Heterosexually." *Science News* 136 (Sept. 1989): 151.

20. Miller, J.A. "Oral Interactions." *Science News* 130 (July 1986): 12-13.

21. Miller, J.A. "Oral Ecology." *Science News* 129 (June 1986): 396-397.

22. Moog, F. "The Lining of the Small Intestine." *Scientific American* 245 (Nov. 1981): 154.

23. Moser, P.W. "It Must Have Been Something You Ate." *Discover* 8 (Feb. 1987): 94.

24. Raloff, J. "Housecleaning Cells May Become Assassins." *Science News* 143 (May 1993): 277.

25. Raloff, J. "Live Cancer: Homing in on the Risks." *Science News* 142 (Nov. 1992): 308.

26. Raloff, J. "Successful Hepatitus A Vaccine Debuts." *Science News* 142 (Aug. 1992): 103.

27. Rios, J. (ed). "Cholesterol Kills." *Cardiac Alert* (Apr. 1984).

19. Turner, C.G. "Teeth and Prehistory in Asia." *Scientific American* 260 (Feb. 1989): 88-96.

28. Stroh, M. "Exposing Salmonella's Gutsy Moves." *Science News* 141 (June 1992): 420.

29. Travis, J. "Surprising New Target Found for Anti-Ulcer Drugs." *Science* 258 (Dec. 1992): 1579.

30. Uvnas-Moberg, K. "The Gastrointestinal Tract in Growth and Reproduction." *Scientific American* 261 (July 1989): 78-83.

31. Weiss, R. "Bulimias Binges Linked to Hormone." *Science News* 134 (Sept. 1988): 182.

32. Wickelgren, I. "What's That Wriggling in My Sushi?" *Science News* 135 (May 1989): 300.

25

Nutrition, Metabolism, and Body Temperature Regulation

Chapter Preview

This chapter deals with metabolism—the chemical reactions that are carried on in the body. When food is metabolized, it is broken down into usable forms: Carbohydrates into simple sugars; lipids into simple fats, fatty acids, and glycerol; proteins into amino acids. These end products of digestion are then absorbed into the body, carried to the liver, and metabolized. The chemical reactions involved include cellular respiration, beta-oxidation, deamination, and other such reactions. The cellular respiration reaction provides the energy for the cell—the energy necessary for the other cellular reactions to occur. This chapter deals with those chemical reactions.

INTEGRATING THE PACKAGE

Suggested Lecture Outline

I. Nutrition (pp. 846-857; Fig. 25.1, p. 847)

    A. Carbohydrate (pp. 847-848)

        1. Dietary Sources

        2. Uses in the Body

        3. Dietary Requirements

    B. Lipids (p. 848)

        1. Dietary Sources

        2. Uses in the Body

        3. Dietary Requirements

    C. Proteins (pp. 848-851)

        1. Dietary Sources (Fig. 25.2, p. 849)

        2. Uses in the Body

            a. All-or-None Law

            b. Adequacy of Calorie Intake

            c. Nitrogen Balance of the Body

            d. Hormonal Controls

        3. Dietary Requirements

    D. Vitamins (p. 851; Table 25.2, pp. 852-855)

        1. Fat-Soluble

        2. Water-Soluble

    E. Minerals (p. 851; Table 25.3, pp. 855-858)

II. Metabolism (pp. 858-881; Fig. 25.3, p. 859)

    A. Overview (pp. 858-860)

        1. Anabolism and Catabolism

        2. Oxidation-Reduction Reactions and the Role of Coenzymes

        3. Mechanisms of ATP Synthesis

    B. Carbohydrate Metabolism (pp. 861-869)

        1. Oxidation of Glucose

            a. Glycolysis (Fig. 25.6, p. 863)

            b. Krebs Cycle (Fig. 25.7, p. 864)

            c. Electron Transport Chain and Oxidative Phosphorylation (Fig. 25.8, p. 866)

            d. Summary of ATP Production (Fig. 25.10, p. 868)

        2. Glycogenesis and Glycogenolysis (Fig. 25.11, p. 868)

        3. Gluconeogenesis

    C. Lipid Metabolism (pp. 869-871)

        1. Oxidation of Glycerol and Fatty Acids

            a. Beta Oxidation

        2. Lipogenesis and Lipolysis

            a. Lipogenesis

            b. Lipolysis

    D. Protein Metabolism (pp. 871-872)

        1. Oxidation of Amino Acids

            a. Transamination (Fig. 25.14, p. 871)

            b. Oxidative Deamination (Fig. 25.14, p. 871)

            c. Keto Acid Modification

        2. Synthesis of Proteins

    E. Catabolic-Anabolic Steady State of the Body (pp. 872-873)

    F. Absorptive and Post-Absorptive States: Events and Controls (pp. 873-878)

        1. Absorptive States

            a. Carbohydrates

            b. Triglycerides

            c. Amino Acids

            d. Hormonal Control

        2. Postabsorptive States

            a. Sources of Blood Glucose

                1. Glycogenolysis in the Liver

                2. Glycogenolysis in Skeletal Muscle

                3. Lipolysis in Adipose Tissues and the Liver

                4. Catabolism of Cellular Protein

            b. Glucose Sparing

            c. Hormonal and Neural Controls

G. Role of the Liver in Metabolism (pp. 879-881)

    1. General Metabolic Functions (Table 25.5, p. 880)

    2. Cholesterol Metabolism and Regulation of Plasma Cholesterol Levels

    3. Lipoproteins and Cholesterol Transport (Fig. 25.21, p. 881)

        a. HDLs

        b. LDLs

        c. VLDLs

    4. Factors Regulating Plasma Cholesterol Levels

III. Body Energy Balance (pp. 881-890)

A. Regulation of Food Intake (pp. 882-884)

    1. Nutrient Signals Related to Energy Stores

    2. Hormones

    3. Body Temperature

    4. Psychological Factors

B. Metabolic Rate and Body Heat Production (pp. 884-995)

C. Regulation of Body Temperature (pp. 885-890; Fig. 25.25, p. 889)

    1. Core and Shell Temperatures

    2. Mechanisms of Heat Exchange

        a. Radiation

        b. Conduction/Convection

        c. Evaporation

    3. Role of the Hypothalamus

    4. Heat-Producing Mechanisms

        a. Vasoconstriction of Cutaneous Blood Vessels

        b. Increase in Metabolic Rate

        c. Shivering

        d. Enhancement of Thyroxine Release

    5. Heat-Loss Mechanisms

        a. Vasodilation of Cutaneous Blood Vessels

        b. Enhanced Sweating

        c. Fever

IV. Developmental Aspects of Nutrition and Metabolism (pp. 890-891)

A. Embryological

B. Aging

Review Items

1. Chemical bonding (Chapter 2, p. 3)

2. Carbohydrates (Chapter 2, pp. 43-46)

3. Lipids (Chapter 2, pp. 45-47)

4. Proteins (Chapter 2, pp. 47-56)

5. Water (Chapter 2, pp. 39-40)

6. ATP (Chapter 2, pp. 55-56)

7. Oxidation/reduction (Chapter 2, p. 30)

8, Chemical equations (Chapter 2, p. 36)

9. Patterns of chemical reactions (Chapter 2, p. 36)

10. Reversibility of reactions (Chapter 2, p. 38)

11. Enzymes (Chapter 2, pp. 50-53)

12. Membrane transport (Chapter 3, pp. 65-73)

13. Cytoplasm (Chapter 3, pp. 75-84)

14. Mitochondria (Chapter 3, p. 77)

15. Hypothalamus (Chapter 12, p. 395)

16. Receptors (Chapter 13, pp. 426-432)

17. Prostaglandins (Chapter 17, p. 548)

18. Growth Hormone (Chapter 17, pp. 555-558)

19. Sex steroids (Chapter 17, pp. 569, 575-576)

20. Glucocorticoids (Chapter 17, pp. 568-569)

21. Diabetes (Chapter 17, pp. 574-575)

22. Insulin (Chapter 17, pp. 573-575)

23. Glucagon (Chapter 17, pp. 572-573)

24. Thyroxine (Chapter 17, pp. 562-564)

25. Blood flow regulation (Chapter 20, pp. 651-664)

26. Chylomicrons (Chapter 24, pp. 835-836)

27. Bile formation (Chapter 24, p. 821)

Cross References

1. Ketone bodies as abnormal urine constituents mentioned in Chapter 26, p. 917.

Transparencies Index

INSTRUCTIONAL AIDS

Lecture Hints

1. A complete understanding of the metabolic pathways necessitates a review of the basic concepts of chemistry in Chapter 2 and cellular structure in Chapter 3. Refer the class to specific sections related to the lecture topic being discussed.

2. Point out that fatty acids perform several functions: structural (membranes); functional (prostaglandins); and as an energy source (enters the Krebs cycle as acetyl CoA).

3. Mention that cholesterol is responsible for membrane fluidity and is the structural basis of the steroid hormones.

4. As a point of interest, mention the logical use of the name anabolic steroid.

5. Point out the difference between vitamins and minerals.

6. The chemist's approach to the metabolic pathways is often very different than that of a biologist. One of the most effective methods for presenting the biochemical pathways of ATP synthesis (from the perspective of a physiologist) is to start with a quick review of cell structure related to the process (membranes, cytoplasm, mitochondria, etc.). Then give the overall outcome of each step (glycolysis, Krebs, electron transport), followed by a more detailed examination of each step. It is essential that students see the "overall picture" in order to understand the significance of the metabolic pathways. Many students have previously studied the pathways and have never realized the relevance to life. You will often receive comments to the effect of: "I've memorized these pathways before, but never understood why until now!"

7. Point out that the Krebs cycle is often considered part of aerobic respiration, but that this step in the pathway does not use oxygen directly.

8. Emphasize that glycolysis occurs whether or not oxygen is present, so the term *anaerobic* must be used with caution.

9. Draw and project a diagram of the cell with a disproportionately large mitochondrion. Label the diagram with the locations of glycolysis, Krebs, and electron transport. Give a brief summary of each step.

10. Mention the different possible names for Krebs cycle: citric acid cycle (citrate is the first substrate in the cycle); tricarboxylic acid cycle (several intermediates have three carboxyl groups).

11. A diagram of the chemiosmotic mechanism of ATP synthesis is very helpful in presenting electron transport. Draw (on a blank acetate) a diagram of the phospholipid bilayer, and fill in electron carriers, ATP synthetase complex, and trace the pathway of electron flow. It is helpful for the instructor to physically draw the diagram (if time allows), especially if the students are required to draw and analyze diagrams on the exam.

12. To measure student understanding, ask plenty of questions during discussion of the pathways. For example, ask: "What would happen to ATP production if $NADH + H^+$ reduced the cytochrome oxidase complex instead of the NADH dehydrogenase complex?"

13. Remind the class of the different ways that nutrients are absorbed and transported (lipids into lacteals, carbohydrates and proteins into blood capillaries). Small reminders help keep the chemistry tied to biological processes.

14. Remind the class that deamination of amino acids is necessary for the carbon "skeletons" to enter catabolic pathways. The nitrogenous compounds are metabolic waste products, the elimination of which is discussed in Chapter 26.

15. When discussing mechanisms of heat control, point out that one can think of heat flowing down its "concentration gradient."

16. Reinforce the concept of the reflex arc when presenting material on hypothalamic control of body temperature.

Demonstrations/Activities

1. Audio-visual materials of choice

2. Use a small portable fan and a container of water to demonstrate the mechanics of cooling the body.

3. Use laminated posters showing the various metabolic pathways as you discuss them.

Critical Thinking/Discussion Topics

1. Discuss the need for a balanced diet.

2. Discuss the idea that more vitamins and minerals are lost in the urine of Americans than are in the diets of people in many other countries. Discuss overdosing from vitamins, etc., from health food store products and other sources, rather than getting vitamins and minerals only from the food we eat.

3. Discuss the various metabolic disorders and relate each one to the dietary deficiency that causes the disorder.

4. Discuss the differences between the lipid and water-soluble vitamins and why care should be taken when using vitamins as a food supplement.

5. Why are there so many steps in the complete oxidation of glucose (i.e., why not just one step)?

6. Discuss the consequences (in terms of ATP production) if $NADH + H^+$ reduced the cytochrome b-c_1 complex instead of the NADH dehydrogenase complex.

Library Research Topics

1. Research the differences between and significance of low-density and high-density cholesterol.

2. Research the effects of the inability to sweat.

3. Research the effects of a liquid protein diet on the body.

4. Research the various types of popular diets currently being publicized. Note differences, similarities, and adverse effects, if any.

Audio-Visual Aids/Computer Software

Slides

1. Metabolism: Structure and Regulation (EIL, 42 Slides)

2. Cellular Respiration (EIL, Slides). A selection of art work combined with live photographs depicting aspects of energy release in biochemical cycles.

Videotapes

1. Respiration (EIL, VHS). The glycolytic system, citric acid cycle, and electron transport are analyzed; student learns to distinguish between aerobic and anaerobic respiration, to describe and explain the main reaction blocks of respiration and the inputs, outputs, and processes of respiration.

2. Metabolism: Structure and Regulation (EIL, VHS). Examines the organization of metabolism from single reactions to complex groupings of pathways and explains principles of metabolic compartmentation and regulatory enzymes.

3. The Living Body: Hot and Cold (FHS, VHS, or BETA). Using the extremes of temperature that occur in a day's skiing, as an example, this program shows the range of mechanisms through which the human body maintains a steady internal temperature and protects its vital organs: shivering, hair erection, and rerouting of blood supplies to conserve heat, increased blood flow to the body surface, sweating, and panting to lose heat.

4. The Chemistry of Life (EI, 525-2066 V, VHS). Introduces the basic concepts of chemistry as related to life processes.

5. Chemistry of Nucleic Acids (EI, 525-2073 V, VHS). Structure and chemistry of DNA and RNA are examined.

6. The Chemistry of Carbohydrates and Lipids (EI, 525-2068 V, VHS). Introduction to the structure and biology of carbohydrates and lipids.

7. The Chemistry of Proteins (EI, 525-2067 V, VHS). General protein function, amino acids, peptide bonds, and protein function.

8. Metabolism: The Fire of Life (EI, 600-2361 V, VHS). Describes the biological pathways.

9. Respiration (EI, 525-2077 V, VHS). Glycolysis, citric acid cycle, and electron transport.

10. Cell Respiration: Anaerobic and Aerobic (EI, 472-2356 V, VHS). A detailed discussion of oxidation-reduction reactions and ATP production.

11. Respiration: The Energy for Life (EI, 475-2355 V, VHS). Relates the processes of breathing and cell respiration.

Computer Software

1. Metabolism and the Production of ATP (EIL, Apple II). All on one disk: how organisms metabolize carbohydrates, lipids, and proteins to synthesize ATP.

2. Experiments in Metabolism (CDL, Apple). Utilizing numerous color graphics and animation, this program guides the student step by step in calculating the Basal Metabolic Rate of "Mr. Faust," the laboratory mouse. These experiments use the indirect method of oxygen consumption to compare the effects of cold and caffeine to the normal resting BMR.

3. Nutritionist I (PLP, Apple, IBM). A diet and nutritional analysis program designed to create and analyze menus and diets for individual needs and preferences.

4. Nutritionist II (PLP, Apple, IBM). An Interactive Graphics Diet Analysis Program.

5. Vitamins, Minerals and Health Foods (PLP, Apple). Defines and describes, identifies the properties of and dispels myths about vitamins and minerals.

6. Dieting (PLP, Apple). Describes the risks of being overweight, the basic concept of nutrition, fallacies about dieting, tips on weight loss, sports nutrition, junk food.

7. Cell Respiration: Basic Unit (CDL, Apple). Features a comprehensive discussion of ATP-ADP, oxidation/reduction and carrier molecules.

8. Cell Respiration: Advanced Unit (CDL, Apple). Presents an integrated lesson on the chemical pathways for aerobic and anaerobic respiration.

9. The Food Processor (CDL, Apple, IBM). A sophisticated system capable of analyzing individual foods and whole menus for nutritional content, and comparing these values to RDA recommendations.

10. Advanced Cell Respiration (EI, C 4077, Apple). Color and graphics as a reinforcement tool for the metabolic pathways.

11. Pathways of Biosynthesis (EI, C 4559, Apple). Details biosynthesis of the major substances of the cell.

12. Basic Biology Computer Series: (EI, Apple)
    Lipids (C 4071)
    Proteins (C 4072)
    Nucleic Acids (C 4073)
    Carbohydrates (C 4074)
    Biochemistry Test (C 4075)

    See Guide to Audio-Visual Resources on p. 339 for Key to AV distributors.

LECTURE ENHANCEMENT MATERIAL

Clinical and Related Terms

1. Anorexia—a loss of appetite.

2. Anorexia nervosa—a serious nervous condition in which the patient loses appetite and becomes greatly emaciated.

3. Bulimia—an eating disorder in which the individual alternately eats excessively and then induces vomiting to purge the stomach of the food.

4. Emaciation—excessive leanness of the body due to tissue wasting.

5. Hyperalimentation—prolonged intravenous nutrition provided for patients with severe digestive disorders.

6. Marasmus—an extreme form of protein and calorie malnutrition.

7. Pica—a hunger for substances that are not suitable foods.

8. Polyphagia—excessive intake of foods.

Disorders/Homeostatic Imbalances

Malnutrition

1. Malnutrition—refers to a defect or abnormal functioning anywhere along the nutritional chain. The nutrition chain includes intake of food, mastication of food, digestion, absorption of digested end products and nutrients, and uptake and utilization of nutrients by cells.

Nutrition Disorders

1. Defective Digestion—can be due to failure to produce digestive juices.

2. Disturbed Intake—can be due to psychosis; alcoholism; food faddism; inability to chew, swallow, or retain food in the stomach.

3. Impaired Utilization—can be due to numerous disorders, i.e., diabetes mellitus; to protein-losing enteropathy, resulting in excretion of proteins from the intestine; and to diarrhea.

4. Malsorption—can be due to absence of digestive juices or to an inflammatory or metabolic disease of the GI tract.

5. Transport Disorder—transport to the liver can be impaired by liver disorders such as hepatitis or cirrhosis.

Common Nutritional Deficiency Syndromes

1. Folate Deficiency (Folate Deficiency Anemia)—see the chapter on blood, Chapter 18, Obesity.

2. Iodine Deficiency (Endemic Goiter)—refers to a condition characterized by the enlargement of the thyroid gland (goiter).

3. Iron Deficiency (Iron Deficiency Anemia)—see the chapter on blood, Chapter 18.

4. Niacin Deficiency (Pellagra)—refers to a condition characterized by dermatitis, diarrhea, and dementia (insanity).

5. Obesity—refers to excessive weight. This is the single most prevalent nutritional disorder in the United States. It is characterized by weight more than 20% over the ideal. It leads to increased mortality, as well as systemic problems (such as hypertension, coronary artery occlusion, etc.).

6. Protein Calorie Malnutrition (Kwashiorkor)—refers to a syndrome resulting from the inadequate intake of protein despite an adequate caloric diet; characterized by apathy or irritability, hypopigmented skin, dermatitis, generalized edema and liver enlargement, atrophy of other organs and tissues, and high mortality due to infectious diseases.

7. Thiamine Deficiency (Beriberi)—refers to a condition characterized by weakness, CNS dysfunction, or cardiac failure; nerves become demyelinated; some individuals may show Wernicke's syndrome consisting of confusion, ataxia, and paralysis of eye movements.

8. Vitamin C Deficiency (Scurvy)—refers to spongy bleeding gums and cutaneous hemorrhage and poor wound healing due to lack of vitamin C.

9. Vitamin D Deficiency (Rickets)—refers to skeletal deformities associated with excessive proliferation of osteoid tissue due to vitamin D lack or to bile duct obstruction resulting in malabsorption of vitamin D; characterized by rosary beading of ribs, bowing of legs, poor maturation of cartilage; deformities of chest and skull in children and softening of bone in adults (osteomalacia).

Hypervitaminosis

Hypervitaminosis—refers to excess intake of vitamins. Acute or chronic toxicity can result from overdoses of specific vitamins.

Common Eating Disorders

1. Anorexia Nervosa—refers to a serious nervous condition in which the patient loses his/her appetite and systematically eats only a small amount of food. The individual, usually a teenage girl, becomes greatly emaciated. Complications include damage to the heart and other organs.

2. Bulimia—an eating disorder characterized by an abnormal increase in the sensation of hunger; the patient, usually a teenage girl, will binge on food and then induce vomiting to try to avoid the caloric intake.

Vitamins, Minerals, and Trace Elements

1. Ascorbic Acid (Vitamin C)—indicated in prevention and treatment of scurvy.

2. Calcium Salts—indicated in treatment of hypocalcemia, osteoporosis, osteomalacia and rickets, and renal osteodystrophy. Calcium is responsible for the mineral content of bone and has numerous physiologic roles, including neuromuscular and cardiac function and blood coagulation.

3. Fluoride—recommended as an aid in the prevention of dental caries.

4. Folic Acid—indicated in macrocytic anemia secondary to folic acid deficiency and as a nutritional supplement in patients with high folic acid demand (as in hemolytic diseases and psoriasis), poor absorption (as in intestinal diseases and diphenylhydantoin therapy), or pregnancy.

5. Niacin (Vitamin B_5)—indicated in prevention and treatment of pellagra, adjuvant in the treatment of hyperlipidemia. Overdose may cause liver failure.

6. Oral Iron Preparations—indicated for oral iron replacement therapy in iron deficiency. Use of this drug never replaces the need to investigate the cause of iron deficiency.

7. Phosphorus—indicated in hypercalcemia and hypophosphatemic rickets or osteomalacia.

8. Piridoxine (Vitamin B_6)—indicated in piridoxine deficiency due to deficient dietary uptake and for drug-induced or inborn errors of metabolism.

9. Potassium—indicated in potassium deficiency, in anticipated calcium deficiency in patients taking diuretics and/or digoxin, in patients with long-term electrolyte replacement regimens, and as routine prophylactic administration after surgery once the urine output has been established. Care must be taken that hyperkalemia does not occur because of its adverse effects on heart activity.

10. Riboflavin (Vitamin B_2)—indicated in B vitamin deficiency states.

11. Vitamin A—indicated in biliary and pancreatic diseases, colitis, sprue, and post gastrectomy.

12. Vitamin B_{12} (Cyanocobalamin)—indicated in vitamin B_{12} deficiency secondary to pernicious anemia or dietary deficiency.

13. Vitamin D (Phytonadione)—indicated for treatment of hypoprothrombinemia resulting from deficient absorption or synthesis of vitamin K, secondary to Coumadin, antibiotic therapy, or salicylates.

14. Vitamin D Preparations—indicated in vitamin D deficiency, rickets and osteomalacia, hypoparathyroidism and pseudohypoparathyroidism, and renal osteodystrophy.

15. Vitamin E—indicated in vitamin E deficiency only.

Nutritional Support Agents

1. Enteral Preps—used to provide partial or complete nutritional support to patients who are unable to meet energy and nutrient requirements on their own. They also provide nutritional support when there are restrictions or limitations for the intake of a high-residue diet (such as inflammatory bowel disease, pancreatitis, gastrointestinal fistulas, and preoperative bowel preparation). Classes include elemental diets (Vivonex, Vivonex-HN and Vivonex-TEN), peptide formulations (Vital, Travasorb-STD, and Travasort-HN), caseinates and whole-protein formulas (Criticare-HN, Ensure, Ensure Plus, Isocal, Magnacal, Nutri-Aid, Osmolite, Sustacal, and others), milk-based preparations (Carnation Instant Breakfast, Meritene Liquid), blenderized preparations (Compleat-B, Formula 2), and selected amino acid formulas (Amin-Aid, Hepatic-Aid, Traum Aid, Stressein).

2. IV Solutions—used to provide nutritional support to patients who cannot, should not, or will not maintain nutritional requirements orally or enterally. Classes include dextrose/amino acid formulation delivered via a central line (Aminosyn 10% and 5%, Vinamine 8%, and others), modified dextrose/amino acid formulations delivered via a central line (Freamine HBC, Hepatamine), dextrose/amino acid formulations delivered via peripheral vein (Freamine III 3%, Travasol-M 3.5%), and fat emulsions (Intralipid 10% and 20%, Soyacal 10% and 20%).

Answers To Textbook Chapter Questions

Multiple Choice/Matching

1.	a	6.	c	11.	b
2.	c	7.	a	12.	d
3.	c	8.	d	13.	c
4.	d	9.	d	14.	d
5.	b	10.	a	15.	a

Short Answer Essay Questions

16. Cellular respiration is a group of reactions that break down (oxidize) glucose, fatty acids, and amino acids in the cell. Some of the energy released is used to synthesize ATP (p. 858). FAD and NAD^+ function as reversible hydrogen acceptors that deliver the accepted hydrogen to the electron transport chain. (p. 863)

17. Glycolysis occurs in the cytoplasm of cells. It may be separated into three major events: (1) sugar activation, (2) sugar cleavage, and (3) oxidation and ATP formation. During sugar activation, glucose is phosphorylated, converted to fructose, and phosphorylated again to yield fructose-1,6-diphosphate; two molecules of ATP are used. These reactions provide the activation energy for the later events of glycolysis. During sugar cleavage, fructose-1,6-diphosphate is split

into two 3-carbon fragments: glyceraldehyde 3-phosphate or dihydroxyacetone phosphate. During oxidation and ATP formation, the 3-carbon molecules are oxidized by the removal of hydrogen (which is picked up by NAD). Inorganic phosphate groups that are attached to each oxidized fragment by high-energy bonds are cleaved off, capturing enough energy to form four ATP molecules. The final products of glycolysis are two molecules of pyruvic acid, two molecules of reduced NAD, and a net gain of two ATP molecules per glucose molecule. (pp. 861-862)

18. Pyruvic acid is converted to acetyl CoA, which enters the Krebs cycle. For pyruvic acid to be converted to acetyl CoA, the following must take place: decarboxylation to remove a carbon, oxidation to remove hydrogen atoms, and combination of the resulting acetic acid with coenzyme A to produce acetyl CoA. (p. 862)

19. Glycogenesis is the process by which glucose molecules are combined in long chains to form glycogen. Glycogenesis is the process by which glycogen is broken down to release its glucose monomers. Gluconeogenesis is the formation of new sugar from noncarbohydrate molecules. Lipogenesis is the term for triglyceride synthesis. Glycogenesis (and perhaps lipogenesis) is likely to occur after a carbohydrate-rich meal. Gluconeogenesis is likely to occur just before waking up in the morning. (pp. 867-868)

20. Metabolic acidosis due to ketosis is the result of excessive amounts of fats being burned for energy. Starvation, unwise dieting, and diabetes mellitus can result in ketosis. (p. 871)

21.

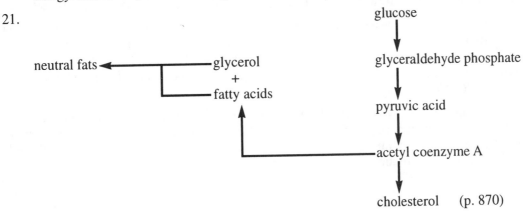

22. HDLs function to transport cholesterol from the peripheral tissues to the liver. LDLs transport cholesterol to the peripheral tissues. (p. 880)

23. Factors influencing plasma cholesterol levels include diet (through intake of cholesterol and/or saturated fatty acids), smoking, drinking, and stress. Sources of cholesterol in the body include the intake of animal foods and production from acetyl coenzyme A in the liver (and intestinal cells). Cholesterol is lost from the body when it is catabolized and secreted in bile salts that are eventually excreted in feces. It is used by body cells in plasma membranes, and in synthesizing vitamin D and steroid hormones. (pp. 880-881)

24. "Body energy balance" means that energy intake is equal to total energy output. If the body is not in exact balance, weight is either gained or lost. (pp. 881-882)

25. Metabolic rate is increased with increased production of thyroxine. Eating increases metabolic rate, an effect called chemical thermogenesis. A higher ratio of body surface area to body volume requires a higher metabolic rate, because heat exchange surface area is greater. Muscular exercise and emotional stress increase metabolic rate. Starvation decreases metabolic rate. (pp. 889-890)

26. The body's core includes organs within the skull and the thoracic and abdominal cavities. The core has the highest temperature. The shell, or skin, has the lowest temperature. Blood serves as the heat transfer agent between the core and shell. (p. 886)

27. Heat-promoting mechanisms to maintain or increase body temperature include vasoconstriction in the shell, which inhibits heat loss via radiation; conduction, and convection; increase in metabolic rate due to epinephrine release; and shivering. Heat-loss mechanisms include vasodilation of blood vessels in the skin and sweating (which enhances heat transfer via evaporation).

Whenever core temperature increases above or decreases below normal, peripheral and central thermoreceptors send input to the hypothalamus. Much like a thermostat, the hypothalamus responds to the input by initiating the appropriate heat-promoting or heat-loss reflex mechanisms via autonomic effector pathways. (pp. 886-890)

Critical Thinking and Application Questions

1. The number of ATP molecules resulting from the complete oxidation of a particular fatty acid can be calculated easily by counting the number of carbon atoms in the fatty acid and dividing by two to determine the number of acetyl CoA molecules produced. For our example, an 18-carbon fatty acid yields 9 acetyl CoA molecules. Because each of these yield 12 ATP molecules per turn of the Krebs cycle, a total of 108 ATP molecules is provided from the oxidative pathways: 9 from electron transport oxidation of 3 NADH + H$^+$, 2 from the oxidation of 1 FADH$_2$, and a net yield of 1 ATP during the Krebs cycle. Also, for every acetyl CoA released during beta oxidation, an additional molecule each of NADH + H$^+$ and FADH$_2$ is produced, which, when reoxidized, yield a total of 5 ATP molecules more. In an 18-carbon fatty acid, this would occur 8 times, yielding 40 more ATP molecules. After subtracting the ATP needed to get the process going, this adds up to a grand total of 147 ATP molecules from that single 18-carbon fatty acid! (pp. 870-871)

2. Hypothermia is abnormally depressed body temperature. It kills by dropping the body temperature below the relatively narrow range in which biochemical reactions can take place. (p. 891)

3. With a diagnosis of high cholesterol and severe arteriosclerosis, he should avoid foods containing saturated fatty acids and avoid eating eggs and large amounts of red meat. He should substitute foods containing unsaturated fatty acids and add fish to his diet. He should also stop smoking, cut down on his coffee, avoid stress situations when possible, and increase his amount of aerobic exercise. (pp. 869-871)

4. The chemiosmotic machinery concerns the operation of the electron transport chain and generation of the proton gradient during which most ATP is harvested in the mitochondria. If uncoupled, cells will use more and more nutrients in an effort to generate needed ATP, leaving fewer "calories" for protein synthesis and tissue maintenance. (p. 860)

Key Figure Questions

Figure 25.2 Yes. He has both grain and beans, which together contain all the essential amino acids. **Figure 25.6** 1. If oxygen were unavailable, the reduced coenzymes (NADH + H$^+$) would transfer their picked-up H$^+$ to pyruvic acid to form lactic acid. 2. Glycolysis would come to a halt because the oxidase enzyme cannot hold onto the H$^+$ removed during substrate oxidation. If all the coenzymes are already reduced and can't release hydrogen, the oxidase enzyme couldn't continue to function. **Figure 25.7** Decarboxylation, shown as removal of CO$_2$, and oxidation, shown as the reduction of the coenzymes, e.g., NAD$^+$ → NADH + H$^+$. **Figure 25.8** In substrate-level phosphorylation, a phosphate group is transferred directly from the substrate to ADP, forming ATP. In oxidative phosphorylation, the substrates are oxidized by the removal of hydrogen, which is eventually transferred to molecular oxygen. The removed hydrogens contain the energy which is eventually used to generate the proton gradient which provides the energy source for attaching phosphate to ADP, making ATP. **Figure 25.12** Acetyl CoA. **Figure 25.19** Insulin.

SUGGESTED READINGS

1. Ames, B.N. "Dietary Carcinogens and Anticarcinogens." *Science* 221 (1983): 1256.

2. Brown, Michael S., and Joseph L. Goldstein. "How LDL Receptors Influence Cholesterol and Arteriosclerosis." *Scientific American* 251 (Nov. 1984) : 58.

3. Cerami, A., H. Vlassara, and M. Brownlee. "Glucose and Aging." *Scientific American* 236 (May 1987): 90-96.

4. Cohen, L.A. "Diet and Cancer." *Scientific American* 257 (Nov. 1987): 42.

5. Fackelman, K.A. "Japanese Stroke Clues: Are There Risks to Low Cholesterol?" *Science News* 135 (Apr. 1989): 250, 253.

6. Fleck, H. *Introduction to Nutrition,* 4th ed. New York: Macmillan, 1981.

7. Hinckle, P.C., and Richard E. McCarty. "How Cells Make ATP." *Scientific American* 238 (Mar. 1978).

8. Jackson, D.D. "Up in the 'Cold Lab' Human Guinea Pigs Shiver for Science." *Smithsonian* 19 (Dec. 1989): 100-110.

9. Kluger, M.J. "The Evolution and Adaptive Value of Fever." *American Scientist* 66 (Jan.-Feb. 1978): 38.

10. Lowe, C. "Diet in a Glass." *Health* 21 (July 1989): 78-83.

11. McKenzie, A. "A Tangle of Fibers." *Science News* 136 (Nov. 1989): 344-345.

12. Raloff, J. "Beyond Oat Bran." *Science News* 137 (May 1990): 330-332.

13. Raloff, J. "Fish Oil Lowers Even Normal Blood Pressure." *Science News* 136 (Sept. 1989): 181.

14. Redgrave, T.G. "A New Approach to the Physiology of Lipid Transport." *NIPS* 3 (Feb. 1988): 10-12.

15. Rowand, A. "Trace Metals Leave More Than Trace Effects." *Science News* 125 (June 1984): 373.

16. Silberner, J. "High Cholesterol = High Cancer Risk?" *Science News* 131 (Jan. 1987): 4.

17. Stockton, W. "Balancing the Equation of Fat and Exercise." *New York Times* (Dec. 12, 1988): C11.

18. Uvnas-Moberg, K. "The Gastrointestinal Tract in Growth and Reproduction." *Scientific American* 26 (July 1989): 78-83.

19. Wurtman, R.J. and J.J. Wurtman. "Carbohydrates and Depression." *Scientific American* 260 (Jan. 1989): 68-75.

20. "Harnessing Fatty Acids to Fight Cancer." *Science News* 133 (May 1988): 332.

26

The Urinary System

Chapter Preview

This chapter deals with the body's filtration system, the urinary system. This system filters blood to remove the nitrogen waste products and other toxic wastes. Other functions of this system include producing the enzyme renin, which helps regulate blood pressure and kidney function; producing the hormone erythropoietin, which stimulates red blood cell production in bone marrow; and metabolizing vitamin D to its active form.

INTEGRATING THE PACKAGE

Suggested Lecture Outline

I. Introduction (p. 896)
    A. Functions of the System
    B. Organs (Fig. 26.1, p. 897)
        1. Kidneys
        2. Ureters
        3. Urinary Bladder
        4. Urethra

II. Kidney Anatomy (pp. 897-903)
    A. Location and External Anatomy (p. 897)
        1. Location (Fig. 26.2, p. 898)
        2. External Anatomy
    B. Internal Anatomy (p. 898; Fig. 26.3, p. 899)
        1. Regions
            a. Cortex
            b. Medulla
            c. Pelvis
        2. Inflammations
    C. Blood and Nerve Supply (p. 899)
        1. Renal Arteries and Veins
        2. Renal Plexus
    D. Nephrons (pp. 901-903; Fig. 26.5, p. 902)
        1. Structure (Fig. 26.4, p. 901)
            a. Glomerulus

           b. Renal Corpuscle
               1. Visceral Layer
               2. Parietal Layer
           c. Renal Tubule
               1. Proximal Convoluted Tubule
               2. Loop of Henle
                   a. Descending Limb
                   b. Ascending Limb
               3. Distal Convoluted Tubule
        2. Capillary Beds
           a. Afferent Arterioles
           b. Efferent Arterioles
           c. Vascular Resistance in the Microcirculation
        3. Juxtaglomerular Apparatus
           a. Juxtaglomerular Cells
           b. Macula Densa
III. Kidney Physiology (pp. 903-917)
    A. Glomerular Filtration (pp. 904-908)
        1. The Filtration Membrane (Fig. 26.8, p. 905)
        2. Net Filtration Pressure (Fig. 26.9, p. 906)
        3. Filtration Rate
        4. Regulation of Glomerular Filtration
           a. Intrinsic Controls: Renal Autoregulation (Fig. 26.10, p. 907)
               1. Myogenic Mechanism
               2. Tubuloglomerular Feedback
               3. Renin-Angiotensin Mechanism
           b. Extrinsic Controls: Sympathetic Nervous System Stimulation
    B. Tubular Reabsorption (pp. 908-911)
        1. Active Tubular Reabsorption
           a. Transport Maximum
        2. Passive Tubular Reabsorption
           a. Obligatory Water Reabsorption
           b. Solvent Drag
        3. Nonreabsorbed Substances
        4. Absorptive Capabilities of Different Regions of the Renal Tubules (Table 26.1, p. 911)
    C. Tubular Secretion (pp. 911-912)
    D. Regulation of Urine Concentration and Volume (pp. 912-916)
        1. The Countercurrent Mechanism and the Medullary Osmotic Gradient (Fig. 26.12, p. 913)
           a. Descending Limb
           b. Ascending Limb
           c. Collecting Tubules
           d. Vasa Recta

Review Items

11. Antidiuretic hormone (Chapter 17, pp. 560-561)

12. Atrial natriuretic factor (Chapter 17, p. 568)

13. Epinephrine (Chapter 14, p. 470; Chapter 17, p. 570)

14. Erythropoietin (Chapter 18, pp. 590-591)

15. Plasma (Chapter 18, p. 598)

16. Fenestrated capillaries (Chapter 20, p. 646)

17. Arterioles (Chapter 20, p. 646)

18. Autoregulation of blood flow (Chapter 20, pp. 660-667)

19. Vascular resistance (Chapter 20, p. 652)

20. Fluid dynamics (Chapter 20, pp. 664-665)

21. Renin-angiotensin (Chapter 20, p. 656)

Cross References

1. The renin-angiotensin mechanism as involved in the control of extracellular fluid volume is examined in Chapter 27, p. 936.

2. Electrolyte balance is explained in greater detail in Chapter 27, pp. 934-940.

3. The male urethra and delivery of semen are described in Chapter 28, p. 958.

4. Glomerulonephritis is mentioned in Chapter 27, p. 933.

5. H^+ and HCO_3^- and kidney function are further described in Chapter 27, pp. 940-941.

6. Hypoaldosteronism is explained in Chapter 27, p. 936.

Laboratory Correlations

1. Marieb, E. N. *Human Anatomy and Physiology Laboratory Manual: Cat and Fetal Pig Versions.* 3rd. ed. Benjamin/Cummings, 1989.

   Exercise 41: Anatomy of the Urinary System

   Exercise 42: Urinalysis

   Exercise 43: Kidney Regulation of Fluid and Electrolyte Balance

2. Marieb, E. N. *Human Anatomy and Physiology Laboratory Manual: Brief Version.* 3rd. ed. Benjamin/Cummings, 1992.

   Exercise 38: Anatomy of the Urinary Tract

   Exercise 39: Urinalysis

   Exercise 40: Kidney Regulation of Fluid and Electrolyte Balance

Transparencies Index

26.1   Organs of the urinary system

26.3   Internal anatomy of the kidney

26.4   Location and structure of nephrons

26.7   Kidney shown schematically as a single, uncoiled nephron

26.11 Directional movement of reabsorbed substances

26.13 Summary of nephron functions

Bassett Atlas Figures Index

INSTRUCTIONAL AIDS

Lecture Hints

1. Emphasize the retroperitoneal location of urinary structures.

2. Stress the importance of maintaining kidney in normal position. Project slides showing kidneys in place and point out the difference in encasement between kidneys and other abdominal organs.

3. Use the analogy of a cone-shaped filter in a glass funnel to illustrate how a pyramid fits into its calyx. This gives students a 3-D structure to relate to kidney anatomy.

4. Emphasize that the "lines" pointing to the papilla are due to the large numbers of microscopic tubules oriented in that direction.

5. Poke a finger into a partially inflated balloon to illustrate how the glomerular capsule forms around the glomerulus.

6. Emphasize the unique microvasculature of the kidney: arterioles feed and drain the glomerulus. During the discussion of anatomy (before physiology has been covered), ask the students if they can come up with a logical function for this design (remind then of the function of arterioles).

7. Students will often confuse the different capillary beds of the kidney. Stress the difference between glomerulus, peritubular capillaries, and the vasa recta.

8. Figure 26.7 on p. 904 is an excellent schematic diagram of overall kidney function. Use this diagram to introduce the class to renal physiology.

9. Emphasize that the filtration membrane is actually composed of three layers, not a single phospholipid bilayer as some students imagine.

10. Introduce the control of glomerular filtration by presenting the overall logic of the system: Filtration is driven by hydrostatic pressure; therefore, control of the afferent arteriole (and possibly efferent arteriole) is the obvious means of pressure control to the nephron, and thereby control of glomerular filtration is achieved. Both extrinsic and intrinsic mechanisms of glomerular filtration regulation work toward the same goal: control of pressure in the glomerulus.

11. Establishing a point of reference is essential for student understanding of renal system terminology. Students often have problems with reabsorption versus secretion: "Is the tubule secreting into the blood?" Make sure the class establishes the epithelial cells of the tubule (or blood) as a reference point when using the terms *secretion* and *reabsorption*. Another possible source of confusion is secretion versus excretion.

12. Clearly distinguish between the effects of aldosterone and antidiuretic hormone. Although both result in decreased urine output (and increased blood volume), the mechanisms are different.

13. Emphasize that the backflow of urine into the ureters is prevented by means of a physiological sphincter; the ureters enter the bladder wall at an angle so that volume and pressure in the bladder increase; pressure forces the openings to the ureter to collapse, preventing retrograde movement of urine. Use a diagram to illustrate this mechanism since it is difficult for most students to visualize.

14. Mention the similar function of the rugae in the bladder to the rugae in the stomach.

15. Emphasize that the urinary system is one of the few locations in the body that contains transitional epithelium.

Demonstrations/Activities

1. Audio-visual materials of choice.

2. Use a torso model and/or dissected animal model to exhibit urinary organs.

3. Use a funnel and filter paper to demonstrate the filtration process in the renal corpuscle.

4. Set up a dialysis bag to show the exchange of ions based on osmolarity.

5. Use a model of a longitudinally sectioned kidney to identify the major anatomical features. If the nephron is part of the model or if one is available, demonstrate the anatomical regions of the nephron and describe the specific functions of each area.

6. Display a hydrometer and other materials used to perform a urinalysis and discuss the importance of the urinalysis in routine physicals and in pathological diagnosis.

7. If possible, arrange for someone from a local renal dialysis center to come and talk to the class about how the artificial kidney works and other aspects of the dialysis process.

Critical Thinking/Discussion Topics

1. Discuss the link between emotions and kidney function.

2. Discuss the effect of certain drugs on kidney function.

3. Explain why physicians tell a sick individual to drink plenty of fluids and why fluid intake and output is so carefully monitored in hospital settings.

4. Discuss how kidney stones are formed, why they are formed, and how they can be treated.

5. Discuss the thirst mechanism and relate it to renal physiology.

6. Discuss the role of the kidneys in blood pressure regulation.

7. Discuss the different types of renal inflammation/infection and the consequences to other body systems.

Library Research Topics

1. Research the effects of common drugs such as penicillin, the myceins, etc., on kidney function.

2. Research the effect of hypertensive drugs on kidney function.

3. Research the effect of circulatory shock on kidney function and explain why the kidneys are affected.

4. Research the link between the emotions and kidney function.

5. Research the available treatments for kidney stones.

6. Research the process of dialysis.

7. Research the latest treatments for incontinence.

8. Research the latest updates in renal physiology.

9. Describe recent advances in the role of atrial natriuretic factor in fluid/electrolyte balance.

Audio-Visual Aids/Computer Software

Slides

1. Urinary System and Its Function (EIL, Slides). Human water balance and its control are discussed with a definite functionalist viewpoint.

Videotapes

1. The Living Body: Water (FHS, VHS or BETA, 26 min.). Shows the crucial part water plays in the body's functioning and the system for keeping it in balance.

2. The Mammalian Kidney (EIL, VHS or BETA). Details the anatomy and function of the mammalian kidney as it (1) regulates the excretion of metabolic wastes, (2) controls concentrations of salts, and (3) preserves overall water balance.

Computer Software

1. The Kidney: Structure and Function (EIL, Apple). Studies the relation between renal structure and function via several laboratory investigations. Students are presented with data that they must collect, organize, and interpret.

2. Dynamics of the Human Urinary System (EIL, Apple II, IBM, 1988). Combines straightforward, concise text with engaging animated graphics, sound effects, and an interactive quiz to instruct and challenge.

3. Body Language: Study of Human Anatomy — Urinary System (PLP, CH-182019, Apple; CH-182018, IBM). Covers the human urinary system from micro- to gross anatomy.

   See Guide to Audio-Visual Resources on page 339 for key to AV distributors.

LECTURE ENHANCEMENT MATERIAL

Clinical and Related Terms

1. Agenesis — absence of one or both kidneys.

2. Anuria — an absence of urine; may be due to kidney failure or to an obstruction in a kidney pathway.

3. Azotemia — presence of urea or other nitrogenous elements in the blood.

4. Cystitis — inflammation of the urinary bladder.

5. Diuresis — an increased production of urine.

6. Dysuria — painful urination.

7. Ectopic kidney — a kidney displaced from its normal position.

8. Enuresis — uncontrolled urination; bed-wetting.

9. Hypoplastic kidney — an undersized kidney.

10. Nephrotic syndrome — refers to a group of abnormalities characterized by severe loss of protein in the urine resulting in low blood protein levels.

11. Oliguria — a scanty output of urine.

12. Polyuria — an excessive output of urine.

13. Uremia — refers to a toxic level of urea in the blood resulting from kidney malfunction.

Disorders/Homeostatic Imbalances

1. Anemia of Kidney Origin — arises from severe urinary tract disease due to hematuria or to suppression of bone marrow by chronic renal failure.

2. Arteriolar Nephrosclerosis — occurs in hypertensive patients. Their renal arteries undergo thickening, with the glomeruli and tubules undergoing secondary degenerative changes.

3. Congenital Polycystic Kidney Disease — transmitted by a mendelian dominant trait; results from faulty development of the renal tubules and collecting ducts. Cysts develop, enlarge, and destroy renal function. This is a relatively common cause of renal failure.

4. Developmental Abnormalities — occur more frequently in the kidneys than in any other organ. Kidneys can be absent (agenesis) or undersized (hypoplastic). They can be displaced (ectopic) from their normal position and extra ureters or renal arteries can be found. The kidneys can fuse giving a horseshoe shape or can be filled with cysts (polycystic renal disease).

5. Diabetic Nephropathy — renal damage, usually diffuse thickening of the glomerular basement membranes, called diabetic glomerulosclerosis. There is usually severe sclerosis of the glomerular arterioles. This is a progressive disorder leading to renal failure.

6. Edema of Renal Failure — results when albumin is passed from the blood into urine, allowing fluid from blood to flow into the tissues due to loss of osmotic pressure in blood. Edema of renal failure is generalized.

7. Glomerulonephritis — an inflammation of the glomeruli, caused by a hypersensitivity antigen-antibody reaction in the glomeruli. Immune complex or acute glomerulonephritis usually occurs as a complication of a beta streptococci infection or other bacterial or viral infections. Anti-GBM nephritis (chronic glomerulonephritis) occurs when autoantibodies are directed against glomerular basement membranes (GBM). Renal failure can result from this condition.

8. Hematuria — presence of blood in the urine, which may be due to lesions in the urinary tract. Blood originating from the upper portion of the tract might be due to glomerulonephritis (an immunologic reaction), a malignancy, infection, trauma, or a stone. Lower tract hemorrhage might be due to infection, malignancy, a stone, or urethritis.

9. Nephrotic Syndrome — refers to a group of abnormalities characterized by severe loss of protein in the urine resulting in low blood protein levels. Patients show marked leg edema, fluid in the abdomen (ascites), and sometimes fluid in the pleural cavities (hydrothorax).

10. Renal Failure (Uremia) — implies loss of renal function. Acute renal failure (sudden loss of renal function) is caused by shock due to prolonged hypertension or hemorrhage. Acute renal tubular necrosis is caused by thrombosis, trauma, or acute nephritis. Chronic renal failure is due to infectious, traumatic, vascular, or immunologic diseases. Symptoms are nonspecific but can include anemia, salt and water retention, hypertension, and toxicity due to retained waste products. Treatment is dialysis or transplantation.

11. Renal Tubule Injury — may be due to toxic chemicals or to reduced renal blood flow. Symptoms include oliguria oranuria. With treatment, tubular function gradually returns.

12. Tumors — include adenomas which arise from epithelium or renal tubules that are small and asymptomatic. Carcinomas are more common. Transitional cell tumors arise from transitional epithelium lining the urinary tract, are of low-grade malignancy, and have a good prognosis.

13. Urinary Tract Calculi (Stones) — due to increased concentration of salts in urine (uric acid in gout and calcium salts in hyperparathyroidism). Clinical symptoms include renal colic, obstruction, pain, and possibly infection.

14. Urinary Tract Infection — usually caused by Gram-negative bacteria ascending the urethra. Cystitis (bladder infection) causes pain and burning during urination. Pyelonephritis (infection of upper urinary tract) is usually due also to ascending bacteria. Stagnation of urine and/or obstruction predisposes individuals to the problem.

15. Urinary Tract Obstruction — blocks urine outflow and leads to progressive dilatation of the urinary tract proximal to the obstruction and eventually causes compression atrophy of the kidneys. Common causes include obstruction of bladder neck by enlarged prostate or urethral stricture, calculus, or tumor. Complications include stone formation and/or infection.

16. Wilm's Tumor (Nephroblasoma) — an uncommon, highly malignant renal tumor of infants and children. It arises from embryonic cells in the kidney.

ANSWERS TO TEXTBOOK CHAPTER QUESTIONS

Multiple Choice/Matching

1. d	4. d	7. a
2. b	5. c	8. c
3. b	6. b	9. a

Short Answer Essay Questions

10. The adipose capsule helps to hold the kidney in place against the posterior trunk muscle and cushions it against blows. (p. 897)

11. A creatine molecule travels the following route from a glomerulus to the urethra. The creatine molecule first passes through the glomerular filtration membrane, which is a porous membrane made up of a fenestrated capillary endothelium, a thin basement membrane and the visceral membrane of the glomerular capsule formed by the podocytes. The creatine molecule then passes through the proximal convoluted tubule, the loop of Henle, and the distal convoluted tubule. Then the molecule passes into the collecting duct in which it travels into the medulla through the renal pyramids. From the medulla the molecule enters the renal pelvis where it then enters a ureter. Then it travels to the urinary bladder and then to the urethra. (pp. 898-903, 919)

12. Renal filtrate is a solute-rich fluid without blood cells or plasma proteins because the filtration membrane is permeable to water and all solutes smaller than plasma proteins. The capillary endothelium restricts passage of formed elements whereas the anion-rich basement membrane holds back most protein and some smaller anionic molecules. (pp. 903-904)

13. The mechanisms that contribute to renal autoregulation are the myogenic mechanism, the tubulo-glomerular feedback mechanism, and the renin-angiotensin mechanism. The myogenic mechanism reflects the tendency of vascular smooth muscle to contract when it is stretched. An increase in systemic blood pressure causes afferent arterioles to constrict, which impedes blood flow into the glomerulus and prevents glomerular blood pressure from rising to damaging levels. Conversely, a decline in systemic blood pressure causes dilation of afferent arterioles and an increase in glomerular hydrostatic pressure. Both responses help maintain a normal GFR.

The tubuloglomerular mechanism reflects the activity of the macula densa cells in response to a slow filtration rate. When so activated they release chemicals that cause vasodilation in the afferent arterioles.

The renin-angiotensin mechanism, triggered by the juxtaglomerular cells, reflects their ability to release the enzyme renin in response to various stimuli. Renin acts on angiotensinogen, a plasma globulin made by the liver, to form angiotensin I, which is, in turn, converted to angiotensin II by converting enzymes present in capillaries, particularly in the lungs. Angiotensin II activates vascular smooth muscle throughout the body, causing systemic blood pressure to rise, and stimulates the adrenal cortex to release aldosterone, which causes the renal tubules to reclaim more sodium ions from the filtrate. Because water follows sodium osmotically, blood volume and blood pressure rise. Angiotensin II may also cause vasoconstriction of the efferent arteriole, thereby increasing glomerular blood pressure and the GFR. Renin release is triggered by several factors acting independently or collectively: (1) reduced stretch of the juxtaglomerular cells; (2) stimulation of the juxtaglomerular cells by macula densa cells, which have been activated by osmotic signals or by slowly flowing filtrate; and (3) stimulation of the juxtaglomerular cells by the sympathetic nervous system.

Renal autoregulation maintains a relatively constant kidney perfusion over an arterial pressure range from about 80 to 180 mm Hg, preventing large changes in water and solute excretion. (pp. 904-908)

14. In active tubular reabsorption substances are usually moving against electrical and/or chemical gradients. The substances usually move from the filtrate into the tubule cells by diffusion but require the services of an ATP-dependent carrier to move across the basolateral membrane of the tubule cell into the interstitial space. Most such processes involve cotransport with sodium.

    Passive tubular reabsorption encompasses diffusion, facilitated diffusion, and osmosis. Substances move along their electrochemical gradient without the use of metabolic energy. (pp. 908-911)

15. The peritubular capillaries are low-pressure, porous capillaries that readily absorb solutes and water from the tubule cells. They arise from the efferent arteriole draining the glomerulus. (p. 903)

16. Tubular secretion is important for the following reasons: (a) disposing of substances not already in the filtrate; (b) eliminating undesirable substances that have been reabsorbed by passive processes; (c) ridding the body of excessive potassium ions; and, (d) controlling blood pH. Tubular secretion moves materials from the blood of the peritubular capillaries through the tubule cells or from the tubule cells into the filtrate. (pp. 911-912)

17. Aldosterone modifies the chemical composition of urine by enhancing sodium ion reabsorption so that very little leaves the body in urine. (pp. 910-911)

18. The filtrate becomes hypotonic as it flows through the ascending limb of the loop of Henle, because it is impermeable to water, and sodium and chloride are being actively pumped to the interstitial fluid, thereby decreasing solute concentration in the tubule. The interstitial fluid at the tip of the loop of Henle and the deep portions of the medulla are hypertonic because: (1) the loop of Henle serves as a countercurrent multiplier to establish the osmotic gradient, a process that works due to the characteristics of tubule permeability to water in different areas of the tubule and ion transport to the interstitial areas, and (2) the vasa recta which acts as a countercurrent exchanger to maintain the osmotic gradient by serving as a passive exchange mechanism which removes water from the medullary areas but leaves salts behind. The filtrate at the tip of the loop of Henle is hypertonic due to the passive diffusion of water from the descending limb to the interstitial areas. (pp. 912-914)

19. The bladder is very distensible. An empty bladder is collapsed and shows rugae. Expansion of the bladder to accomodate increased volume is due to the ability of the transitional epithelial cells lining the interior of the bladder to slide across one another, thinning the mucosa, and the ability of the detrusor muscle to stretch. (p. 919)

20. Micturition is the act of emptying the bladder. The micturition reflex is activated when distension of the bladder wall activates stretch receptors. Afferent impulses are transmitted to the sacral region of the spinal cord and efferent impulses return to the bladder via the parasympathetic pelvic splanchnic nerves, causing the detrusor muscle to contract and the internal sphincter to relax. (p. 920)

21. In old age the kidneys become smaller, the nephrons decrease in size and number, and the tubules become less efficient. By age 70, the rate of filtrate formation is only about one half that of middle-aged adults. This slowing is believed to result from impaired renal circulation caused by arteriosclerosis. The bladder is shrunken, with less than half the capacity of a young adult. Problems of urine retention and incontinence occur. (pp. 921-923)

Critical Thinking and Application Questions

1. Diuretics will remove water from the blood and eliminate it in the urine. Consequently, water will move from the peritoneal cavity into the bloodstream reducing her ascites.

    (1) Osmotic diuretics are substances that are not reabsorbed or that exceed the ability of the tubule to reabsorb it; (2) loop diuretics inhibit chloride reabsorption in the loop of Henle by diminishing sodium chloride uptake. They reduce the normal hyperosmolality of the medullary interstitial fluid, reducing the effects of ADH, resulting in loss of NaCl and water; and (3) thiazide diuretics act mainly on the early segments of the distal tubule to inhibit NaCl reabsorption. The increased Na^+ load in the distal tubule stimulates Na^+ exchange with K^+ and H^+, increasing their excretion and causing hypokalemia and a metabolic alkalosis.

    Her diet is salt-restricted because if salt content in the blood is high, it will cause her to retain water, rather than allowing her to eliminate it. (pp. 914-916)

2. A fracture at the lumbar region will stop the impulses to the brain, so there will be no voluntary control of micturition and he will never again feel the urge to void. There will be no dribbling of urine between voidings as long as the internal sphincter is undamaged. Micturition will be triggered by a reflex arc at the sacral region of the spinal cord as it is in an infant. (p. 920)

3. Cystitis is bladder inflammation. Women are more frequent cystitis sufferers than men because the female urethra is very short and its external orifice is closer to the anal opening. Improper toilet habits can carry fecal bacteria into the urethra. (p. 919)

4. Hattie has a renal calculus, or kidney stone, in her ureter. Predisposing conditions are frequent bacterial infections of the urinary tract, urinary retention, high concentrations of calcium in the blood, and alkaline urine. The woman's pain comes in waves because waves of peristalsis pass along the ureter at intervals. The pain results when the ureter walls close in on the sharp kidney stone during this peristalsis. (p. 918)

Key Figure Questions

Figure 26.4 It would go from the glomerular blood through the filtration membrane into Bowman's capsule, then through the renal tubule (PCT → loop of Henle → DCT), then into the collecting duct that carries the urine through the cortex and medulla, and into the minor calyx, into the major calyx, and into the pelvis. **Figure 26.7** The proximal convoluted tubule. The glomerulus. **Figure 26.9** With liver disease, there probably wouldn't be a normal supply of plasma proteins. Hence, more filtrate would be formed due to a lower osmotic pressure in the blood. **Figure 26.12** The countercurrent multiplier is the loop of Henle. The countercurrent exchanger is the vasa recta.

SUGGESTED READINGS

1. Amatnik, J. "Kidney Hormone May Limit Osteoporosis." *Science News* 124 (Dec. 1983): 373.

2. Anderson, B. "Regulation of Body Fluids." *Annual Review of Physiology* 39 (1977): 185.

3. Ash, S.R., et al. "Peritoneal Dialysis for Acute and Chronic Renal Failure: An Update." *Hospital Practice* (Jan. 1983).

4. Bauman, J.W., and F.P. Chinard. *Renal Function: Physiological and Medical Aspects*. St. Louis: C.V. Mosby, 1975.

5. Beeuwkes, R. "The Vascular Organization of the Kidney." *Annual Review of Physiology* 42 (1980): 531.

6. Franklin, D. "Pain Killer Abuse Spells Renal Damage." *Science News* 125 (Mar. 1984): 151.

7. Harvey, R.J. *The Kidneys and the Internal Environment*. New York: Halsted Press, 1976.

8. Malti, J., and Debra Wellons. "CAPD: A Dialysis Breakthrough with Its Own Burdens." *RN* (Jan. 1988): 46-53.

9. Pollie, R. "Comprehending Kidney Disease." *Science News* 122 (Oct. 1982): 218.

10. Reily, N.J. "The New Wane in Lithotrypsy." *RN* (Mar. 1988): 44-51.

11. Rivers, R. "Nursing the Kidney Transplant Patient." *RN* (Aug. 1987): 46.

12. Simonson, M.S. "Endothelins: Multifunctional Renal Peptides." *Physiological Reviews* 73 (Jan. 1993): 375.

13. Smith, H.W. "The Kidney." *Scientific American* (Jan. 1953).

14. Thomsen, D.E. "Plasma Physics Breaks Stones." *Science News* 130 (Sept. 1986): 157.

15. Vander, A.J. *Renal Physiology*. 2nd ed. New York: McGraw-Hill, 1980.

27

Fluid, Electrolyte, and Acid-Base Balance

Chapter Preview

This chapter deals with the balance of fluids and electrolytes and the action of acids and bases in the body. The fluids of the body are either extracellular or intracellular. The balance of those fluids and the electrolytes within them is a major process of the body. Without homeostatic balance, the cells of the body are not maintained.

Acid-base balance is also important. Body metabolism results in the production of numerous hydrogen ions. Those ions must be buffered in order to prevent acidosis from occurring. Lack of hydrogen ions can result in alkalosis. Therefore, the body has numerous buffer systems designed to maintain hydrogen ion homeostasis. This chapter deals with those homeostatic mechanisms as well as the mechanisms designed to maintain fluid/electrolyte homeostasis.

INTEGRATING THE PACKAGE

Suggested Lecture Outline

I. Body Fluids (pp. 928-930)
    A. Body Water Content (p. 928)
    B. Fluid Compartments (p. 928)
        1. Intracellular Fluid
        2. Extracellular Fluid
            a. Plasma
            b. Interstitial Fluid
    C. Composition of Fluids (pp. 928-930)
        1. Solutes
            a. Electrolytes
            b. Nonelectrolytes
        2. Comparison of Extracellular and Intracellular Fluids
    D. Movement of Fluids Between Compartments (p. 930)

II. Water Balance (pp. 930-933)
    A. Regulation of Water Intake: The Thirst Mechanism (pp. 931-932)
        1. The Thirst Mechanism (Fig. 27.6, p. 932)
    B. Regulation of Water Output (p. 932)
        1. Obligatory Water Losses
    C. Disorders of Water Balance (Fig. 27.7, p. 933)
        1. Dehydration
        2. Hypotonic Hydration
        3. Edema

III. Electrolyte Balance (pp. 934-940; Table 27.1, p. 935)
    A. Central Role of Sodium in Fluid and Electrolyte Balance (p. 934)
    B. Regulation of Sodium Balance (pp. 934-937)
        1. Influence and Regulation of Aldosterone (Fig. 27.8, p. 936)
        2. Cardiovascular System Baroreceptors
        3. Influence and Regulation of ADH (Fig. 27.9, p. 937)
        4. Influence and Regulation of Atrial Natriuretic Factor
        5. Influence of Other Hormones
            a. Female Sex Hormones
            b. Glucocorticoids
    C. Regulation of Potassium Balance (pp. 937-939)
        1. Tubule Cell Potassium Content
        2. Aldosterone Levels
        3. pH of the Extracellular Fluid
    D. Regulation of Calcium Balance (pp. 939-940)
        1. Effects of Parathyroid Hormone
            a. Bones
            b. Small Intestine
            c. Kidneys
        2. Effect of Calcitonin
    E. Regulation of Magnesium Balance (p. 940)
    F. Regulation of Anions (p. 940)
IV. Acid-Base Balance (pp. 940-950; Table 27.2, p. 943)
    A. Introduction (pp. 940-941)
        1. Alkalosis
        2. Acidosis
    B. Chemical Buffer Systems (pp. 941-942)
        1. Bicarbonate Buffer System
        2. Phosphate Buffer System
        3. Protein Buffer System
    C. Respiratory System Regulation of Hydrogen Ion Concentration (pp. 942-944)
    D. Renal Mechanisms of Acid-Base Balance (pp. 944-946)
        1. Regulation of Hydrogen Ion Secretion
        2. Conservation of Bicarbonate Ions
        3. Buffering of Excreted Hydrogen Ions (Fig. 27.13, p. 945)
            a. Ammonia-Ammonium Ion Buffer Systems
    E. Abnormalities of Acid-Base Balance (pp. 946-950)
        1. Respiratory Acidosis or Alkalosis
            a. Respiratory Acidosis
            b. Respiratory Alkalosis
        2. Metabolic Acidosis or Alkalosis
            a. Metabolic Acidosis
            b. Metabolic Alkalosis

3. Effects of Acidosis or Alkalosis

4. Respiratory and Renal Compensation

V. Developmental Aspects (p. 950)

  A. Early Development

    1. Water Content

    2. Problems with Fluid, Electrolyte, Acid-Base Balances

  B. Aging Problems

Review Items

1. Ions (Chapter 2, pp. 32-33, 39)

2. Water (Chapter 2, pp. 39-40)

3. Acid-base reactions and pH (Chapter 2, pp. 40-42)

4. Sodium-potassium pump (Chapter 3, p. 71)

5. Membrane transport (osmosis, diffusion) (Chapter 3, pp. 67-70)

6. Hypothalamus (Chapter 12, p. 395)

7. ADH (water conservation) (Chapter 17, pp. 560-561; Chapter 26, pp. 914-916)

8. Diabetes (mellitus, insipidus) (Chapter 17, pp. 574-575)

9. Aldosterone (sodium conservation) (Chapters 17, p. 567; 26, p. 911)

10. Atrial natriuretic factor (Chapters 17, p. 568; 26, p. 911)

11. Estrogens (Chapter 17, pp. 575-576)

12. Glucocorticoids (Chapter 17, pp. 568-569)

13. Parathyroid hormone/calcitonin (Chapter 17, p. 565)

14. Plasma (Chapter 18, p. 598)

15. Baroreceptors (Chapter 20, pp. 654-655)

16. Capillary exchange (Chapter 20, pp. 664-665)

17. Carbon dioxide and bicarbonate (Chapter 23, pp. 770-772)

18. Hemoglobin and pH control (Chapter 23, pp. 769, 772)

19. Ketone bodies and metabolism (Chapter 25, p. 871)

20. Control of renal blood flow (Chapter 26, pp. 906-908)

21. Glomerular filtration (Chapter 26, pp. 904-907)

22. Glomerulonephritis (Chapter 26, p. 923)

23. Juxtaglomerular apparatus (Chapter 26, p. 907)

24. Renin-angiotensin mechanism (Chapter 26, p. 908)

25. Potassium reabsorption (Chapter 26, p. 911)

26. H^+ and HCO_3^- and the kidney (Chapter 26, p. 912)

Laboratory Correlations

1. Marieb, E. N. *Human Anatomy and Physiology Laboratory Manual: Cat and Fetal Pig Versions.* 3rd. ed. Benjamin/Cummings, 1989.

   Exercise 43: Kidney Regulation of Fluid and Electrolyte Balance

2. Marieb, E. N. *Human Anatomy and Physiology Laboratory Manual: Brief Version.* 3rd. ed. Benjamin/Cummings, 1992.

    None in this chapter

Transparencies Index

27.1   Major fluid compartments of the body
27.5   Sources of water intake and water output
27.8   Mechanism and consequence of Aldosterone release
27.9   Mechanism and consequence of ADH release

27.11   Dissociation of acids
27.12   Reabsorption of filtered HCO_3^-
27.13   Buffering of excreted hydrogen ions in urine

INSTRUCTIONAL AIDS

Lecture Hints

1. Stress that fluid compartments are an abstract idea; draw a schematic diagram depicting the cell and its external environment. Stress that the plasma membrane is the structure separating these fluid compartments.

2. Point out the difference between plasma and interstitial fluid and that these two divisions of the extracellular compartment are not the only examples in the body. CSF, lymph, and other serous fluids are also extracellular fluids.

3. Refer students to a review of osmosis and diffusion. A thorough understanding of the movements of solute and solvent are crucial for comprehension of fluid/electrolyte balance.

4. Stress the different solute compositions of intracellular and extracellular compartments. Remind students of the physiology of the action potential as an example of the importance of maintaining cellular boundaries (therefore the relative compositions of the fluid compartments).

5. Get the class involved by asking students questions: "What are some of the ways we lose water (water output)?" "What are the sources of body water (water intake)?"

6. Emphasize that water will always move with solutes whenever possible. Water cannot be actively transported, so balance is achieved by controlling solute movement and water permeability.

7. Remind the class of blood pressure control by nervous, renal, and hormonal mechanisms. All of these control systems are highly integrated, and this is an opportunity to illustrate the cooperative nature of body systems in maintaining homeostasis. Although we treat the body as having individual systems, one must not forget that all systems are part of a functional individual.

8. Stress the importance of acid-base balance and levels of intracellular potassium, especially in excitable cells. Also point out that potassium is the major ion in the intracellular fluid and therefore is the major control of water balance within the cell. As a point of interest, mention the consequences of excessively high or low levels of potassium.

9. Emphasize the importance of acidity or basicity on all chemical reactions.

10. Start the discussion of acid-base balance by mentioning that the respiratory and renal systems are powerful pH control mechanisms.

11. Emphasize the difference between using strong acid-base combinations versus weak acid-base combinations as buffering systems. Relating the two makes it easier for students to realize the need for the latter.

12. Clearly distinguish between metabolic and respiratory acids and bases.

Demonstrations/Activities

1. Audio-visual materials of choice.

2. Demonstrate the principles of osmosis and diffusion as a reminder to students about those processes.

3. Perform a simple titration to demonstrate how strong acids and bases can be neutralized by weaker acids and bases.

Critical Thinking/Discussion Topics

1. Discuss the effects of IV therapy on the fluid and electrolyte balance in the body; distinguish between the infant or small child and the adult.

2. Discuss why a sodium bicarbonate IV is used in cases of cardiac arrest or circulatory shock.

3. Discuss the effects of alcoholism on acid-base balance.

4. Discuss the effects of prolonged use of antacids on acid-base balance.

Library Research Topics

1. Research the rationale behind taking arterial blood gas values to help determine acid-base balance.

2. Research the reasons for and effects of IV therapy in cases of heart attack, surgery, chemotherapy, etc.

3. Research the roles in the body of the more common electrolyte such as Na^+, Mg^{++}, Ca^{++}, etc.

LECTURE ENHANCEMENT MATERIAL

Clinical and Related Terms

1. Acetonemia — presence of abnormal amounts of acetone in the blood.

2. Acetonuria — presence of abnormal amounts of acetone in the urine.

3. Albuminuria — presence of albumin in the urine.

4. Antacid — an alkaline substance used to neutralize an acid.

5. Azotemia — an accumulation of nitrogenous wastes in the blood.

6. Glycosuria — excessive sugar in the urine.

7. Hyperkalemia — excessive potassium in the blood.

8. Hypernatremia — excessive sodium in the blood.

9. Hyperuricemia — excessive uric acid in the blood.

10. Hypoglycemia — an abnormally low level of sugar in the blood.

11. Ketonuria — presence of ketone bodies in the urine.

12. Ketosis — acidosis due to presence of excessive ketone bodies in body fluids.

Disorders/Homeostatic Imbalances

1. Dehydration — lack of fluids in the body caused by inadequate water intake or excessive water loss. This is the most common disturbance of water balance. Excessive fluid loss may be due to vomiting and diarrhea. Fluid intake is also decreased, which contributes to the dehydration.

2. Electrolyte Balance Disturbances — abnormal concentrations of ions (electrolytes) in the body. This condition will generally be caused by the same conditions that result in water imbalances. Diuretics, administered to heart failure, cirrhosis of the liver, and kidney disease patients, often will cause electrolyte imbalance.

3. Metabolic Acidosis — occurs when excess endogenous acid depletes bicarbonate stores. Causes include renal failure, ketosis, and overproduction of lactic acid. Compensation involves hyper-ventilation (lowers PCO_2), kidney excretion of more hydrogen ions and formation of more bicarbonate.

4. Metabolic Alkalosis — occurs in the presence of excess plasma bicarbonate. Usual causes include loss of gastric juice, chloride depletion, excess corticosteroid hormones, ingestion of excessive bicarbonate or other antacids. The body does not compensate for this problem.

5. Overhydration — presence of excessive fluids in the body. It is less common than dehydration. It can be due to excessive intake of fluids in an individual with renal dysfunction or to excessive administration of IV fluids.

6. Respiratory Acidosis — occurs when there is inefficient excretion of carbon dioxide by the lungs. The usual cause is chronic pulmonary disease. Compensation is by formation of additional bicar-bonate by the kidneys.

7. Respiratory Alkalosis — occurs when hyperventilation lowers the PCO_2 levels. Usual causes include severe anxiety with hyperventilation, stimulation of respiratory centers by drugs, and CNS disease. Compensation is by increased excretion of bicarbonate by the kidneys.

ANSWERS TO END-OF-CHAPTER QUESTIONS

Multiple Choice/Matching

1. a	5. f, h, i	8. b	11. b, d, f
2. c	6. c, g	9. a	12. a
3. b	7. e, h	10. j	13. c
4. b			

Short Answer Essay Questions

14. The body fluid compartments include the intracellular fluid compartment, located inside the cells with fluid volume approximately 25 liters, and the extracellular fluid compartment, located in the external environment of each cell, with fluid volume approximately 15 liters. (p. 928)

15. It is believed that a decrease in plasma volume of 10% or more and/or an increase in plasma osmolality of 1% to 2% results in a dry mouth and excites the hypothalamic thirst or drinking center. Hypothalamic stimulation occurs because the osmoreceptors in the thirst center become

irritable and depolarize as water, driven by the hypertonic ECF, moves out of them by osmosis. Collectively, these events cause a subjective sensation of thirst. Thirst is quenched almost as soon as the required amount of water is drunk. The quenching of thirst begins as the mucosa of the mouth and throat are moistened and continues as stretch receptors in the stomach and intestine are activated, providing feedback signals that inhibit the hypothalamic thirst center. (pp. 928-930)

16. Sodium is pivotal to fluid and electrolyte balance and to the homeostasis of all body systems since it is the principal extracellular ion. While the sodium content of the body may be altered, its concentration in the ECF remains stable because of immediate adjustments in water volume. The regulation of the sodium-water balance is inseparably linked to blood pressure and entails a variety of neural and hormonal controls: (1) Aldosterone — increases the reabsorption of sodium from the filtrate; water follows passively by osmosis, increasing blood volume (and pressure). The renin-angiotensin mechanism is an important control of aldosterone release; the juxtaglomerular apparatus responds to: (a) decreased stretch (due to decreased blood pressure), (b) decreased filtrate osmolality, or (c) sympathetic nervous system stimulation resulting ultimately in aldosterone release from the adrenal cortex. (2) ADH — osmoreceptors in the hypothalamus sense solute concentration in the ECF: increases in sodium content stimulates ADH release, resulting in increased water retention by the kidney (and increasing blood pressure). (3) Atrial natriuretic factor — released by cells in the atria during high-pressure situations, has potent diuretic and natriuretic (sodium-excreting) effects; the kidneys do not reabsorb as much sodium (therefore water) and blood pressure drops. (pp. 934-937)

17. Respiratory system regulation of acid-base balance provides a physiological buffering system. Falling pH, due to hydrogen ion concentration or rising PCO_2 in plasma, excites the respiratory center (directly or indirectly) to stimulate deeper, more rapid respirations. When pH begins to fall, the respiratory center is depressed. (pp. 942-943)

18. Chemical acid-base buffers prevent pronounced changes in H^+ concentration by binding to hydrogen ions whenever the pH of body fluids drops and releasing them when pH rises. (pp. 941-942)

19. The rate of H^+ secretion rises and falls directly with CO_2 levels in the ECF. The higher the content of CO_2 in the peritubular capillary blood, the faster the rate of H^+ secretion. The rate of ammonia secretion is coupled to the rate of H^+ secretion. Each time an ammonia molecule combines with a hydrogen ion, another ammonia molecule diffuses into the filtrate. The dissociation of carbonic acid in the tubule cells liberates HCO_3^- as well as H^+. HCO_3^- is shunted into the peritubular capillary blood. The rate of reabsorption of bicarbonate depends on the rate of secretion of H^+ in the filtrate.

20. Factors that place newborn babies at risk for acid-base imbalances include very low residual volume of infant lungs, high rate of fluid intake and output, relatively high metabolic rate, high rate of insensible water loss, and inefficiency of the kidneys. (p. 950)

Critical Thinking and Application Questions

1. This patient has diabetes insipidus caused by insufficient production of ADH by the hypothalamus. The operation for the removal of the cerebral tumor has damaged the hypothalamus and thus interfered with the function of the hypothalamohypophyseal tract leading to the posterior pituitary. Because of the lack of ADH, the collecting tubules and possibly the convoluted part of the distal convoluted tubule are not absorbing water from the glomerular filtrate. The large volume of very dilute urine voided by this man and the intense thirst that he experiences are the result. (p. 936)

2. Problem 1: pH 7.63, PCO_2 19 mm Hg, HCO_3^- 19.5 m Eq/L

   a. The pH is elevated = alkalosis.

   b. The PCO_2 is low and is the cause of the alkalosis.

   c. The HCO_3^- is also low = compensating. This is a respiratory alkalosis, possibly due to hyperventilation, being compensated by metabolic acidosis.

Problem 2: pH = 7.22, PCO_2 30 mm Hg, HCO_3^- 12.0 mEq/L

a. The pH is below normal = acidosis.

b. The PCO_2 is low, therefore not the cause of the acidosis, but is compensating.

c. The HCO_3^- is very low and is the cause of the acidosis. This is a metabolic acidosis. Possible causes include ingestion of too much acid (drinking too much alcohol), excessive loss of bicarbonate ion (diarrhea), accumulation of lactic acid during exercise, or shock, or by the ketosis that occurs in diabetic crises or starvation. (pp. 946-950)

3. Emphysema impairs gas exchange or lung ventilation, leading to retention of carbon dioxide and respiratory acidosis. Congestive heart failure produces oxygenation problems as well as edema and causes metabolic acidosis due to an increase in lactic acid. (pp. 942-943)

4. The patient has a normal sodium ion concentration, CO_2 is slightly low as is Cl^-. The potassium ion concentration is so abnormal that the patient has a medical emergency. The greatest danger is cardiac arrhythmia and cardiac arrest. (p .935)

Key Figure Questions

Figure 27.2 Na. K. Protein anions and SO_4^{2-}. **Figure 27.6** It would increase thirst because body fluids would become more concentrated. **Figure 27.9** An increased water absorption. **Figure 27.11** A weak base, because the weak base would act as a buffer.

SUGGESTED READINGS

---

1. Anderson, B. "Regulation of Body Fluids." *Annual Review of Physiology* 39 (1977): 185.

2. Barta, M.A. "Correcting Electrolyte Imbalances." *RN* (Feb. 1988).

3. Beauchamp, Gary K. "The Human Preference for Excess Salt." *American Scientist* 75 (Jan.-Feb. 1987): 27.

4. Folk-Lighty, Marie. "Solving the Puzzles of Patient's Fluid Imbalance." *Nursing* 84 (14) (Feb. 1984).

5. Friedman, P.A. "Renal Calcium Transport: Sites and Insights." *NIPS* 3 (Feb. 1988): 17-20.

6. Halperin, M.L., and M Goldstein. *Fluid, Electrolytes, and Acid-Base Emergencies*. Philadelphia: W.B. Saunders Co., 1988.

7. Hills, A.G. *Acid-Base Balance*. Baltimore: Williams and Wilkins, 1973.

8. Hinton, B.T., and T.T. Turner. "Is the Epididymis a Kidney Analog?" *NIPS* 3 (Feb. 1988): 28-31.

9. Horne, M.M., and P.L. Swearingen. *Pocket Guide to Fluids and Electrolytes*. St. Louis: C.B. Mosby Co., 1989.

10. Miller, J.A. "At the Heart of Blood Pressure Control." *Science News* 126 (July 1984): 4.

11. Sawka, M.M., et al. "Influence of Hydration Level and Body Fluids on Exercise Performance in the Heart." *Journal of the American Medical Association* (Sept. 1984).

12. Schwartz, Michele W. "Potassium Imbalances." *American Journal of Nursing* 87 (Oct. 1987): 1292.

13. Simonson, M.S. "Endothelins: Multifunctional Renal Peptides." *Physiological Reviews* 73 (Jan. 1993): 375.

28

The Reproductive System

Chapter Preview

This chapter deals with the male and female reproductive systems. These systems are designed differently, but their common purpose is to produce offspring. The male reproductive system produces gametes or sex cells and delivers the gametes (spermatozoa) to the female reproductive tract, where fertilization can occur. The female reproductive system also produces gametes (ova), receives the male gametes, and if fertilization occurs, provides the organ (uterus) in which the developing offspring will grow. In addition, both the male and female systems produce hormones that play vital roles in the development and function of the reproductive organs and in sexual behavior and drives.

INTEGRATING THE PACKAGE

Suggested Lecture Outline

Review Items

1. Cell division (Chapter 3, pp. 88-95)
2. Tight junctions (Chapter 3, p. 64)
3. Organelles (Chapter 3, pp. 75-84)
4. Microvilli (Chapter 3, p. 64)
5. Pseudostratified epithelium (Chapter 4, p. 107)
6. Tubuloalveolar glands (Chapter 4, p. 111)
7. Peristalsis (smooth muscle contraction) (Chapter 9, pp. 278-280)
8. Male perineum (Chapter 10, p. 311)
9. Female perineum (Chapter 10, p. 311)
10. Muscles of the pelvic floor (Chapter 10, p. 311)
11. Testosterone and brain anatomy (Chapter 12, p. 420)
12. Reflex activity (Chapter 13, pp. 450-456)
13. Sympathetic and parasympathetic effects (Chapter 14, p. 472)
14. Brain-testicular axis (Chapter 17, p. 558)
15. Prostaglandins (Chapter 17, p. 548)
16. Testosterone (Chapter 17, p. 575)
17. FSH and LH (Chapter 17, p. 558)
18. Ovaries and estrogen (Chapter 17, pp. 575-576)
19. Male urethra (Chapter 26, pp. 919-920)

Cross References

1. Fertilization (union of egg and spermatozoon) is described in detail in Chapter 29, pp. 1000-1002.
2. The vaginal environment and sperm viability are presented in Chapter 29, p. 1000.
3. The passage of sperm through the female reproductive tract in preparation of fertilization is covered in Chapter 29, pp. 1000-1001.
4. The relationship of spermatozoon and oocyte structure related to the process of fertilization is described in Chapter 29, pp. 1002-1003.
5. Uterine function in reproduction is described in detail in Chapter 29, pp. 1017-1019.
6. Interruption of uterine and ovarian cycles by pregnancy is presented in Chapter 29, pp. 1004-1005.
7. The completion of meiosis II is examined in Chapter 29, p. 1003.
8. The process of meiosis as related to genetics is presented in Chapter 30, pp. 1029-1030.
9. The importance of tetrad formation and recombination is detailed in Chapter 30, p. 1029.

Laboratory Correlations

1. Marieb, E. N. *Human Anatomy and Physiology Laboratory Manual: Cat and Fetal Pig Versions.* 3rd. ed. Benjamin/Cummings, 1989.

    Exercise 44: Anatomy of the Reproductive System

    Exercise 45: Physiology of Reproduction: Gametogenesis and the Female Cycles

2. Marieb, E. N. *Human Anatomy and Physiology Laboratory Manual: Brief Version*. 3rd. ed. Benjamin/Cummings, 1992.

    Exercise 41: Anatomy of the Reproductive System

    Exercise 42: Physiology of Reproduction: Gametogenesis and the Female Cycles

Transparencies Index

28.1   Reproductive organs of the male, sagittal view

28.2   Relationship of the testis to the scrotum and spermatic cord

28.3   Internal structure of the testis

28.4   Longitudinal and transverse sections of the penis

28.9   Relationship between sustentacular and spermatogenic cells

28.10 Hormonal regulation of testicular function

28.11 Internal organs of the female reproductive system

28.13 Structure of an ovary

28.14 Photo — anterior view of internal female reproductive organs

28.17 Structure of lactating mammary glands

28.18 Mammograms

28.20 The ovarian cycle

28.21 Feedback interactions involved in the regulation of ovarian function

Bassett Atlas Figures Index

Slide Number	FigureNumber	Description
44	5.1A,B	Left inguinal area
45	5.2	Right inguinal area from within
47	5.4A,B	Sagittal section, male pelvis
48	5.5A,B	Looking down into the female pelvis
49	5.6A,B	Sagittal section, female pelvis
51	5.8	Female perineum

INSTRUCTIONAL AIDS

Lecture Hints

1. Emphasize that sperm are not capable of fertilizing an egg immediately, but must first be naturally or artifically capacitated.

2. The ducting system of the male reproductive system is difficult for most students to visualize. Use models and diagrams so that 3-D structure becomes apparent.

3. Emphasize the different secretions (and their functions) in the male reproductive tract.

4. Use cross-sectional and longitudinal diagrams of the anatomy of the penis. Both sections are necessary to establish the correct 3-D internal structure.

5. Emphasize the difference between mitosis and meiosis. Mitosis involves a single round of DNA synthesis followed by a single cytokinetic event to result in two diploid cells. Meiosis involves a single round of DNA synthesis followed by two successive cytokinetic events resulting in four haploid structures (four spermatozoa in the male, or two or three polar bodies and an egg in the female).

6. Students will have a clearer understanding if you draw a schematic diagram of spermatogenesis and relate it to a cross section of a seminiferous tubule.

7. Be sure to indicate the reasoning behind the terms *reduction* and *equatorial division*. Students often have difficulty with the concept of chromatid versus chromosome, and therefore have difficulty with these terms.

8. Clearly distinguish between spermatogenesis and spermiogenesis. Students are often confused by the similar-sounding names.

9. Point out that erection is a parasympathetic response and ejaculation is due to sympathetic reflex action.

10. Stress the importance of the blood-testis barrier in preventing immune action against sperm antigens.

11. Emphasize that testosterone has somatic effects as well as those involving reproductive functions.

12. Mention that the term *germinal epithelium* has nothing to do with ovum formation.

13. Show a diagram or slide of a mature follicle and primordial follicle to illustrate the size difference between the two.

14. It is often of benefit to compare and contrast spermatogenesis and oogenesis side-by-side to emphasize the similarities and differences.

15. It is important to stress that the secondary oocyte (even when initially ovulated) does not complete meiosis II until fertilized by the spermatozoon, therefore should technically not be called an ovum until fertilization has occurred.

16. Mention that the polar bodies are actually tiny nucleate haploid cells (that are not fertilizable) and that the size difference between polar bodies and oocyte is due to the amount of cytoplasm present.

17. A great deal of confusion can be generated during the presentation of ovarian and menstrual cycles. Regardless of the sequence of presentation, emphasize that both cycles occur concurrently and that both should be visualized as one continuous process even though different events occur at different times. At the end of the discussion, present a plot of hormone concentration versus day in the cycle, and review the function of each hormone as the level rises and falls. Be sure to indicate sources of hormones (follicle vs. corpus luteum) so that students will understand maintenance of pregnancy in Chapter 29.

18. Emphasize the difference between menstrual phase and menstrual cycle. Students will often confuse the two.

Demonstrations/Activities

1. Audio-visual materials of choice.

2. Use a torso model, reproductive model, and/or dissected animal model to exhibit reproductive organs.

3. Display a large wall chart of the hormone levels during ovarian and uterine (menstrual) cycles.

4. Project a series of 2 x 2 color slides of the process of cell reproduction to refresh memories of the sequence of events.

5. Obtain and project slides of the effects of various sexually transmitted diseases.

6. Obtain or prepare a display of the various methods of birth control.

7. Use models showing the process of meiosis in spermatogenesis and oogenesis.

Critical Thinking/Discussion Topics

1. Discuss the need for mammograms and self-examination for early diagnosis of breast cancer.

2. Discuss the need for self-exam for testicular cancer.

3. Discuss the current treatments available for breast cancer.

4. Discuss the signs and symptoms of premenstrual syndrome and menopause.

5. Discuss the various treatments available for infertility.

6. Describe the consequences of a lack of the blood-testis barrier.

7. Would the injection of a man's own sperm be a reasonable method of birth control? Why or why not?

8. Discuss the possible consequences of a lack of estrogen (or any other reproductive hormone) during the ovarian/menstrual cycles.

Library Research Topics

1. Research the current treatments for breast cancer.

2. Research the current prostate gland disorder treatments.

3. Research the disorders associated with the menstrual cycle.

4. Research the various causes of infertility in males and females and how fertility can be enhanced.

5. Research the latest advances in birth control. How soon will a male oral contraceptive be available? What about the five-year female contraceptive implant?

Audio-Visual Aids/Computer Software

Slides

1. Reproductive System and Its Function (EIL, Slides). This graphic depiction of human male and female sexual anatomy introduces a discussion of the hormonally directed physiological changes associated with the reproductive system.

2. Histology of the Reproductive System (EIL, Slides). Views of the male and female tissues, including sections of the forming gametes within the gonads, ducts of the ovaries and testes, auxiliary glands associated with the systems, organs of copulation, and the relationships of tissues within each system.

Videotapes

1. Human Embryology Series: Reproductive or Sexual Cycles in the Female I (TFI, VHS or BETA, 11 min.)

2. Human Embryology Series: Reproductive or Sexual Cycles in the Female II (TFI, VHS or BETA, 11 min.)

3. Safe Sex (FHS, VHS or BETA, 28 min., C). Discusses safe sex, especially in reference to AIDS.

4. Preventing Teen Pregnancy (FHS, VHS or BETA, 28 min., C). Discusses sex education, sexual abstinency, education about contraceptives.

5. PMS and Endometriosis (FHS, VHS or BETA, 19 min., C). Explains the nature of PMS and its various treatments. Also explains the pertinent facts about endometriosis.

6. Rape: An Act of Hate (FHS, VHS or BETA, 30 min., C). Seeks to determine why people rape and to help potential victims protect themselves.

7. The Sexual Brain (FHS, VHS or BETA, 28 min., C). Provides some of the answers that separate cultural and social differences between men and women from physiological ones.

8. Shares in the Future (FHS, VHS or BETA, 26 min., C). Shows the characteristics of sperm and ova and how each contains a partial blueprint for the future offspring. The mechanism of cell division is shown through exceptional microphotography; the mechanisms of heredity are carefully described.

9. Meiosis, Reduction Division (PLP, CH-100023, VHS)

Computer Software

1. Describing Patterns in Reproduction, Growth, and Development (QUE, MVM4008A, Apple; MVM4008H, IBM)

2. Introduction to General Biology Series: Disk XI — The Human Body, Part II (QUE, COM4240A, Apple; COM4240B, IBM)

3. Reproduction, Growth, and Development (QUE, INT4158A, Apple; INT4158B, IBM)

4. Biology Achievement Series, Series II: Reproduction and Development (QUE, MIC4000A, Apple). Over 10,000 problems, some with graphics.

5. Dynamics of the Human Reproductive System (EI, C 3055A, Apple, C3055M, IBM)

6. Body Language: Study of Human Anatomy, Reproductive Systems (PLP, CH-182019, Apple; CH-182018, IBM)

7. The Human Systems: Series 3 (PLP, CH-140220, Apple)

8. Reproductive Systems (PLP, CH-378013, Apple)

    See Guide to Audio-Visual Resources on page 339 for key to AV distributors.

LECTURE ENHANCEMENT MATERIAL

Clinical and Related Terms

1. Amenorrhea — absence of menstrual flow.

2. Cryptorchidism— a developmental defect characterized by failure of the testes to descend into the scrotum.

3. Dysmenorrhea — painful menstruation.

4. Endometritis — inflammation of the uterine lining.

5. Hermaphroditism — presence of both male and female sex organs in one individual.

6. Hysterectomy — surgical removal of the uterus.

7. Mastitis — inflammation of a mammary gland.

8. Monorchism — condition characterized by having only one testis in the scrotum.

9. Oophorectomy — surgical removal of an ovary.

10. Orchiectomy — surgical removal of a testis.

11. Prostatectomy — surgical removal of all or part of the prostate gland.

12. Salpingectomy — surgical removal of a uterine tube.

13. Vaginitis — inflammation of the vaginal lining.

Disorders/Homeostatic Imbalances

1. Benign Prostatic Hyperplasia — refers to non-cancerous enlargement of the prostate gland. The enlarged prostate obstructs urine flow, possibly resulting in infection, calculi, and hydronephritis. Treatment is by transurethral resection.

2. Breast Cancer — the most common malignant tumor in women. Early diagnosis and treatment improves cure rate. The disorder tends to occur later in life. Clinical signs may include a lump in a breast, a retracted nipple or skin, skin edema due to plugging of lymphatics, or fixation of a breast to the chest wall. Early diagnosis usually is due to the finding of a lump or to a mammogram discovery of the cancer. Most widely used treatment is a modified radial mastectomy, examination of axillary lymph nodes for metastasis, and chemotherapy or radiation if indicated. Some tumors require estrogen and other hormones for continual growth. Estrogen-positive tumors can be treated with estrogen antagonists which may cause tumor regression. Breast sarcoma is rare. The tumor is large and bulky and treated by surgical resection.

3. Breast Lump — tissue growth which presents a diagnostic problem. It may be due to cystic disease, fibroadenoma, carcinoma, or some other less common disease of the breast. Positive diagnosis requires a surgical excision of the mass. Treatment depends on the diagnosis.

4. Cancer of the Prostate — a relatively common malignancy of the prostate gland causing enlargement and therefore obstruction of urine flow. The cancer often metastasizes to pelvic and vertebral bones, if untreated. Treatment includes surgery and hormone use to suppress tumor growth.

5. Carcinoma of the Penis — an uncommon disorder almost never seen in the circumcised male.

6. Carcinoma of the Testes (Testicular Cancer) — a relatively uncommon problem. Types include seminoma (a malignant tumor of semen-producing epithelium), malignant teratoma (tumor derived from cells that have the potential to differentiate into many different types of tissue), and choriocarcinoma (a tumor resembling placental trophoblastic tissue). Treatment is usually surgical.

7. Cervical Dysplasia — refers to abnormal growth and maturation of cervical squamous epithelium. The abnormal growth can lead to cervical carcinoma. This type of abnormality can be detected in its early stages by Pap smears. Treatment is usually surgical.

8. Condylomas — benign, warty, tumor-like growths of squamous epithelium caused by a sexually transmitted virus. Treatment consists of destruction of the growth.

9. Cryptorchidism — condition characterized by a failure of the testes to descend into the scrotum, resulting in sterility.

10. Dysmenorrhea — refers to painful menstruation that occurs prior to the beginning of menstruation and for the first few days of the cycle, due to cramping. Primary dysmenorrhea results when pelvic organs are normal. Secondary dysmenorrhea results from various diseases of the pelvic organs, such as endometriosis. Treatment of primary dysmenorrhea includes use of prostaglandins, birth-control pills, and aspirin or other painkillers to ease the pain. Treatment of secondary dysmenorrhea involves treating the basic cause of the problem.

11. Dysplasia of Breast Tissue — refers to abnormal growth of breast tissue. It is the most common benign tumor that produces a mass in the breast. Dysplastic breast tissue includes fibrosis, adenosis, and cystic disease. Fibrosis results from a proliferation of the fibroblasts that produce collagen between the ducts and lobules. This lesion does not become malignant. Cystic disease (fibrocystic disease) is associated with cancer development. Cysts are formed from dilated ducts surrounded by fibrous tissue. The mucosal lining of the cysts is often hyperplastic. Adenosis presents a rubbery or hard lesion that can mimic cancer. It consists of tightly compressed glands which are two cell layers in thickness and surrounded by dense fibrous tissue. Adenosis involves no increased risk of cancer.

12. Endometriosis — refers to the presence of endometrium in any location outside the endometrial cavity. The tissue responds to hormonal stimuli and undergoes the same changes as the uterine lining. Secondary scarring may occur, obstructing the fallopian tubes, and causing infertility.

13. Fibroadenoma — a common benign tumor that occurs in women under the age of 35. It consists of both fibrous and adenomatous ductal tissue. It has no association with cancer.

14. Gonorrhea — a sexually transmitted disease caused by the *Neisseria gonorrhoea* (gonococcus) bacterium. Clinical symptoms include urethritis, cervicitis, pharyngitis, and proctitis (inflammation of the rectal mucosa). Major complications include disseminated bloodstream infection, tubal infection with impaired fertility, and spread (in the male) of infection to prostate and epididymides. Treatment involves use of antibiotics.

15. Herpes — a sexually transmitted disease caused by a herpes virus. This disease causes superficial vesicles and ulcers on the external genitalia and in the genital tract. Regional lymph nodes are often enlarged and tender. The major complication in females is a possible predisposition to cervical cancer.

16. Monorchidism — condition characterized by having only one testicle in the scrotum, and may result in sterility.

17. Ovarian Tumors and Cysts — develop from several parts of the tract. Follicle or corpus luteum-derived cysts occur frequently and usually spontaneously regress. Endometrial cysts, filled with blood and old debris, may form in the ovary. Dermoid cysts (benign cystic teratomas) are also common. They arise from unfertilized ova and often contain skin, hair, teeth, bone, brain tissue, etc. Granulosa cell tumors produce sex hormones and may induce endometrial hyperplasia. Some produce male sex hormones and cause masculinization.

18. Pelvic Inflammatory Disease — refers to any infection that involves the fallopian tubes and adjacent tissue. Causes range from bacterial infection to the use of an IUD. Symptoms include pain, tenderness, fever, and elevated WBCs. Even with treatment the affected tissue may be so scarred that fertilization is difficult to impossible, resulting in sterility.

19. Prostatitis — inflammation of the prostate that occurs when an acute inflammation of the bladder or urethra spreads to the prostate. It may be secondary to gonococcal infection or be a chronic mild inflammation with few symptoms.

20. Syphilis — a sexually transmitted disease caused by the bacterium *Treponema pallidum*. Symptoms occur in stages. Primary symptoms include presence of chancre on the reproductive organ. Secondary stage symptoms include a systemic infection with a rash and enlarged lymph nodes. The tertiary stage shows late destructive lesions in internal organs. Major complications include damage to the cardiovascular system and nervous system (tertiary stage) which may be fatal. Antibiotics are used for treatment.

21. Toxic Shock Syndrome (TSS) — a disorder caused by a toxin produced by a penicillin-resistant *Staphylococcus aureus* bacterium. It occurs almost exclusively in menstruating women who use high-absorbency tampons. Clinical signs include high fever, vomiting, diarrhea, muscular aches and pains, a fall in blood pressure, and other symptoms including a characteristic rash. Treatment is symptomatic until the effects of the toxin wear off. Prevention involves not using tampons. TSS can occur in non-menstruating women and in men due to staphylococcal infections in other parts of the body.

22. Trichomoniasis — a sexually transmitted disease caused by the flagellated protozoan *Trichomonas vaginalis*. This disease causes an inflammation of the mucous membrane of the vagina in females and urethra in males. Men may be asymptomatic but can still transmit the disease.

23. Uterine Tumors — various types occur. A fibroid (leiomyoma) is a benign tumor of smooth muscle. Fibroids, the most common form of uterine tumor, cause hemorrhage, infertility, and can produce pressure on pelvic organs. Fibroids are estrogen dependent and tend to atrophy with onset of menopause.

24. Vaginitis — an infection of the vagina. This common problem can be caused by a fungus (*Candida albicans*), a protozoan parasite (*Tricomona vaginalis*), and a small gram-negative bacterium. Treatment depends on the organism producing the problem.

ANSWERS TO END-OF-CHAPTER QUESTIONS

Multiple Choice/Matching

1. a, b
2. d
3. a
4. d
5. b
6. d
7. d

8. c
9. a, c, e, f
10. (1) c, f; (2) e, h; (3) g; (4) a; (5) b, g, (and perhaps e); (6) f
11. a
12. c

13. d
14. b
15. a
16. b
17. c

Short Answer Essay Questions

18. In males, the urethra serves both the urinary and reproductive systems; in females, the two systems are structurally and functionally separate. (p. 958)

19. A spermatid is converted to a motile sperm by a process called spermiogenesis, during which most of the superfluous cytoplasmic "baggage" is sloughed off and a tail is fashioned. The sperm regions are the head: the genetic (DNA delivering) region; the midpiece: the metabolizing (ATP producing) region; and the tail: the locomotor region. (pp. 963-965, Fig. 28.8, p. 965)

20. Three tiny polar bodies, nearly devoid of cytoplasm, assure that the fertilized egg has enough nutrient reserves to support it during its journey to the uterus. (pp. 977-980)

21. Secondary sexual characteristics of the female include breasts; deposits of subcutaneous fat, especially in the hips and breasts; appearance of pubic hair; and widening and lightening of the pelvis. (pp. 982-984)

22. The events of menopause include a decline in estrogen production, an anovulatory ovarian cycle, and erratic menstrual periods that are shorter in length and eventually cease entirely. Possible consequences of menopause include atrophy of the reproductive organs and breasts, dryness of the vagina, painful intercourse, vaginal infections, irritability and mood changes, intense vaso-dilation of the skin's blood vessels ("hot flashes"), gradual thinning of the skin, loss of bone mass, and slowly rising blood cholesterol levels. (pp. 991-994)

23. Menarche is the first menstrual cycle, occurring when the adult pattern of gonadotropin cycling is achieved. (p. 980)

24. The most used contraceptive in the U.S is the birth control pill. It is a preparation containing minute amounts of estrogens and progestins. It "tricks" the hypothalamic-pituitary axis, because the relatively constant blood levels of the ovarian hormones make it appear that the woman is pregnant. The incidence of failure is less than 6 pregnancies per 100 women per year. Some women cannot tolerate the changes caused by the pill; they become nauseated and/or hypertensive.

    The diaphragm, a barrier method, is quite effective. Many avoid it because it can reduce the spontaneity of sexual encounters.

    Coitus interruptus is withdrawal of the penis just before ejaculation; however, control of ejaculation is never assured. (pp. 986-987)

25. The pathway of a sperm from the male testes to the uterine tubule of a female is as follows: testis, epididymis, ductus deferens, urethra, vagina, uterus, and uterine tube. (pp. 956-959, 966, 972-975)

Critical Thinking and Application Questions

1. This patient has a prolapsed uterus, no doubt caused by the stress on the pelvic floor muscles during her many pregnancies. Since she also has keloids, one can assume that the central tendon to which those muscles attach has been severely damaged and many vaginal tears have occurred. (pp. 972-973)

2. The patient probably has a gonorrhea infection caused by the *Neisseria gonorrhoea* bacterium. It is treated with penicillin and other antibiotics. If untreated, it can cause urethral constriction and inflammation of the entire male duct system. (pp. 984-985)

3. No, she will not be menopausal, because the ovaries will not be affected; they will continue to produce hormones. Tubal ligation is the cutting or cauterizing of the uterine tubes. (p. 972, 980, 987)

4. The man would be asked questions such as does he have difficulty in urination or problems with impotence. The major test to be run would be sperm count. (p. 959)

Key Figure Questions

Figure 28.6 1. To reestablish the normal diploid number of the species when sperm and egg fuse. 2. Random alignments of tetrads at metaphase I and crossover. **Figure 28.7** Spermatids. **Figure 28.20** One through five contain primary oocytes; six contains secondary oocytes.

SUGGESTED READINGS

1. Bennett, D.D. "Pelvic Inflammatory Disease." *Science News* 127 (Apr. 1985): 263.

2. Cohen, R.J. "Diagnosis: Breast Cancer." *Hospital Medicine* (July 1984).

3. Edwards. D.D. "Beating Breast Cancer." *Science News* 133 (20) (May 1988): 314-315.

4. Fackelmann, K.A. "Preventing Pregnancy with the Cervical Cap." *Science News* 136 (4) (July 1989): 52.

5. Federman, D.D. "Impotence: Etiology and Management." *Hospital Practice* (Mar. 1982).

6. Franklin, D. "Menstrual Hiatus Can Prompt Bone Loss in Female Athletes." *Science News* 126 (Aug. 1984): 69.

7. Franklin, D. "The Anti-Sperm Vaccine." *Hippocrates* 3 (5) (Sept.-Oct. 1989): 16.

8. Garnick, M.B. "The Dilemmas of Prostate Cancer." *Scientific American* 270 (Apr. 1994): 72.

9. Goldfinger, S.E. (ed.). "Lumpy Breasts." *The Harvard Medical School Health Letter* (Nov. 1984).

10. Goldman, L. "Premenstrual Syndrome: Is It Oversold or Over-diagnosed?" *Modern Medicine* (June 1985).

11. Grabowski, C.T. *Human Reproduction and Development*. Philadelphia: Saunders, 1983.

12. Hafez, E.S.E. *Human Reproduction,* 2nd ed. New York: Harper & Row, 1980.

13. Johnson, G.T. (ed.). "Sexually Transmitted Diseases." *The Harvard Medical School Health Letter* (Apr. 1981).

14. Jones, R.E. *Human Reproduction and Sexual Behavior*. Englewood Cliffs, NJ: Prentice-Hall, 1984.

15. Loscalzo, J., D.J. Singel, and J.S. Stamler. "Biochemistry of Nitric Oxide and its Redox-Activated Forms." *Science* 258 (Dec. 1992): 189.

16. Lutz, Ronald. "Stopping the Spread of Sexually Transmitted Disease." *Nursing* 86 (16) (Mar. 1986): 47.

17. Miller, J.A. "X Chromosomes: Too Few and Too Many." *Science News* 129 (23) (June 1986): 358.

18. Moniwe, M., and M. Laird. "Contraceptives: A Look at the Future." *American Journal of Nursing* 89 (4) (Apr. 1989): 496-499.

19. Nimmons, D. "Sex and the Brain." *Discover* 15 (Mar. 1994): 64.

20. Piziak, V., and B.L. Shull. "Menopausal Hormone Replacement." *Hospital Practice* (Feb. 1985).

21. Roberts, L. "Sex and Cancer." *Science* 867 (6) (July-Aug. 1986): 30-33.

22. Snyder, S.H. "Nitric Oxide: First in a New Class of Neurotransmitters?" *Science* 257 (July 1992): 494.

23. Stahl, F.W. "Genetic Recombination." *Scientific American* 256 (2) (Feb. 1987): 90-101.

24. Stark, M. "The Private Struggles of Endometriosis and Infertility." *Mount Holyoke Alumnae Quarterly* (Spring 1987): 18-21.

25. Stolzenberg, W. "PMS Study Pans Popular Prescription." *Science News* 138 (3) (July 1990): 37.

26. The Cancer and Steroid Hormone Study of the Centers for Disease Control and the National Institute of Child Health and Human Development. "Oral Contraceptive Use and the Risk of Breast Cancer." *New England Journal of Medicine* 315 (1986): 405-411.

27. Ulmann, A., G. Teutsch, and D. Philbert. "RU 486." *Scientific American* 262 (6) (June 1990): 42-48.

28. Williams, *Textbook of Endocrinology* 7th ed. Philadelphia, W.B. Saunders Co. 1985.

29. Wingfield, J.C., et al. "Testosterone and Aggression in Birds." *American Scientist* 75 (6) (Nov.-Dec. 1987): 602-608.

30. "Estrogen Use Raises Questions." *Science News* 128 (Nov. 1985): 279.

31. "Studies Shed Light on Impotence." *Science News* 121 (Mar. 1982): 201.

32. "Surgery for Breast Cancer: Preserve and Protect." *The Harvard Medical School Health Letter* (July 1985).

29

Pregnancy and Human Development

Chapter Preview

This chapter introduces developmental events from conception to birth. The process starts with fertilization and continues with the pre-embryonic, embryonic, and fetal development of the conceptus, the developing offspring. The events that occur immediately after birth are also discussed.

INTEGRATING THE PACKAGE

Suggested Lecture Outline

I. From Egg to Embryo (pp. 1000-1008; Fig. 29.1, p. 1000)
   A. Accomplishing Fertilization (pp. 1000-1003)
      1. Sperm Transport and Capacitation
      2. Acrosomal Reaction and Sperm Penetration
      3. Blocks to Polyspermy (Fig. 29.2, p. 1001)
      4. Completion of Meiosis II and Fertilization
   B. Pre-embryonic Development (pp. 1003-1007)
      1. Cleavage and Blastocyst Formation (Fig. 29.4, p. 1003)
      2. Implantation (Fig. 29.5, p. 1004; Fig. 29.6, p. 1005)
      3. Placentation (Fig. 29.7, p. 1009)
II. Events of Embryonic Development (pp. 1008-1014)
   A. Formation and Roles of the Embryonic Membranes (p. 1008)
      1. Amnion
      2. Yolk Sac
      3. Allantois
   B. Germ Layers (p. 1010; Fig. 29.8, p. 1010)
      1. Gastrulation: Germ Layer Formation
      2. Organogenesis: Differentiation of the Germ Layers
         a. Specialization of the Ectoderm
         b. Specialization of the Endoderm
         c. Specialization of the Mesoderm
      3. Development of Fetal Circulation
III. Events of Fetal Development (p. 1014; Table 29.2, p. 1016)
IV. Effects of Pregnancy on the Mother (pp. 1017-1019)
   A. Anatomical Changes (p. 1017; Fig. 29.15, p. 1018)
   B. Metabolic Changes (p. 1018)

Review Items

Laboratory Correlations

1. Marieb, E. N. *Human Anatomy and Physiology Laboratory Manual: Cat and Fetal Pig Versions.* 3rd. ed. Benjamin/Cummings, 1989.

    Exercise 46: Survey of Embryonic Development

2. Marieb, E. N. *Human Anatomy and Physiology Laboratory Manual: Brief Version.* 3rd. ed. Benjamin/Cummings, 1992.

    Exercise 43: Survey of Embryonic Development

Transparencies Index

29.3 Events following sperm penetration	29.10 Folding of the embryonic body that produces the tubular body trunk
29.4 Mitotic divisions	
29.7 Events of placentation, early embryonic development, and formation of the embryonic membranes	29.13 Circulation in the fetus and newborn
	29.15 Relative size of the uterus
	29.16 Process of positive feedback mechanism during birth
29.8 Gastrulation: formation of the three primary germ layers	

Bassett Atlas Figures Index

Slide Number	Figure Number	Description
50	5.7A,B	Fetus and placenta

INSTRUCTIONAL AIDS

Lecture Hints

1. Emphasize the difference between the terms *conceptus*, *embryo*, and *fetus* (and the associated periods).

2. Stress that the early cell divisions of the conceptus increase total cell number but do not result in cell size increase. Cells become increasingly smaller until the zona ruptures.

3. When discussing the maintenance of the corpus luteum (which initially maintains pregnancy and is under hormonal control by the trophoblast cells) mention that a measurement of HCG levels would be an ideal indicator of pregnancy (EPT home tests, blood tests).

4. Point out that the blastocyst is actually embedded into the endometrial wall, not attached to the surface as some students first imagine.

5. Use plenty of models and/or wall charts illustrating the different stages of embryonic and fetal development. Students are often overwhelmed by the terminology of development.

6. Point out the dual origin of the placenta.

7. Emphasize that from the mother's point of view, the placenta is just another organ drawing resources from the mother's blood supply. This idea is helpful to establish the placenta as an exchange organ.

8. Stress that embryonic/fetal blood does not come into contact with maternal blood under normal circumstances but that most substances (including the gamma class of immunoglobulins) cross the placental barrier to embryonic/fetal circulation.

9. Spend time emphasizing the embryonic/fetal membranes since the anatomical orientation of these membranes is difficult for many students to visualize.

10. Mention that the yolk sac is not the source of nutrients for the egg as it is in birds and reptiles, but instead is an early site of blood formation.

11. As a point of interest, reveal that "eating for two" is a popular belief that has no physiological basis.

12. Review hypothalamic and pituitary control of the ovarian cycle.

13. Point out the logic behind the various modification of fetal circulation and how those shunts must be redirected when the umbilical cord is cut.

14. When discussing the changes that occur in the mother during pregnancy, emphasize each in a commonsense way: urinary output must increase because the fetus is adding a considerable amount of waste to the mother's blood; blood volume increases due to the excess draw on the mother's resources, etc.

15. The Miracle of Life video is one of the best available depicting developmental events from fertilization to birth. It is worthwhile to take time to present this videotape, especially if you do not use any other.

Demonstrations/Activities

1. Audio-visual materials of choice.

2. Use a pregnancy model to exhibit fetal development, placement, and birth.

3. Use a doll and a drawstring sack to illustrate the placement of the fetus for vaginal delivery and the turning movements that result in delivery.

4. Use a doll and a drawstring sack to illustrate abnormal placements for delivery.

5. Obtain a fresh placenta from a local hospital to demonstrate the anatomical features of this vital structure.

6. Use a series of models to illustrate embryonic and fetal development.

7. If available, display embryos and fetuses in different stages of development.

8. Have the students bring in a recent article that deals with the effects of maternal drug-taking or disease (e.g., AIDS, herpes, etc.) on the well-being of the fetus.

Critical Thinking/Discussion Topics

1. Discuss the need for all females of childbearing age who are sexually active, whether practicing contraception or not, to maintain a lifestyle which will enhance, not harm, the developing embryo and fetus if the woman should become pregnant.

2. Discuss the drastic changes the fetus must undergo at birth and how those changes might be minimized.

3. Discuss the pros and cons of determining when life begins (be sure to advance both sides of the issue).

4. Discuss the methods available to produce pregnancy (i.e., artificial insemination, in vitro fertilization, etc.) and what cautions should be given to those choosing these methods.

5. Student assignment for class discussion:

    a. Define episiotomy and explain why this procedure is performed.

    b. Define Down syndrome (trisomy 21) and indicate in which maternal age group it is most common.

Library Research Topics

1. Research the types of birth presentations and note the symptoms, prognosis, and difficulties encountered in each type.

2. Research several types of birth defects by symptom category, i.e., skeletal system, circulatory system, etc.

3. Research various environmental effects on embryological and fetal development, i.e., alcohol, drugs (legal and nonlegal), infectious diseases, etc.

4. Research the various methods of contraception, including those currently being used as well as projected methods.

5. Research the fetal and infant problems associated with the mother's lifestyle, i.e., sexually transmitted disease infections, infectious disease infections, alcoholism and DTs, AIDS, etc.

6. Research the pros and cons concerning traditional delivery procedures versus underwater births or births in warm, dimly lit rooms, etc.

Audio-Visual Aids/Computer Software

Videotapes

1. Human Embryology Series (TFI, 12-18 min.). Covers all areas of human prenatal development, both normal and abnormal. Titles include: Highlights of Reproduction and Prenatal Development (16 min.); Formation of Sex Cells and Chromosomal Abnormalities (14 min.); Formation of Sex Cells and Chromosomal Abnormalities (14 min.); Reproductive or Sexual Cycles in the Female (11 min.); Reproductive or Sexual Cycles in the Male (11 min.); and Fertilization, Cleavage, and Implantation (17 min.).

2. The Living Body: Shares in the Future (FHS, 26 min., C). Looks at how the male and female bodies are prepared for their task of increasing the human race. The program shows the characteristics of sperm and ova. The mechanism of cell division is also shown.

3. The Living Body: Coming Together (FHS, 26 min., C). Attraction, desire, and sexual coupling lead to conception. This program covers the physiological events that underlay the process of reproduction.

4. The Living Body: A New Life (FHS, 26 min., C). This program looks at the events that lead from the fertilized cell to a human baby. Using film of living fetuses in the womb, it explains how the familiar human shape is sculpted out of the basic cell mass, what controls the timing of the various stages of fetal development, and what life is like for a fetus.

5. The Living Body: Into the World (FHS, 26 min., C). This program covers the tumultuous events of birth, using fetoscopy and specially constructed models to show what happens from the baby's viewpoint. It also shows the physiological events immediately following the birth.

6. The Discovery Year (FHS, 52 min., C). Looks into the first year of human life.

7. High-Risk Pregnancy (FHS, 19 min., C). Smoking, alcohol consumption, and drug abuse are among the factors which may make pregnancy risky for millions of women, despite the advances in medical skills and technology. The program profiles a young woman for whom a sudden episode of high blood pressure resulted in the loss of her first child.

8. Contemporary Childbirth (FHS, 19 min., C). New trends in childbirth include a strong emphasis on preparation even before conception.

9. Saving Premature Infants (FHS, 19 min., C). This program highlights revolutionary technologies in peri- and neonatology, which now save thousands of babies each year.

10. The Placenta and Fetal Membranes (TFI, 24 min., C, 1988). This program demonstrates the early development of the embryo and its associated membranes, and describes how the placenta develops from a combination of embryonic and maternal tissues. The placental circulation is also discussed.

11. Infertility: Nature's Heartache (FHS, 28 min., C). Examines some of the treatments, including drug therapy, surgery, and in vitro fertilization, for infertility.

12. Options for Infertility (FHS, 19 min., C). The causes and treatments of male and female infertility are the subject of this program, which focuses on two non-traditional means of having children: in vitro fertilization and artificial insemination with donor sperm.

LECTURE ENHANCEMENT MATERIAL

Clinical and Related Terms

1. Abortion — the spontaneous or deliberate termination of a pregnancy; a spontaneous abortion is usually termed a miscarriage.

2. Amniocentesis — a diagnostic procedure in which a sample of amniotic fluid is withdrawn from the amniotic cavity.

3. Ectopic pregnancy — development of an embryo outside of its normal location within the uterus.

4. Hydraminos — excessive amniotic fluid.

5. Hydatid mole — uterine tumor that originates from placental tissue.

6. Hermaphroditism — presence of both male and female sex organs in one individual.

7. Meconium — the first fetal discharge following childbirth; initially consists of blood and later of serous fluid.

8. Preeclampsia — a pregnancy-related syndrome characterized by sudden hypertension, large amounts of protein in urine, and generalized edema.

9. Tubal pregnancy — development of an embryo in the fallopian tube; causes pain and eventual rupturing of the tube.

Disorders/Homeostatic Imbalances

1. Anencephaly — condition in which the brain fails to develop; results from a failure of the cranial neural folds to develop, fuse, and form the front portion of the brain.

2. Atresia — failure of any part of the GI tract to form a channel for the passage of food.

3. Cleft Lip — condition in which the upper lip is separated by a notch.

4. Cleft Palate — condition due to failure of the mesodermal masses of the maxillae to fuse in the midline.

5. Clubfoot — general term for abnormalities of the foot.

6. Cyanotic Congenital Heart Disease — refers to cardiac abnormalities that result in inadequate oxygenation of the blood.

7. Ectopic Pregnancy — the development of an embryo outside of its normal location within the uterus. Most ectopic pregnancies occur in the fallopian tubes. Predisposing factors of tubal pregnancy include previous tubal infection and disturbed tubal motility. Consequences of tubal pregnancy include rupture of the tube with profuse bleeding and extreme pain. This condition may be fatal.

8. Fetal Jaundice — occurs because of inefficient excretion of bilirubin by a newborn infant's liver. It is a common occurrence and usually corrects itself within a short period of time.

9. Gestational Trophoblast Disease — a term used to describe the neoplasm produced by the trophoblast when masses of abnormal proliferating trophoblastic tissue invade the uterus and other areas. The more benign form of this condition is called a hydatidiform mole. The villi invade and over-distend the uterus. Treatment involves emptying the uterus and then chemotherapy if the mole recurs. The more malignant form of this disease is choriocarcinoma. It consists of neoplastic trophoblast without villi. It is treated by chemotherapy.

10. Gonadal Dysengesis — refers to errors in development of sex organs. Types include infertility without ambiguity in genital development, infertility with ambiguity in genital development, and ambiguity without infertility. Infertility without genital ambiguity occurs in individuals with an abnormal number of X chromosomes as in Klinefelter's syndrome (XXY sex chromosomes) or Turner's syndrome (XO sex chromosomes).

11. Hemolytic Disease of the Newborn — occurs when an infant sensitizes the mother to a blood-group antigen. When the mother produces antibodies to the antigen, fetal red blood cells can be damaged. The condition varies in its severity.

12. Hydrocephalus — condition where cerebrospinal fluid accumulates in the brain under increased pressure; may result from an overproduction of cerebrospinal fluid, from failure of resorption of the fluid, or from obstruction of fluid flow.

13. Imperforate anus — failure to form an anus.

14. Meningoencephalocele — condition where the brain protrudes through a bony defect in the skull.

15. Meningomyelocele — condition where a portion of the spinal cord and meninges protrude through defective vertebrae.

16. Microcephaly — uncommon condition in which the cranium is small and the face is normal size; usually marked mental retardation occurs.

17. Placenta Previa — occurs when the placenta blocks the exit from the uterus by attaching in the lower part of the uterus. This condition will cause bleeding late in pregnancy. It is hazardous to both the mother and infant. Cesarean section delivery is required.

18. Renal Agenesis — failure of one kidney to form.

19. Rh Hemolytic Disease (Erythroblastosis Fetalis) — occurs when there are blood-group differences between the mother and the infant. Most cases involve an Rh- mother and an Rh+ child. The mother has formed antibodies against the infant's blood groups; the antibody coats the infant's blood cells, causing blood destruction. Treatment is by exchange transfusion, fluorescent light therapy, or intrauterine fetal transfusion. The problem can be prevented by administration of Rh immune globulin to the mother to eliminate antigenic fetal cells.

20. Rubella-Induced Malformations — occur when the mother has rubella (German measles) during the first trimester. Eye, ear, heart, and brain malformations are most common. Fewer malformations occur when rubella infection occurs in later parts of the pregnancy.

21. Septal Cardiac Defects — failure of partitions to form between atria or ventricles.

22. Spina Bifida — failure of the neural tube and the bones that encase the CNS to fuse. The opening to the spinal cord may be pinpoint or greatly enlarged. The severity of the disorder is determined by the size of the opening.

23. Spontaneous Abortion (Miscarriage) — usually occur early in pregnancy due to chromosomal abnormalities, maldevelopment of the embryo that prevented its survival, or some other problem. Miscarriage later during the pregnancy is usually due to partial detachment of the placenta from the uterine wall or obstruction of blood flow through the umbilical cord. If a dead fetus is retained in the uterus, products of degenerated fetal tissue diffuse into the maternal circulation. This material may produce thrombi in the mother.

24. Tetralogy of Fallot — common cardiac abnormality consisting of pulmonary stenosis (narrowing), a ventricular septal defect, an overriding aorta, and hypertrophy (thickness of the right ventricle).

25. Toxoplasmosis-Induced Malformation — occurs when the pregnant mother eats raw or poorly cooked meat and comes in contact with an animal infected by the protozoan *Toxoplasma gandii*. Common malformations include microcephaly and hydrocephaly.

26. Tracheoesophageal Fistula — an abnormal passage connecting the trachea and esophagus.

ANSWERS TO END-OF-CHAPTER QUESTIONS

Multiple Choice/Matching

1. (1) a; (2) b; (3) b; (4) a
2. b
3. c
4. b
5. a

6. d
7. d
8. c
9. a
10. b

11. c
12. a
13. c
14. a
15. (1) e; (2) g; (3) a; (4) f; (5) i; (6) b; (7) h; (8) d; (9) c

Short Answer Essay Questions

16. The process of fertilization involves numerous steps. First, the sperm must reach the ovulated secondary oocyte. The sperm must travel from the vagina to the uterine tubes through the cervical mucus of the uterus. The sperm deposited in the vagina must be capacitated; that is, their membranes must become fragile so that the hydrolytic enzymes in their acrosomes can be released. The acrosomal reaction is the release of acrosomal enzymes (hyaluronidase, acrosin, proteases, and others) that occurs in the immediate vicinity of the oocyte. Hundreds of acrosomes must rupture to break down the intercellular cement that holds the granulosa cells together and to digest holes in the zona pellucida. Once a path has been cleared and a single sperm makes contact with the oocyte membrane receptors, its nucleus is pulled into the oocyte cytoplasm.

As soon as the plasma membrane of one sperm makes contact with the oocyte membrane, sodium channels open and ionic sodium moves into the oocyte from the extracellular space, causing its membrane to depolarize. The depolarization causes ionic calcium to be released into the oocyte cytoplasm. This surge in intracellular calcium levels initiates the cortical reaction and activates the oocyte. The activated secondary oocyte completes meiosis II to form the ovum nucleus and ejects the second polar body. The ovum and sperm nuclei swell, becoming the female and male pronuclei, and approach each other as a mitotic spindle develops between them. The pronuclei membranes then rupture, releasing their chromosomes into the immediate vicinity of the spindle. Combination of the maternal and paternal chromosomes constitutes the act of fertilization and produces the diploid zygote. The effect of fertilization is the formation of a single cell (zygote) with chromosomes from the egg and sperm, and determination of the offspring's sex. The zygote undergoes cleavage, forming first a morula, then a blastocyst. Then it undergoes gastrulation, neurulation, and final development of the fetus. (pp. 1000-1003)

17. In cleavage, daughter cells become smaller and smaller, resulting in cells with high surface-to-volume ratio and providing a larger number of cells to serve as building blocks for constructing the embryo. (pp. 1003-1004)

18. a. Viability of the corpus luteum is due to human chorionic gonadotropin secreted by trophoblast cells of the blastocyst which bypasses the pituitary-ovarian controls, prompting continued production of estrogen and progesterone to maintain the endometrium.

    b. The corpus luteum must remain functional following implantation until the placenta can assume the duties of hormone production, otherwise the endometrium will not be maintained. (pp. 1004-1005)

19. The placenta is formed from embryonic (trophoblastic) and maternal (endometrial) tissues. When the trophoblast acquires a layer of mesoderm it becomes the chorion. The chorion sends out chorionic villi, which come in contact with maternal blood. Oxygen and nutrients diffuse from the maternal to the embryonic blood; embryonic wastes diffuse from the embryo to the mother's circulation. (pp. 1005-1007)

20. As soon as the plasma membrane of one sperm makes contact with the oocyte membrane, sodium channels open and ionic sodium moves into the oocyte from the extracellular space, causing the membrane to depolarize. This "fast block to polyspermy" prevents other sperm from fusing with the oocyte. This is followed by the cortical reaction which constitutes the "slow block to polyspermy." (p. 1002)

21. The gastrulation process gives rise to the three primary germ layers, the ectoderm, mesoderm, and endoderm, and the embryonic membranes, the amnion, yolk sac, chorion, and allantois. (pp. 1010-1011)

22. a. Breech presentation is buttocks-first presentation.

    b. Two problems of breech presentation include a more difficult delivery and difficulty for the baby breathing. (p. 1021)

23. The factors that bring about uterine contractions include high levels of estrogen, production of oxytocin by the fetus, which acts on the placenta to stimulate the production and release of prosta-glandins, and activation of the hypothalamus to produce oxytocin for release by the posterior pituitary. (p. 1019)

24. The flat embryonic disc achieves a cylindrical body shape as its sides fold inward and it lifts up off the yolk sac into the amniotic cavity. At the same time, the head and tail regions fold under. All this folding gives the month-old embryo a tadpole shape. The sequence is illustrated in Figure 29.10 on p. 1012.

Critical Thinking and Application Questions

1. (C). Most major developmental events occur during the first three months of pregnancy, and events that are blocked for whatever reason never occur because development has a precise timetable. Assessment should be done to analyze possible problems. (pp. 1008-1009)

2. An episiotomy is a midline incision from the vaginal orifice laterally or posteriorly toward the rectum. It is performed to reduce tissue tearing as the baby's head exits from the perineum. (p. 1021)

3. The woman was in labor, the expulsion stage. She probably would not have time to get to the hospital. Typically it takes 50 minutes in the first birth and 20 minutes in subsequent births for birth to occur once the expulsion stage has been reached. A 60-mile drive would take over an hour. (p. 1020)

4. Mary's fetus might have respiratory problems or even congenital defects due to her smoking, since smoking causes vasoconstriction, which would hinder blood delivery to the placenta. (p. 1007)

5. Segmentation is the presence of multiple, repeating units, lined up from head to tail along the axis of the body. The body's segmented structures, such as vertebrae, ribs, and the muscles between the ribs, are primarily derived from somites of mesoderm. (p. 1012)

Key Figure Questions

Figure 29.4 Because virtually no growth occurs between the successive divisions of cleavage. **Figure 29.9** The ectoderm. **Figure 29.15** As the pregnancy continues, the uterus pushes higher in the abdomen, eventually pressing on the diaphragm, inhibiting the downward movement.

SUGGESTED READINGS

1. Balinsky, B.I. *An Introduction to Embryology*. 5th ed. Philadelphia: Saunders, 1981.

2. Begley, S., and J. Carey. "How Human Life Begins." *Newsweek* (Jan. 1982).

3. Cooke, J. "The Early Embryo and the Formation of Body Pattern." *American Scientist* 76 (1) (Jan.-Feb. 1988): 35-41.

4. Fischer, K., and A. Lazerson. *Human Development: From Conception Through Adolescence*. San Francisco: W.H. Freeman, 1984.

5. Gehring, W.J. "The Molecular Basis of Development. " *Scientific American* (Oct. 1985).

6. Grady, D. "Pregnancies That Can Kill." *Discovery* 4 (June 1983): 80-83.

7. Grobstein, C. "External Human Fertilization. " *Scientific American* 240 (6) (June 1979): 57-67.

8. Hall, Stephen S. "The Fate of the Egg." *Science* 85 (Nov. 1985): 40-49.

9. Hayflick, L. "The Cell Biology of Human Aging." *Scientific American* (Jan. 1980).

10. Hynes, Richard O. "Fibronectins." *Scientific American* 254 (June 1986): 42-51.

11. Langercrantz, Hugo, and Theodore A. Slothkin. "The "Stress of Being Born." *Scientific American* 254 (4) (Apr. 1986): 100-107.

12. Langman, J. *Medical Embryology*. 4th ed. Baltimore: Williams & Wilkins, 1981.

13. Miller, J.A. "Sexual Ambiguity: Getting Down to the Gene." *Science News* 125 (15) (Apr. 1984): 230.

14. Mills, J.A. "Switching-On Genes in Development." *Science News* 128 (Nov. 1985): 295.

15. Moore, K.L. *The Developing Human*. 3rd ed. Philadelphia: Saunders, 1982.

16. Morse, G. "Why is Sex?" *Science News* 126 (Sept. 1984): 154-157.

17. Poole, W. "The First 9 Months of School." *Hippocrates* 1 (2) (July-Aug. 1987): 68-73.

18. Radetsky, P. "Wire Tapping the Womb." *Science* (Oct. 1984): 58-63.

19. Short, R.V. "Breast Feeding." *Scientific American* (4) 250 (Apr. 1984): 35.

20. Tollefsbol, T.O., and R.W. Gracy. "Premature Aging Diseases: Cellular and Molecular Changes." *BioScience* (Nov. 1983).

21. Treichel, J.A. "Embryo Transfers Achieved in Humans." *Science News* 124 (July 1983): 69.

22. Wasserman, P.M. "Fertilization in Mammals." *Scientific American* 259 (6) (Dec. 1988): 78-84.

23. Wechsler, R. "Hostile Womb." *Discover* (Mar. 1988): 82-87.

24. Weiss, R. "Breast Milk May Stimulate Immunity." *Science News* 133 (13) (Mar. 1988): 19.

25. Weiss, R. "Simple Sugars Are Key to Complex Sex." *Science News* 133 (Feb. 1988): 133.

26. Weiss, R. "New Therapy Blocks Newborn Jaundice." *Science News* 133 (16) (Apr. 1988): 247.

27. Weiss, R. "Uneven Inheritance." *Science News* 138 (1) (July 1990): 8-11.

28. Young, P. "Mom's Mitochondria May Hold Mutation." *Science News* 134 (5) (July 1988): 70.

30

Heredity

Chapter Preview

This chapter deals with genetics, the study of the mechanism of heredity. The study of the human body is not complete without a brief look at the science of genetics. Genetics explains how traits are passed from one generation to another. Recent advances in genetics have enabled geneticists to manipulate and engineer genes in order to examine their expression and to treat or cure disease. This chapter concentrates on the basic principles of heredity.

INTEGRATING THE PACKAGE

Suggested Lecture Outline

I. The Vocabulary of Genetics (pp. 1027-1030)
    A. Introduction (p. 1027)
        1. Homologous Chromosomes
        2. Karyotype (Fig. 30.1, p. 1028)
    B. Gene Pairs (Alleles) (p. 1028)
        1. Alleles
        2. Homozygous
        3. Heterozygous
        4. Dominant
        5. Recessive
    C. Genotype and Phenotype (p. 1028)
    D. Sexual Sources of Genetic Variation (p. 1028)
    E. Segregation and Independent Assortment of Chromosomes (p. 1029; Fig. 30.2, p. 1029)
    F. Crossover of Homologues and Gene Recombination (p. 1029; Fig. 30.3, p. 1030)
    G. Random Fertilization (p. 1030)
II. Types of Inheritance (pp. 1030-1034)
    A. Dominant-Recessive Inheritance (pp. 1031-1032)
        1. Dominant Traits
        2. Recessive Traits
    B. Incomplete Dominance (Intermediate Inheritance) (pp. 1032-1033)
    C. Multiple-Allele Inheritance (p. 1033; Table 30.2, p. 1033)
    D. Sex-Linked Inheritance (p. 1033)
    E. Polygene Inheritance (Fig. 30.5, p. 1034)
III. Influence of Environmental Factors on Gene Expression (pp. 1034-1035)

IV. Genetic Screening and Counseling (pp. 1035-1038)

    A. Carrier Recognition (pp. 1035-1036)

    B. Fetal Testing (Fig. 30.10, p. 1037)

Review Items

1. Mitosis (Chapter 3, p. 89)

2. Chromatin (Chapter 3, p. 86)

3. Meiosis (Chapter 28, pp. 960-963)

4. Tetrad formation (Chapter 28, pp. 960, 962)

Laboratory Correlations

1. Marieb, E. N. *Human Anatomy and Physiology Laboratory Manual: Cat and Fetal Pig Versions.* 3rd. ed. Benjamin/Cummings, 1989.

   Exercise 47: Principles of Heredity

2. Marieb, E. N. *Human Anatomy and Physiology Laboratory Manual: Brief Version.* 3rd. ed. Benjamin/Cummings, 1992.

   Exercise 45: Principles of Heredity

Transparencies Index

30.2 Production of variability in gametes by independent assortment of homologous chromosomes during metaphase of meiosis I

30.3 Crossover and genetic recombination occurring during meiosis I introduces genetic variability in the gametes formed

---

INSTRUCTIONAL AIDS

Lecture Hints

1. Stress that an individual receives a member of an allele from each parent.

2. Students often confuse the terms genotype and phenotype. Mention that the genotype is the entire genetic complement (geno) of an individual. It is sometimes easier for students to remember the difference between genotype and phenotype by this simple association.

3. Emphasize the importance of segregation and independent assortment.

4. Review the process of recombination and use diagrams to reinforce the concepts.

5. Remind the class of the structure and function of DNA.

6. Refer the class to a review of basic probability problem solving.

Demonstrations/Activities

1. Audio-visual materials of choice.

2. Use pipe cleaners and craft balls (ones with holes in them) to form chromosomes. Then use those "chromosomes" to demonstrate various genotypes and other genetic patterns.

3. Use ice cream sticks with looped tape to form chromosomes and use them to demonstrate how dominant and recessive genes can be combined.

Critical Thinking/Discussion Topics

1. Discuss the tests available to detect various genetic and/or development problems prior to birth.

2. Discuss the moral dilemma concerning terminating a pregnancy due to genetic disorders.

3. Discuss the rationale for increasing possibility of birth defects with age of the mother.

Library Research Topics

1. Research several types of birth defects by system category, i.e., skeletal system, circulatory system, etc.

2. Research the chromosomal aberrations that result in congenital disorders.

3. Research the multiple-allele inheritance disorders.

Audio-Visual Aids/Computer Software

Videotapes

1. Human Embryology Series: Formation of Sex Cells and Chromosomal Abnormalities (TFI, 14 min., C, VHS). Common congenital malformations are illustrated against a background of normal embryology.

2. Birth Defects (FHS, 19 min., C, VHS or BETA). This program looks at the genetic and environmental causes of birth defects and reviews such risk factors for birth defects as family history, maternal age, ethnic background, and noninherited factors. The program profiles a couple with a son who has Down syndrome and interviews a genetic counselor who discusses the reactions of parents when they learn they have a child with birth defects.

3. Regulation of Gene Expression in Eucaryotes (EIL, VHS). Introduces chromatin structure, chromosome puffs, and steroid hormone-binding proteins to describe some current ideas on how nuclear genes in eucaryotic cells may be turned on and off. Compares and contrasts gene regulation in eucaryotes with gene regulation in bacteria; discusses the complex molecular interactions required for correct, coordinated development of higher organisms.

4. Heredity, Health and Genetic Disorders: Parts 1-2 (CDL, VHS). A two-part program: Part I — The Causes of Genetic Disorders; and Part II — Genetic Screening and Counseling. Illustrates how genetic traits are passed from generation to generation and examines, through case studies, such methods of genetic screening as amniocentesis and ultrasound.

Computer Software

1. Genetics (CDL, Apple, IBM). Examines the DNA molecule and progresses to applied genetics. Explores how selected traits are used by breeders to improve plant and animal organisms. Human genetics and related problems are presented.

2. Gene Regulation (EIL, Apple). This four-part program illustrates the basic principles of gene regulation in procaryotes, with emphasis on the lactose operon.

3. Genetic Engineering (EIL, Apple). Discusses RNA, DNA, enzymes, and linkers. Tutorial highlights cloning, replica plating, how to identify the genes to isolate, and the many uses of genetic engineering. Covers the use of genetic engineering in diagnosing diseases such as sickle-cell anemia.

4. Heredity Dog (QUE, Apple, Commodore 64). This program allows students to breed dogs of different colors and patterns to demonstrate graphically the relationship of genotypes and phenotypes of puppies to those of the parent dogs.

5. Genetic Engineer's Toolbox (QUE, Apple). This program provides students with an introduction to the principles and techniques of genetic engineering or recombinant DNA technology, and to the application of this technology to the manufacture of biologically important proteins.

6. Gene Machine (QUE, Apple, Commodore 64, TRS-80). With this program, students investigate some of the basic ideas concerning the structure and function of DNA and RNA.

7. Human Genetic Disorders (QUE, Apple). Provides software, activities, an extensive Teaching Guide, and student worksheets to allow students to investigate inherited disorders.

8. Heredity Pattern Generator (CDL, Apple). Lets students apply fundamental concepts of heredity that are related to single-trait crosses and allele interactions.

9. Gene Structure and Function (PLP, CH-600005, Apple). A comprehensive introduction to molecular biology.

10. Genetics Counselor (PLP, CH-790127, Apple). This classroom tool explains the basic concepts of inheritance.

LECTURE ENHANCEMENT MATERIAL

Clinical and Related Terms

1. Achondroplasia — refers to dwarfism due to failure of cartilage to develop properly.

2. Albinism — absence of skin, hair, eye pigment.

3. Brachydactyly — abnormal shortness of the fingers and toes.

4. Chorea — ceaseless occurrence of a wide variety of rapid, highly complex, jerky movements that appear to be well coordinated but are performed involuntarily.

5. Color blindness — a deviation from normal perception of color.

6. Fragilitas ossium — also called osteogenesis imperfecta or osteitis fragilitas; an inherited autosomal dominant trait in which the bones are abnormally brittle and subject to fractures.

7. Friedreich's ataxia — an inherited disease showing sclerosis of the dorsal and lateral columns of the spinal cord; characterized by ataxia, speech impairment, lateral curvature of the spinal column and peculiar swaying and irregular movements, with paralysis of the muscles, especially of the lower extremities.

8. Hemorrhagic telangiectasia — refers to formation of multiple small angiomas of the skin and mucous membranes; Rendu-Osler-Weber disease.

9. Huntington's chorea — a rare hereditary disease characterized by chronic progressive chorea and mental deterioration terminating in dementia; usually occurs after 40 years of age with death in about 15 years after occurrence; inherited as an autosomal recessive trait.

10. Marfan's syndrome — a congenital disorder of connective tissue characterized by abnormal length of the extremities, especially of fingers and toes; subluxation of the lens; cardiovascular abnormalities and other deformities; inherited as an autosomal dominant trait.

Disorders/Homeostatic Imbalances

Sex Chromosome Abnormalities (Female)

1. Multiple X Syndrome — extra X chromosome (XXX, XXXX, etc., genotype). Embryos with triple X chromosome syndrome are relatively common. Usually there are specific abnormalities. Sex development is usually normal. Fertility and intelligence are either normal, slightly decreased, or the individual may be sterile and mentally retarded (depending on the number of extra X chromosomes).

2. Turner's Syndrome — absence of one X chromosome (XO genotype). Embryos with Turner's syndrome are usually aborted early in the pregnancy. If the infant survives, the syndrome characteristics include short stature, broad neck with prominent lateral skin folds, broad chest lacking breast development, widely spaced nipples, small uterus, and nonfunctional ovaries.

Sex Chromosome Abnormalities (Male)

1. Klinefelter's Syndrome — extra X chromosome (XXY genotype). Individuals with Klinefelter's syndrome show a feminine body configuration including breast formation but with external genital organs of a male. Testicles are atrophic. Intelligence seems to be subnormal.

2. XYY Syndrome — extra Y chromosome. These individuals are usually taller than normal, show no specific abnormalities of body configuration, show some reduction of fertility and intelligence, and in many cases exhibit aggressive and antisocial behavior.

Autosomal Abnormalities

1. Down Syndrome (mongolism, trisomy 21) — the most common chromosomal abnormality. Characteristics include mental retardation, upward slanting of eyes, prominent skin folds extending from the base of the nose to the inner aspect of the eyebrows, large protruding lips, and congenital cardiac malformations.

2. Trisomy 13 — a disorder associated with severe developmental abnormalities including cleft lip and palate, abnormal development, congenital heart defects, and polydactyly. This condition is usually fatal.

3. Trisomy 18 — associated with severe congenital malformations and is usually fatal.

Autosomal-Dominant Disorders

1. Achondroplasia — refers to dwarfism with disproportionally short limbs due to distorted growth of bones due to defective cartilage development.

2. Colorectal Polyposis — refers to formation of multiple polyps in the rectum; may be neoplastic.

3. Congenital Polycystic Kidney Disease — refers to formation of multiple cysts in kidneys leading to renal failure.

4. Hemorrhagic Telangiectasia — refers to formation of multiple small angiomas of the skin and mucous membranes, often with epistaxis or GI bleeding and sometimes with pulmonary or hepatic arteriovenous fistula; Rendu-Osler-Weber disease.

5. Medullary Thyroid Cancer — refers to congenital cancer of the thyroid gland.

6. Neurofibromatosis — refers to disfigurement and deformities due to multiple tumors in peripheral nerves.

Autosomal Recessive Disorders

1. Albinism — refers to absence of pigment in the skin, hair, and eyes due to a complete defect of melanin precursors.

2. Cretinism (Familial) — refers to a chronic condition due to congenital lack of thyroid secretion, marked by arrested physical and mental development, dystrophy of the bone and soft parts, and lowered basal metabolism.

3. Cystic Fibrosis — refers to dysfunction of the exocrine glands; characterized by signs of chronic pulmonary disease, pancreatic deficiency, abnormally high levels of electrolytes in sweat, and biliary cirrhosis.

4. Galactosemia — an inborn error of metabolism attributable to a deficiency of the enzyme galactose-1-phosphate uridyl transferase, which digests galactose. Characteristics include mental retardation, cataracts, and cirrhosis of the liver. Treatment is a diet free of galactose.

5. Gaucher's Disease — refers to accumulation of damaging glycosphinogolipid in phagocytic cells of the spleen, liver, and bone marrow. This disease is due to lysosomal enzyme deficiency of glucocerebrosidase. Characteristics include spleen enlargement, increased susceptibility to disease, and fatal hemorrhage.

6. Neiman-Pick Disease — refers to a lysosomal storage disease common in Ashkenazi (European) children. It is caused by a hexosaminidase deficiency. It is rapidly fatal because of an accumulation of gangliosides in brain neurons.

7. Phenylketonuria (PKU) — an inborn error of metabolism attributable to a deficiency of or a defect in phenylalanine hydroxylase. Characteristics include mental retardation, neurological manifestations, eczema, and a mousy odor. Treatment is a diet low in phenylalanine.

8. Tay-Sachs Disease — refers to an inherited lysosomal storage disease of children of Jewish ancestry, characterized by mental retardation, motor weakness, and blindness due to hexosaminidase A deficiency.

Codominant Disorders

1. Sickle-Cell Anemia — refers to a condition of blood where the red cells contain no normal hemoglobin. It occurs almost exclusively in blacks. It is characterized by severe anemia and obstruction of blood flow to organs by masses of sickled red cells.

2. Sickle-Cell Trait — refers to a condition of blood where the red cells contain a mixture of normal (A) and sickle (B) hemoglobin.

X-Linked Recessive Disorder

1. Color Blindness — refers to a deviation from normal perception of hues. This is a "catchall" term describing numerous perception disorders.

2. Duchenne's Muscular Dystrophy — a muscular disorder characterized by weakness and possibly atrophy of muscular tissue. Skeletal muscle is particularly affected; however, all muscular tissue can be involved.

3. Hemophilias A and B — refers to disorders causing uncontrolled bleeding into joints and internal organs following minor injuries.

4. Various Immunodeficiencies — (See Chapter 22).

Hereditary Disorders

1. Eye Diseases — retinitis pigmentosa, hereditary optic atrophy or Leber's disease, color blindness, and some forms of night blindness.

2. Hereditary Blood Disorders — hemophilia, pernicious anemia, sickle-cell anemia.

3. Mental Diseases — schizophrenia; Huntington's chorea; some forms of feeblemindedness, including amaurotic familialidiocy and mongolian idiocy.

4. Metabolic Disorders — PKU, fructose intolerance, cystinuria, diabetes mellitus.

5. Neuromuscular Disorders — progressive muscular atrophy, pseudohypertrophic muscular dystrophy, Friedreich's ataxia, peroneal atrophy, amyotonia congenita, and myotonia congenita.

6. Skeletal Defects — brachydactyly (short fingers), multiple cartilaginous exostoses, fragilitas ossium (multiple fractures), Marfan's syndrome (extremities are long and thin, face is long and narrow, individual is "double jointed").

7. Skin Diseases — baldness, xeroderma pigmentosum, von Recklinghausen's disease of multiple neurofibromatosis.

ANSWERS TO END-OF-CHAPTER QUESTIONS

Multiple Choice/Matching

1. (1) d; (2) g; (3) b; (4) a; (5) f; (6) c; (7) e; (8) h
2. (1) e; (2) a; (3) b; (4) d; (5) c; (6) d

Short Answer Essay Questions

3. The mechanisms that lead to genetic variations in gametes are segregation and independent assortment of chromosomes, crossover of homologues and gene recombination, and random fertilization. Segregation implies that the members of the allele pair determining each trait are distributed to different gametes during meiosis. Independent assortment of chromosomes means that alleles for the same trait are distributed independently of each other. The net result is that each gamete has a single allele for each trait, but that allele represents only one of the four possible parent alleles. Crossover of homologues and gene recombination implies that two of the four chromatids in a tetrad take part in crossing over and recombination, but these two may make many crossovers during synapsis. Paternal chromosomes can precisely exchange gene segments with the homologous maternal ones, giving rise to recombinant chromosomes with mixed contributions from each parent. Random fertilization implies that a single human egg will be fertilized by a single sperm on a totally haphazard basis. (pp. 1029-1030)

4.   Tt x Tt   T   t

                  T   TT   Tt

                  t   Tt   tt

a. 75% or $^3/_4$ tasters possible — any one or all of the three offspring could be or not be tasters. The chance that all three will be tasters is $^{27}/_{64}$ or approximately 42%. The chance all will not be tasters is $^1/_{64}$ or under 2%. The chance that two will be tasters and one will be a nontaster is $^9/_{64}$ or slightly more than 14%.

   Tt x tt   T   t

              t   Tt   tt

              t   Tt   tt

b. Percentage of tasters is 50%. Nontasters: 50%. Homozygous recessive: 50%. Heterozygous: 50%. Homozygous dominant, none. (pp. 1031-1033)

5. Both nonalbino parents carry the recessive gene for albinism, hence they are heterozygous for the trait. (p. 1033)

6. The mother's genotype is IAi. The father's genotype is IBi; his phenotype, B. Child number one has an ii genotype; child number two, IBi. (p. 1033)

7. a. AABBCC x aabbcc

    (very dark) (very light)

    offspring genotype: AaBbCc

    offspring phenotype: medium range of color

    b. AABBCC x AaBbCc

    (very dark) (medium color)

offspring:	genotype	phenotype
	AABBCC	very dark
	AaBbCc	medium color

    c. AAbbcc x aabbcc

    (light) (very light)

    offspring: genotype Aabbcc

    offspring: phenotype lighter than AAbbcc parent, but not as light as aabbcc parent

    This is an example of polygenic inheritance. (p. 1033)

8. Amniocentesis is done after the 14th week. A needle is inserted through the mother's abdominal wall to remove fluid (or fetal cells) to be tested. Chorionic villi sampling is done at eight weeks. A tube is inserted through the vagina and cervical os. It is guided by ultrasound to an area where a piece of placenta can be removed. (p. 1037)

Critical Thinking and Application Questions

1. Maternal grandfather XcY
    Mother with normal vision XCXc

    Color-blind father XcY

	XC	Xc
Xc	XCXc	XcXc
Y	XCY	XcY

    a. One chance in two of the first child being a son. A son will either be color-blind or have normal vision. To determine the probability of two events happenings in succession, we must multiply together the probabilities of the separate events happening. The probability of getting a son as the first child is $1/2$. The probability of a son being color-blind is $1/2$, hence the probability of the first child being a color-blind son is $1/4$ ($1/2 \times 1/2$). The combined probability will be the same for the first child being a color-blind daughter ($1/4$).

    b. The probability that there will be one color-blind son is $1/4$. The probability of two color-blind sons is $1/4 \times 1/4 = 1/16$, or slightly more than 6%.

    The production of each child is an independent event that does not influence the others. (p. 1008)

2.

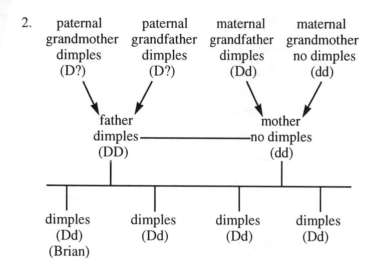

paternal grandmother dimples (D?) paternal grandfather dimples (D?) maternal grandfather dimples (Dd) maternal grandmother no dimples (dd)

father dimples (DD) ——————— mother no dimples (dd)

dimples (Dd) (Brian)   dimples (Dd)   dimples (Dd)   dimples (Dd)

(pp. 1028-1030)

3. Mrs. Lehman should have testing, since Tay-Sachs is a recessive disorder. The baby would have to get both recessive genes for the disease. If there is no incidence of the disease in her family, the recessive gene could be there but would always be masked by the dominant gene. If her husband carries one recessive gene and she carries a recessive gene, the baby would have a chance of getting both recessives and having the disease. (pp. 1028-1030)

Key Figure Questions

Figure 30.2 Crossover and exchange of chromosomal parts. **Figure 30.3** You would get a different selection of linked genes in the resulting gametes.

SUGGESTED READINGS

---

1. Anderson, W.F. "Beating Nature's Odds." *Science* 85 (Nov. 1985): 49.

2. Baskin, Y. "Gene Bank: Storehouse for Life's Secret Code." *Science Digest* 91 (May 1983): 94.

3. Brady, R.O. "Inherited Metabolic Diseases of the Nervous System." *Science* 193 (Aug. 1976): 733.

4. Cambon, P. "Split Genes." *Scientific American* 244 (May 1981).

5. Capecchi, M.R. "Targeted Gene Replacement." *Scientific American* 270 (Mar. 1994): 52.

6. Conner, J.M., and M.A. Ferguson-Smith. *Essential Medical Genetics*. St. Louis: C.V. Mosby, 1984.

7. DeRobertis, E.M., G. Oliver, and C.V.E. Wright. "Homeobox Genes and the Vertebrate Body Plan." *Scientific American* 263 (1) (July 1990): 46-52.

8. Diamond, J. "The Cruel Logic of Our Genes." *Discover* 10 (11) (Nov. 1989): 72-78.

9. Eigen, M., W. Gardiner, P. Schuster, and R. Winkler-Oswatitsch. "The Origin of Genetic Information." *Scientific American* 244 (Apr. 1981).

10. Ferguson-Smith, M.A. "Chromosomal Abnormalities II: Sex Chromosome Defects." *Hospital Practice* (Apr. 1970).

11. Hall, S.S. "James Watson and the Search for Biology's 'Holy Grail.'" *Smithsonian* 20 (11) (Feb. 1990): 40-49.

12. Hartl, D.L. *Human Genetics*. New York: Harper & Row, 1983.

13. Hirschhorn, K. "Chromosomal Abnormalities I: Autosomal Defects." *Hospital Practice* (Feb. 1970).

14. Lawn, R.M., and G.A. Vehar. "The Molecular Genetics of Hemophilia." *Scientific American* (Mar. 1986).

15. Miller, J.A. "New Molecular Analysis for Genetic Disorders." *Science News* 129 (Feb. 1986): 129.

16. Ptashne, M. "How Gene Activators Work." *Scientific American* 260 (1) (Jan. 1989): 40-47.

17. Stahl, F.W. "Genetic Recombination." *Scientific American* 256 (Feb. 1987): 91.

18. Suzuki, D., A. Griffith, J. Miller, and R. Lweontin. *An Introduction to Genetic Analysis*. 4th ed. New York: W.H. Freeman, 1989.

19. Temple, M.J. "Chromosomal Syndromes." *Hospital Practice* (Feb. 1983).

20. Weiss, R. "A Genetic Gender Gap." *Science News* 135 (20) (May 1989): 312-315.

21. White, R., and J. Lalonel. "Chromosome Mapping with DNA Markers." *Scientific American* 258 (Feb. 1988): 40.

22. "X Chromosomes: Too Few and Too Many." *Science News* 129 (June 1986): 358.

Guide to Audio-Visual Resources

ACR — American College of Radiology, 20 N. Wacker Drive, Chicago, IL 60606

ACS — American Cancer Society, 19 West 56th Street, New York, NY 10019 (212) 586-8700

AEF — American Educational Films, 3807 Dickerson Road, Nashville, TN 37207 (615) 868-2040

AF — Academy Films, P.O. Box 1023, Venice, CA 90291

AFI — Association Films, 799 Stevenson Street, San Francisco, CA 94103

AIF — Australian Instructional Films, 39 Pitt Street, Sydney, Australia

AMA — American Medical Association, 535 N. Dearborn, Chicago, IL 60610 (800) 621-8335

APH — Alfred Higgins Productions, 9100 Sunset Blvd., Los Angeles, CA 90069

ASFT — Association-Sterling, 8616 Directors Row, Dallas, TX 75240

BARR — Barr Films, 12801 Schabarum Ave., Irwindale, CA 91706 (818) 338-7878

BFA — BFA Films and Videos, 468 Park Ave. South, New York, NY 10016 (800) 221-1274

BM — Biology Media, P.O. Box 10205, Berkeley, CA 94710

BNF — Benchmark Films, Inc., 145 Scarborough Road, Briarcliff Manor, NY 10510 (914) 762-3838

BYU — Brigham Young University, Audio-Visual Services, 101 Fletcher Building, Provo, UT 84602 (801) 378-4071

CA — Career Aids, 20417 Nordhoff St., Post AH98, Chatsworth, CA 91311 (818) 341-8200

CBS — Carolina Biological Supply Company, 2700 York Road, Burlington, NC 27215 (800) 334-5551

CCM — CCM Films, Inc. Distributed by: Films, Inc., 5547 Ravenswood Ave., Chicago, IL 60640

CCMI — Classroom Consortia Media, Inc.

CDL — Cambridge Development Laboratory, Inc., 42 4th Ave., Waltham, MA 02154

CDR — Center for Devices and Radiological Health, Training Resources Center, 5300 Fishers Lane, Rockville, MD 20857 (301) 443-4647

CF — Churchill Media, 12210 Nebraska Ave., Los Angeles, CA 90025 (800) 334-7830

CFI — Counselor Films, Inc., 1728 Cherry St., Philadelphia, PA 19103

CHM — Cleveland Health Museum, 8911 Euclid Ave., Cleveland, OH 44106

CIBA — Ciba Pharmaceutical Company, Medical Communications Dept., 556 Morris Ave., Summit, NJ 07901

CIF — Coronet/MTI Film and Video, Supplementary Education Group, Simon and Schuster Communications, 108 Wilmot Rd., Deerfield, IL 60015-9990 (800) 621-2131

COND — Conduit, P.O. Box 388, Iowa City, IA 52244

CRM — CRM Films, 2215 Faraday, Carlsbad, CA 92008 (800) 421-0833

DA — Document Associates/The Cinema Guild, 1697 Broadway, Suite 802, New York, NY 10019 (212) 246-5522

EBE — Encyclopedia Britannica Educational Corporation, 310 South Michigan Ave., Chicago, IL 60604 (800) 621-3900

EI — Educational Images (see EIL below)

EIL — Educational Images Limited, P.O. Box 3456, Elmira, NY 14905 (607) 732-1090

EL — Eli Lilly and Company, Medical Division, Indianapolis, IN 46206 (317) 261-2000

FAD — F.A. Davis Company, 1915 Arch St., Philadelphia, PA 19103 (215) 568-2270

FHS	Films for the Humanities and Sciences, Inc., P.O. Box 2053, Princeton, NJ 08540 (800) 257-5126
HP	Hoechst-Roussel Pharmaceuticals, Rt. #202-206N, Somerville, NJ 08876 (201) 685-2648
HR	Harper and Row Publishers. Distributed by: MTI Teleprograms, 108 Wilmot Rd., Deerfield, IL 60015 (212) 207-7000
HRM	Human Relations Media, 175 Tompkins Ave., Pleasantville, NY 10570 (800) 431-2050
IBIS	IBIS Media (now HRM above)
ICI	Imperial Chemicals Industries, Inc., P.O. Box 1274, 151 South St., Stamford, CT 06904
ICIA	ICI America, Inc., Concord Pike & Murphy Rd., Wilmington, DE 19899 (302) 575-3275
IFB	International Film Bureau, 332 S. Michigan Ave., Chicago, IL 60604 (312) 427-4545
IOWA	Iowa State University Media Resources Center, 121 Pearson Hall, Ames, IA 50011 (515) 294-1540
IP	Iwanami Productions, Inc., 22-2 Kanda Misakicho, Chiyoda-Ku, Tokyo, Japan
IU	Indiana University Audio-Visual Center, Bloomington, IN 47405-5901 (800) 552-8620
JBL	J.B. Lippincott, East Washington Square, Philadelphia, PA 19105 (800) 523-2945
JJ	Johnson and Johnson, Grandview Road, Skillman, NJ 08558
JW	John Wiley and Sons, Inc., 605 Third Ave., New York, NY 10158 (212) 850-6276
LPI	Lawren Productions, Inc., 930 Pitner Ave., Evanston, IL 60202 (800) 323-9084
MAP	Medical Arts Production
MC	Mayo Clinic, Section of Photography, 200 S.W. First St., Rochester, MN 55905 (507) 284-2511
McG	McGraw-Hill Book Co., Inc., Text-Film Division, 1221 Avenue of the Americas, New York, NY 10020 (see CRM above)
MF	Milner-Fenwick, Inc., 2125 Greenspring Drive, Timonium, MD 21093 (800) 638-8652
MG	Media Guild, 11722 Sorrento Valley Rd., Suite E, San Diego, CA 92121 (619) 755-9191
MGHT	McGraw-Hill Films (see CRM above)
MI	Medcom, Inc., 1633 Broadway, New York, NY 10019
NEIF	National Educational and Information Films, Ltd., National House, Tullock Rd., Apollo Bunder, Bombay 1, India
NET	Nebraska Educational Television Council for Higher Education, Inc., P.O. Box 83111, Lincoln, NE 68501 (402) 472-3611
NGS	National Geographic Society, 1145 17th St. N.W., Washington, DC 20036 (800) 368-2728
NTA	National Teaching Aids, 120 Fulton Ave., Garden City Park, NY 11040
NYAM	New York Academy of Medicine, 2 East 103rd St., New York, NY 10029
PAR	Paramount Communications. Distributed by: AIMS Media, 9710 DeSoto Ave., Chatsworth, CA 91311 (818) 773-4300
PE	Perennial Education, Inc., 930 Pitner Avenue, Evanston, IL 60202 (800) 323-9084
PFP	Pyramid Films and Videos, P.O. Box 1048, Santa Monica, CA 90406 (800) 421-2304
PHM	Prentice-Hall Media. Distributed by: Vocational Media Associates, P.O. Box 1000, Mount Kisco, NY 10549 (800) 431-1242
PLP	Projected Learning Programs, Inc., P.O. Box 3008, Paradise, CA 95969 (916) 893-4223
PMR	Peter M. Robeck and Co., Inc. Distributed by: Time-Life Films, Inc. (see TL below)
POLY	Polymorph Films, Inc., 118 South Street, Boston, MA 02111 (617) 542-2004
PSP	Popular Science Publishing Company, 355 Lexington Ave., New York, NY 10017

PYR   Pyramid Films and Videos, P.O. Box 1048, Santa Monica, CA 90406 (800) 421-2304

QUE   Queue, Inc., 562 Boston Ave., Bridgeport, CT 06610

REX   REX Educational Resources Company, P.O. Box 2379, Burlington, NC 27216

RJB   Robert J. Brady Company, 130 Q Street N.E., Washington, DC 20002

RL   Roche Biomedical Lab, Inc., 1447 York Ct., Burlington, NC 27215 (919) 584-5171

SC   Scherring Corporation, 1011 Morris Avenue, Union, NJ 07083 (201) 558-4000

SEF   Sterling Educational Films, 241 E. 34th St., New York, NY 10016 (212) 779-0202

SM   Science and Mankind, Inc., P.O. Box 1000, Mount Kisco, NY 10549 (800) 431-1242

SQ   E.R. Squibb and Sons, Inc., P.O. Box 4500, Princeton, NJ 08543-4500

SU   Syracuse University Film Library, 1455 E. Colvin St., Collendale Campus, Syracuse, NY 13210

TC   Trainex Corp., 12601 Industry St., Garden Grove, CA 92641 (714) 898-2561

TF   Teaching Films, Inc., 930 Pitner Ave., Evanston, IL 60202 (312) 328-6700

TL   Time-Life Films (see Time-Life Video below)

TLV   Time-Life Video. Distributed by: Ambrose Video Publishing, Inc., 381 Park Avenue South, Suite 1601, New York, NY 10016 (212) 696-4545

TNF   The National Foundation • March of Dimes, Professional Film Library, c/o Association, Inc., 600 Grande Ave., Ridgefield, NJ 07657

UI   University of Illinois Film Center (see UIFC below)

UIFC   University of Illinois Film Center, 1325 South Oak St., Champaign, IL 61820 (800) 367-3456

UJ   Unijapan Films, 9-13 Ginza 5-Chome, Chuo-Ku, Tokyo, Japan

UN   University of Nebraska Audio-Visual Instruction, 421 Nebraska Hall, Lincoln, NE 68508 (402) 472-1907

UP   Upjohn Professional Film Library, 7000 Portage Rd., Kalamazoo, MI 49002

USNAC   U.S. Audiovisual Center, General Services Administration, Washington, DC 20409 (301) 763-1896

UT   University of Texas Medical Branch, Galveston, TX 77550 (713) 765-2481

UWF   United World Films, Inc., 221 Park Ave. South, New York, NY 10003

UWM   University of Washington, School of Medicine, Seattle, WA 98105

WNSE   Wards Natural Science Establishment, Inc., P.O. Box 1712, Rochester, NY 14622